UTILIZATION OF FOOD WASTE AND BYPRODUCTS

UTILIZATION OF FOOD WASTE AND BYPRODUCTS

Mudasir Ahmad Malik, PhD
Kshirod Kumar Dash, PhD
Editors

AAP | APPLE ACADEMIC PRESS

First edition published 2027

Apple Academic Press Inc.
1265 Goldenrod Circle, NE,
Palm Bay, FL 32905 USA

760 Laurentian Drive, Unit 19,
Burlington, ON L7N 0A4, Canada

CRC Press
2385 NW Executive Center Drive,
Suite 320, Boca Raton FL 33431

4 Park Square, Milton Park,
Abingdon, Oxon, OX14 4RN, UK

Library and Archives Canada Cataloguing in Publication

CIP data on file with Canada Library and Archives

Library of Congress Cataloging-in-Publication Data

CIP data on file with US Library of Congress

ISBN: 978-1-77964-114-4 (hbk)
ISBN: 978-1-77964-115-1 (ebk)

DOI: 10.1201/9781779641151

About the Editors

Mudasir Ahmad Malik, PhD
Assistant Professor, Department of Food Engineering and Technology, Ghani Khan Choudhury Institute of Engineering and Technology, Malda, West Bengal, India

Mudasir Ahmad Malik, PhD, is an Assistant Professor in the Department of Food Processing Technology at Ghani Khan Choudhury Institute of Engineering and Technology, Malda, West Bengal. He earned his BTech in Food Technology from Islamic University of Science and Technology, Awantipora, and his MTech and PhD in Food Engineering and Technology from Sant Longowal Institute of Engineering and Technology, Longowal.

He has authored several international publications in the field of food technology in reputed peer-reviewed journals. His MTech research focused on buckwheat starch, and his PhD research explored sunflower protein isolates. He has participated in numerous national and international conferences related to food technology.

Currently, he is working on the utilization of food waste from food industries for the development of value-added products. Before joining his current institute, he worked at several universities across the country. His expertise includes food engineering, food chemistry, fruits and vegetables, and cereals and pulses.

Kshirod Kumar Dash, PhD
Professor, Department of Food Engineering and Technology, Ghani Khan Choudhury Institute of Engineering & Technology (GKCIET), Malda, West Bengal, India

Kshirod Kumar Dash, PhD, is a Professor in the Department of Food Processing Technology at Ghani Khan Choudhury Institute of Engineering and Technology (GKCIET), Malda, West Bengal, India. Previously, he served as an Assistant Professor in the Department of Food Engineering and Technology at Tezpur University, Tezpur, Assam.

Dr. Dash earned his BTech in Agricultural Engineering from the College of Agricultural Engineering and Technology, Odisha University of Agriculture and Technology (OUAT), Bhubaneswar; his MTech in Dairy and Food Engineering from the Indian Institute of Technology (IIT) Kharagpur; and his PhD in Food Process Engineering from IIT Kharagpur. He completed his postdoctoral research at The Ohio State University, Columbus, Ohio, USA.

He has guided several postgraduate and doctoral students and has published more than 50 papers in peer-reviewed international journals, as well as book chapters in edited volumes. He is a life member of several professional bodies in India and a member of the Institute of Food Technologists (IFT), USA.

Dr. Dash's research interests include food process engineering, unit operations in food processing, fluid mechanics, and heat and mass transfer in food processing.

Contents

Contributors

Ria Ahuja
Department of Foods and Nutrition, The M.S. University of Baroda, Vadodara, Gujarat, India

Abha Anand
Dr. B. R. Ambedkar State Institute of Medical Sciences, Mohali, Punjab, India

Aneeta Babu
Department of Food Technology, School of Agriculture and Food Technology, Vignan's Foundation for Science, Technology and Research, Vadlamudi, Guntur, Andhra Pradesh, India

Radheshyam B. Bajad
Department of Food Technology, School of Agriculture and Food Technology, Vignan's Foundation for Science, Technology and Research, Vadlamudi, Guntur, Andhra Pradesh, India

Khalid Bashir
Department of Food Technology, Jamia Hamdard, New Delhi, Delhi, India

Kunal Bhardwaj
Public Health Foundation of India, Saidulajab, Sainik Farm, New Delhi, India

Farhan M. Bhat
Department of Food Science, Nutrition and Technology, CSK Himachal Pradesh Krishi Vishvavidyalaya, Palampur, Himachal Pradesh, India

Jithender Bhukya
Department of Agricultural Engineering, School of Agricultural and Bio-Engineering, Centurion University of Technology and Management, Paralakhemundi, Odisha, India

Nilakshi Chauhan
Department of Food Science, Nutrition and Technology, CSK Himachal Pradesh Krishi Vishvavidyalaya, Palampur, Himachal Pradesh, India

Preeti Choudhary
Department of Food Science and Technology, Dr. Yashwant Singh Parmar University of Horticulture and Forestry, Neri, Hamirpur, Himachal Pradesh, India

B. N. Dar
Department of Food Technology, Islamic University of Science and Technology, Awantipora, Jammu and Kashmir, India

Moirangthem Kalpana Devi
Department of Agricultural and Food Engineering, Indian Institute of Technology, Kharagpur, West Bengal, India

Harshavardhan Dhulipalla
Department of Food Technology, School of Agriculture and Food Technology, Vignan's Foundation for Science, Technology and Research, Vadlamudi, Guntur, Andhra Pradesh, India

Olivia Francis
Department of Food Technology, School of Agriculture and Food Technology, Vignan's Foundation for Science, Technology and Research, Vadlamudi, Guntur, Andhra Pradesh, India

Kulsum Jan
Department of Food Technology, Jamia Hamdard, New Delhi, Delhi, India

Shumaila Jan
National Institute of Food Technology Entrepreneurship and Management (NIFTEM), Kundli, Sonipat, Haryana, India

Sweta Joshi
Department of Food Technology, School of Interdisciplinary Sciences, Jamia Hamdard, New Delhi, India

Pir Mohammad Junaid
Department of Post-Harvest Engineering and Technology, Aligarh Muslim University, Aligarh, Uttar Pradesh, India

M. Kamalish
Department of Clinical Nutrition, Sri Ramachandra Institute of Higher Education and Research, Chennai, Tamil Nadu, India

Devinder Kaur
Center of Food Technology, University of Allahabad, Prayagraj, Uttar Pradesh, India

Manmeet Kaur
Department of Microbiology, Punjab Agricultural University, Ludhiana, Punjab, India

Vivek Kumar
Department of Food Processing Technology, Ghani Khan Choudhury Institute of Engineering and Technology, Malda, West Bengal, India

Radha Kushwaha
Center of Food Technology, University of Allahabad, Prayagraj, Uttar Pradesh, India

Palak Mahajan
Department of Biotechnology Engineering and Food Technology, University Institute of Engineering, Chandigarh University, Mohali, Punjab, India

Darakshan Majid
Department of Food Technology, Islamic University of Science and Technology, Awantipora, Jammu and Kashmir, India

Hilal Ahmad Makroo
Department of Food Technology, Islamic University of Science and Technology, Awantipora, Jammu and Kashmir, India

Mudasir Ahmad Malik
Department of Food Engineering and Technology, Ghani Khan Choudhury Institute of Engineering and Technology, Malda, West Bengal, India

Arooj Mattoo
Department of Food Technology, Islamic University of Science and Technology, Awantipora, Jammu and Kashmir, India

Khalid Muzzafer
Department of Food Technology, Islamic University of Science and Technology, Awantipora, Jammu and Kashmir, India

Gazia Nasir
Department of Bioengineering, Faculty of Engineering and Technology, Integral University, Lucknow, Uttar Pradesh, India

Mariya Nayeem
Division of Food Science and Technology, Sher-e-Kashmir University of Agricultural Sciences and Technology, Srinagar, Jammu and Kashmir, India

Ruby Pandey
Department of Agricultural Engineering, School of Agricultural and Bio-Engineering, Centurion University of Technology and Management, Paralakhemundi, Odisha, India

Aasima Rafiq
Krishi Vigyan Kendra, Srinagar, Sher-e-Kashmir University of Agricultural Sciences and Technology, Jammu and Kashmir, India

Sharika Raouf
Department of Food Technology, Islamic University of Science and Technology, Awantipora, Jammu and Kashmir, India

Mrinmoy Roy
Department of Food Technology, School of Agriculture and Food Technology, Vignan's Foundation for Science, Technology and Research, Vadlamudi, Guntur, Andhra Pradesh, India

Pinki Saini
Center of Food Technology, University of Allahabad, Prayagraj, Uttar Pradesh, India

Sakshi
Department of Food Technology, Jamia Hamdard, New Delhi, Delhi, India

Anwesa Sarkar
Department of Food Processing Technology, Ghani Khan Choudhury Institute of Engineering and Technology, Malda, West Bengal, India

Tajamul Rouf Shah
Department of Food Technology, Islamic University of Science and Technology, Awantipora, Jammu and Kashmir, India

Neha Sharma
Department of Microbiology, Punjab Agricultural University, Ludhiana, Punjab, India

P. Shruthi
Department of Community Science, College of Agriculture, Vellayani, Kerala Agricultural University, Thiruvananthapuram, Kerala, India

Darakhshan Tahseen Siddiqui
Department of Food Technology, School of Interdisciplinary Sciences, Jamia Hamdard, New Delhi, India

Anchal Singh
Center of Food Technology, University of Allahabad, Prayagraj, Uttar Pradesh, India

Mohini Prabha Singh
Department of Microbiology, Punjab Agricultural University, Ludhiana, Punjab, India

Sajad Ahmad Sofi
Department of Food Technology, Islamic University of Science and Technology, Awantipora, Jammu and Kashmir, India

Irshaan Syed
Department of Food Technology, School of Agriculture and Food Technology, Vignan's Foundation for Science, Technology and Research, Vadlamudi, Guntur, Andhra Pradesh, India

Nazia Tabassum
Department of Post-Harvest Engineering and Technology, Faculty of Agricultural Sciences, Aligarh Muslim University, Aligarh, Uttar Pradesh, India

Manisha Thakur
Department of Food Technology, Bhaskaracharya College of Applied Sciences, University of Delhi, New Delhi, Delhi, India

Ranjana Verma
Department of Food Science, Nutrition and Technology, CSK Himachal Pradesh Krishi Vishvavidyalaya, Palampur, Himachal Pradesh, India

Sadaf Zaidi
Department of Post-Harvest Engineering and Technology, Aligarh Muslim University, Aligarh, Uttar Pradesh, India

Abbreviations

ACE	angiotensin I-converting enzyme
AD	anaerobic digestion
AME	apparent metabolizable energy
ARF	aleurone rich fraction
ASBR	anaerobic sequencing batch reactors
BDAI	Biodiesel Association of India
BGH	black gram husk water extract
BMP	biochemical methane potential
BOD	biochemical oxygen demand
BRGA	byproduct rich in germ and aleurone
BSE	bovine spongiform encephalopathy
BSG	Brewer's Spent Grain
BUN	blood urea nitrogen
CAGR	compound annual growth rate
CHP	combined heat and power
CMF	cellulose microfibers
CNC	nanocrystals
COD	chemical oxygen demand
CP	crude protein
CS	coffee silverskin
DGCI&S	Directorate General of Commercial Intelligence and Statistics
DHA	docosahexaenoic acid
EAE	enzyme-assisted extraction
EPA	eicosatetraenoic acid
FAO	Food and Agriculture Organization
FBH	faba bean husks
FID	flame ionization detector
FSC	food supply chains
GC	gas chromatography
GDP	gross domestic product
GOS	galacto-oligosaccharides
GRAS	generally recognized as safe
HCA	hydroxycinnamic acids
HF	hydrolyzed feather meal
IARI	Indian Agricultural Institute
IPSP	instant pea soup powder

LCYC	lignocellulosic yeast
LDL	low-density lipoprotein
MAE	microwave-assisted extraction
MEC	microbial electrolysis cells
MFC	microbial fuel cells
MMC	mixed microbial cultures
MSG	monosodium glutamate
MSW	municipal solid waste
MUFA	monounsaturated fatty acids
ODS	ordinary denatured spirit
OLR	organic loading rate
PBR	packed bed reactors
PCM	phase change materials
PLE	pressurized liquid extraction
PRF	plumule rich fraction
PUFA	polyunsaturated fatty acids
RSLD	rapid solid–liquid dynamic
SCG	spent coffee grounds
SCP	single-cell protein
SDG	Sustainable Development Goals
SFE	supercritical fluid extraction
SHF	separate hydrolytic fermentation
SSF	simultaneous saccharification-cum-fermentation
STP	standard temperature and pressure
TNPL	Tamil Nadu Newsprints Limited
TOC	total organic carbon
TSS	total suspended solids
UAE	ultrasound-assisted extraction
UASB	upflow anaerobic sludge blanket
UFA	unsaturated fatty acids
UNEP	United Nations Environment Program
VFA	volatile fatty acid
WHC	water hydration capacity

Preface

In the food processing industry, large quantities of wastes and byproducts are generated, which have significant negative implications for the environment. These wastes and byproducts include pomace from fruit and vegetable processing; meal or oil cake from oilseed processing; hulls and bran from grain milling; and many others. They are rich sources of nutrients such as proteins, dietary fiber, vitamins, and bioactive compounds like polyphenols and antioxidants. Proper utilization of these wastes and byproducts can generate economic gains for the food industry, help reduce nutritional deficiencies, and, most importantly, mitigate environmental problems.

This book was written with the aim of exploring the wide range of byproducts and wastes produced by the food industry for their proper utilization and, if necessary, disposal. It discusses the different sources of waste, their impact on the environment, and methods to convert these byproducts and wastes into valuable products. We have covered almost all segments of the food processing industry that generate waste. This book will serve as a guide for food industries in the proper utilization of byproducts and wastes, ultimately having a positive environmental impact and providing monetary benefits to the industry.

The book comprehensively covers the sources, utilization, and disposal of different byproducts and wastes from both plant and animal sources. The major plant sources of food waste and byproducts discussed include cereals, pulses, oilseeds, fruits, vegetables, citrus fruits, sugarcane, brewer's spent grain, coffee, and spices. The animal sources include milk, meat, fish, poultry, and seafood. Methods used for the proper disposal of waste are also discussed in the final section of this book.

Introduction

In the food processing industry, large quantities of wastes and byproducts are generated, which have significant negative implications for the environment. These wastes and byproducts include pomace from fruit and vegetable processing; meal or oil cake from oilseed processing; hulls and bran from grain milling; and many others. They are rich sources of nutrients such as proteins, dietary fiber, vitamins, and bioactive compounds like polyphenols and antioxidants. Proper utilization of these wastes and byproducts can generate economic gains for the food industry, help reduce nutritional deficiencies, and, most importantly, mitigate environmental problems. This book discusses the different food wastes and byproducts produced from food processing and their proper utilization. It covers wastes and byproducts from both plant and animal sources, as well as various treatment methods for their proper disposal. The major plant sources of food waste and byproducts include cereals, pulses, oilseeds, fruits, vegetables, citrus fruits, sugarcane, brewer's spent grain, coffee, and spices. Animal sources include milk, meat, fish, poultry, and seafood. Treatments such as biochemical, enzymatic, and microbiological methods are also discussed in this book.

CHAPTER 1

Utilization of Food Wastes and Byproducts: An Overview

SAKSHI,[1] MANISHA THAKUR,[2] SHUMAILA JAN,[3] KHALID BASHIR,[1] MUDASIR AHMAD MALIK,[4] and KULSUM JAN[1,*]

[1]*Department of Food Technology, Jamia Hamdard, New Delhi, Delhi, India*

[2]*Department of Food Technology, Bhaskaracharya College of Applied Sciences, University of Delhi, New Delhi, Delhi, India*

[3]*National Institute of Food Technology Entrepreneurship and Management (NIFTEM), Kundli, Sonipat, Haryana, India*

[4]*Department of Food Engineering and Technology, Ghani Khan Choudhury Institute of Engineering and Technology, Malda, West Bengal, India*

*Corresponding author

ABSTRACT

Today's world is facing a serious problem with food waste. According to the FAO, over half of all food produced will never be eaten, making it a severe problem in our society that affects "poor and rich countries" equally. Instead of procrastinating, we have to deal with these issues and take many initiatives to reduce food waste. More research is needed to determine the efficiency of these methods or their potential to reduce food waste. Information is lacking regarding the economic, environmental, and social performance of efforts to reduce food waste, especially with reference to agro byproducts. The effectiveness of these techniques or their

Utilization of Food Waste and Byproducts. Mudasir Ahmad Malik and Kshirod Kumar Dash (Eds.)

DOI: 10.1201/9781779641151-1

potential to reduce food waste requires more research. There is a lack of information on the effectiveness of programs to minimize food waste in terms of economic, environmental, and social performance, particularly with regard to agro byproducts. Therefore, the bifurcation and utilization of food waste has been categorically identified and discussed in this chapter, emphasizing food waste management. This chapter fills in the knowledge gap by exploring the techniques currently trending in nonfiction to assess food waste categorically and understand the various alternatives to utilize them in order to cut the nutrition left over from various byproduct categories, minimize food wastage, and increase food sustainability.

1.1 INTRODUCTION

Since a consequence of increased food demand and stronger environmental laws, agriculture is transforming towards more effective and environmentally friendly methods. However, food waste continues to be a problem on a worldwide scale with negative effects on the atmosphere, society, and the economy. Spang et al. (2019) linked projected yearly losses of almost one-third of all edible food (1.3 billion metric tons, MMT) across the supply chain to this problem (FAO, 2011). Two indices have been used more recently by the FAO (2019) to describe food waste: the Food Waste Index, which describes food that is eventually wasted by customers or retailers, and the Food Loss Index, which describes food that is lost during production or in the supply chain prior to reaching the retail level. Before it reaches the retail level, 14% of the world's food is lost, although the Food Waste Index role in this loss is still being researched (FAO, 2019). According to Dou et al. (2016), the production and processing of fruits, vegetables, meat, poultry, and fish would result in food loss streams totaling 35.9 MMT in the United States alone. However, total retail and consumer waste will be estimated to be 19.5 MMT and 40.8 MMT, respectively. In Canada, there is a total of 35.5 MMT of yearly loss and waste in the food value chain, of which 11.2 MMT (32%) is preventable, valued at $49.5 billion and accounting for 51.8% of the food dollars spent in retail stores in 2016 (Gooch et al., 2019). According to the Global Hunger Index, between 2019 and 2022, there may have been an additional 150 million people suffering from undernourishment. This disaster, which was mostly fueled by war, climate change, and the COVID-19 pandemic, has resulted in a 10% rise in population worldwide. Nevertheless, up to 828 million people remain undernourished. Talking about India, according to various researches, 40% of food manufactured in India is wasted as it cannot be utilized due to lack of food supply chain (FSC) and fragmented food systems (Figure 1.1). Nutrition that is not consumed in the retail, food service, and consumer phases of the FSC is referred to as food waste and is associated with consumer behavior (Dou et al., 2016). When food is lost along the supply chain's stages of manufacturing, processing, and distribution, quality byproducts are often also lost. In developing countries, the production of food results in significant waste and byproduct production; this is pricey and has an adverse effect on the environment (Figures 1.1 and 1.2).

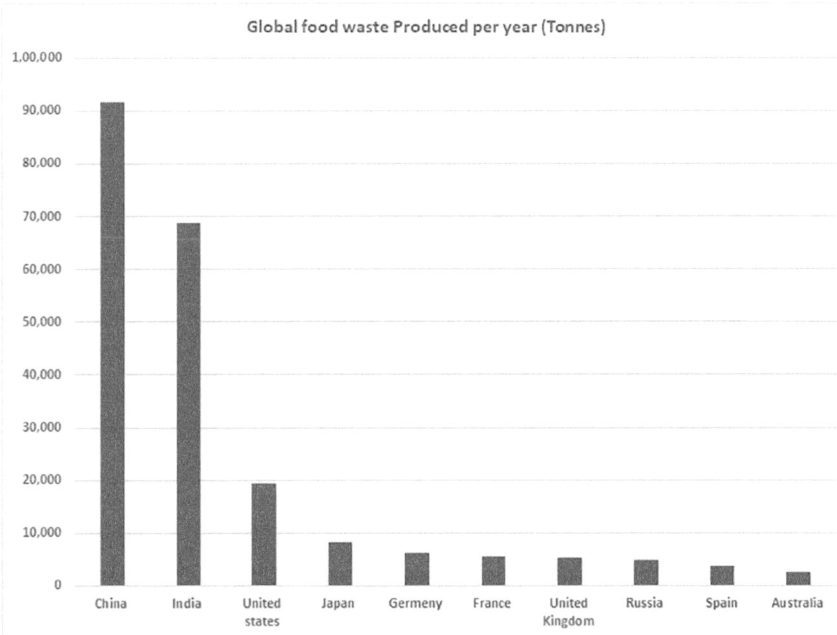

FIGURE 1.1 Global wastage in various countries per year in tons.

Source: https://thewire.in/food/global-food-waste-index-china-india; Tang et al. (2023).

As we are discussing food waste management, all the other steps involved from "farm to fork" should be well maintained and controlled in order to achieve less wastage. Agricultural produce has defined shelf lives, which limits its utilization and processing time.

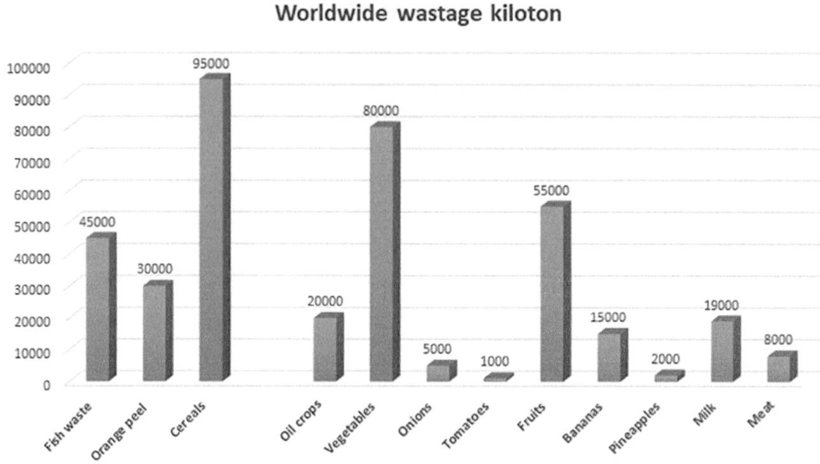

FIGURE 1.2 Category wise annual wastage in kiloton.

Source: https://thewire.in/food/global-food-waste-index-china-india; Tang et al. (2023).

To enlist key reasons for such huge wastages, the following points are discussed to identify and improve the possibilities to enhance utilization of food waste.

1. Improper harvesting time given to the agro products leads to various rejections and unacceptability of products by the consumers for further processing.
2. Improper harvesting techniques delay the harvesting and cause a huge amount of wastage and lack of availability of equipment, which can increase the efficiency of agricultural produce.
3. Fluctuating environmental conditions during harvesting will cause serious damage to the agricultural products, and further inadequacy of an unregimented supply chain will delay the transportation and ultimate utilization of the produce.
4. We are witnessing every year in India, especially, food being thrown away in tons due to improper storage conditions, which can be controlled if we have better facilities to keep them in off seasons.
5. Insect infestation is one of the major spoilers of agricultural produce, leading to agricultural waste if insecticides and pesticides are not provided to the plants on time.

6. After harvesting, both storage and transportation should go hand in hand; otherwise, the waste percentage will increase.
7. Non-efficient processing techniques can produce a higher percentage of waste than normal; therefore, it should also be considered an important drive responsible for food waste management.
8. Since we have labor-dependent work for pre- and post-harvest activities, human mismanagement is quite possible, which can further lead to food wastage.
9. Packaging is the final step for any food product; however, if it is not proper, it might lead to damage or spoilage of the food product. Therefore, improper packaging should be taken care of to avoid such wastages.
10. Improper internal and external conditions during transportation, storage, or processing can spoil the product. For example, temperature, humidity, pH, and so forth. Such changes, especially in climatic products, can spoil the whole produce.
11. Excessive trimming, slicing, grading, and so forth, can increase the percentage of waste in agro products; therefore, proper training and control are required for such activities and practices.
12. Improper knowledge and entitlement amongst the humans handling the produce might also cause a high amount of food wastage. Therefore, small awareness programs and training can help us minimize food waste.

Figure 1.3 depicts the actual scenario of where has the food been wasted and lost before reaching to the consumer.

Biomaterials are used to generate food additives, which will reduce poverty and malnutrition in the poor nations where they are produced. Numerous of these biomaterials are sources of good stuff, including dietary fiber, proteins, lipids, carbs, and micronutrients. Biotechnological methods can also be used to lessen the antinutritional components in certain byproducts, which can then be added to food or used to create balanced meals.

1.2 AGRO BYPRODUCTS

Food wastes are often additive-free byproducts of the conversion of food basic elements. Waste is defined as a product that does not contribute

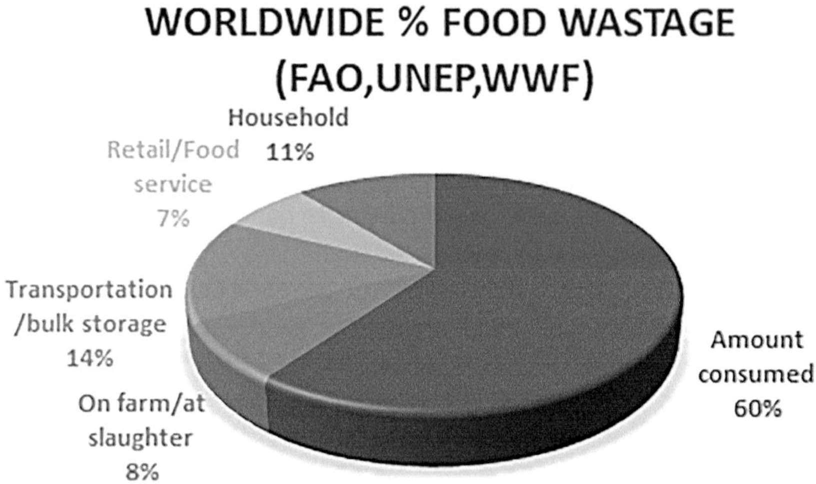

FIGURE 1.3 Global food wastage at various stages of food distribution chain.

worth to a product; however, a byproduct is a secondary product that is produced as a byproduct of the primary product's manufacture and frequently has a market worth. Before being sold, many byproducts need to be processed further (Castro-Muñoz et al., 2018). Therefore, if the necessary technical tools are accessible to produce value that surpasses the cost of reprocessing, the wastes might be termed valuable byproducts. In this situation, residues are no longer deemed to be wastes but rather a more valuable product. Utilizing the byproducts of food processing offers the possibility of turning them into useful applications. Agricultural production, harvesting, and processing in farm regions, as well as agricultural processing businesses including oilseed extraction, brewing, malt production, milling cereal grains, and fruit and vegetable processing, are the foremost sources of agro byproducts or agro wastes. These byproducts have a huge potential as a source of protein for animal feed and can be economically transformed into biofuels, bioenergy, and other goods.

1.2.1 TYPES OF AGRO BYPRODUCTS

A large quantity of trash and byproducts are produced during agricultural production and agro-industrial processing. More than 50% of fresh fruit is made up of byproducts such as bagasse, peels, trimmings, stems, shells, bran, and seeds, some of which occasionally have a higher nutritional or functional value than the finished product (Ayala et al., 2011). Damage that occurs during processing, storage, and transportation also results in wasted fruit and food. In recent years, byproducts and wastes have also increased due to the increasing demand of juices made from fruits, nectars, thawed food, and least-processed goods.

1.2.1.1 BYPRODUCTS OF PLANT ORIGIN

Plant origin is made up of processing companies' leftovers such as bran, leftover flour, trash from cleaning grain, wheat, maize, and rye germs, as well as the hulls of some seeds including pea plant, barleycorn, and buckwheat. Additionally, byproducts of the oil industry, such as cakes from oil-producing plants including soy, rape, sunflower, and flax, as well as lecithin and fatty acids, can be used to make animal feed. Significant quantities of protein and carbs can be found in gluten gums (Ajila et al., 2012). They are byproducts of the starch and sugar industries that consist of beet pulp, molasses, remains after deface-saturation, potato pulp, juice from potato cells, and different seed leftover after starch extraction.

1.2.1.2 BYPRODUCTS OF THE FERMENTATION INDUSTRY

The main ingredients of fermentation sector byproducts are grain, molasses, potato distillers soluble, brewers and wine yeast, as well as leftover grains and malt sprout in breweries. Lactic, citric, and other organic acids are also produced by these organisms as waste. A sticky byproduct called molasses is created when sugarcane or sugar beets are converted into sugar. According to Cheeke et al. (1991), molasses is one of the primary sources of fermentable carbohydrates and a convenient source of sugar and phosphate for the animal body. The urea molasses diet can be used as an addition to a low-protein diet (Ranjhan et al., 1993).

1.2.2 UTILIZATION OF AGRO BYPRODUCTS

Earlier, agriculturalists considered these leftovers as garbage and dumped them into the environment, causing contamination. However, they began to use it as animal feed after realizing the importance of these byproducts, the invulnerable costs of animal feed and fertilizers, and adverse effects on the environment. Crop leftovers can be used to get rid of items that could otherwise pose a risk to your health and the environment. Agricultural byproducts have a vital role in improving the nutritional status of various animal rations and feeds because they are abundant in the macro- and micronutrients needed for body growth and productivity (Gupta et al., 2012). Agricultural byproducts can be used in a variety of ways, such as fuel, livestock feed, agricultural inputs, and rural enterprises (Ayoo et al., 2019). The growth of rural economies may be supported and accelerated by increased use of agricultural byproducts. By being employed as per manure, bio-fertilizers, raw materials for generating energy, and consumer goods, crop residue may boost agricultural firms' profitability, the environment's quality, and energy security.

1.3 BYPRODUCTS AND WASTES OF FRUIT AND VEGETABLE PROCESSING

The farming industry and agro-industrial processing generate a lot of trash and byproducts. Fruit byproducts such as bagasse, peels, trimmings, stems, shells, bran, and seeds make up more than half of fresh fruit, and some of these byproducts may be more valuable nutritionally or functionally than the fruit itself (Ayala-Zavala et al., 2011). Food wastes and byproducts are extremely important in this situation because there are sufficient amounts of proteins, lipids, carbohydrates, micronutrients, bioactive compounds, and dietary fibers. Damage that occurs during processing, storage, and transportation also causes wasted fruit and food. Fruit juices, nectars, frozen foods, and items with little to no processing are becoming more and more popular, and it has caused the production of trash and byproducts to rise during the past few years. The production of garbage and byproducts has an effect on the social, economic, and environmental spheres. These have an impact on the environment by increasing GHG emissions (Girotto et al., 2015).

1.3.1 VEGETABLES BYPRODUCTS

Vegetable byproducts are the secondary products that are commonly wasted or lost during the production of food or other stages of food processing. For instance, most of the components of vegetables that are abandoned in the production process are the hulls, bagasse, and seeds; however, stems and leaves of several vegetables, including broccoli, cauliflower, and pumpkin, are not eaten and are instead discarded. The seed coat, which is the outer covering that covers the seeds or legumes and serves to shield them from outside danger, is one of the vegetable byproducts (Res et al., 2000). Apart from very tiny kernel coats, like peanut seed coats, most seed coats are not eaten because of their texture and flavor. Because 22% of the world's oilseeds and pulses are lost or squandered every year, a significant portion of seed coatings are thrown away in the canned food business (Faostat et al., 2015). In North Africa, West Asia, and Central Asia, oilseed and pulse losses are largest, and the majority of these losses occur during the agricultural stage (Faostat et al., 2015). The hard covering that covers the seeds or grains throughout the growth season, in contrast to the seed coat, is called the hull. Hulls are removed before being used in cooking or manufacturing because of how tough their texture is. Due to the substantial consumption of rice in Asian nations, hulls play a significant role in global food waste, particularly rice hulls (Toriyama et al., 2005). According to research, 30% of whole grains are lost, and post-harvest operations and human consumption account for the majority of these losses (Faostat et al., 2015). Hulls are used as fuel, fertilizers, and building materials, as one method to combat this issue and decrease waste (Kumar et al., 2013).

Vegetable peels are so frequently thrown away as kitchen or production trash in the manufacturing process. The consuming stage of tubers and roots results in the annual loss of up to 5,814,000 tons of roots and tubers in the areas of North America and Oceania. Additionally, 45% of tubers and roots are lost globally throughout the agricultural stage, with the largest losses occurring in Europe, North America, Oceania, and industrialized Asia (Faostat et al., 2015).

1.3.2 MANGO WASTE

Mango (*Mangifera indica* L.), a member of the Anacardiaceous family, is mostly farmed as a crop in tropical and subtropical areas. Due to its wonderful flavor, alluring smell, stunning color, and nutritional content, mango is one of the most significant fruits in the word, earning it the title of "the king of fruits." According to Robles Sánchez et al. (2013), they contain a lot of bioactive substances such as ascorbic acid, lupeol, carotene, and polyphenolic compounds. Mango is traditionally consumed as a dessert fruit, although use of mango products, such as preserved in containers, thawed concentrates, beverages, jams, mashed, dried, and minimally processed items has grown (Dussan-Sarria et al., 2014a, b; Masibo et al., 2009). Mango fruit processing produces a sizable number of byproducts, including peels (13–16%) and seeds (9.5–25%) (Sols et al., 2011; Torres León et al., 2017). According to estimates by O'shea et al. (2012), 60% of the fruit's weight is lost following processing. Mango seeds are an excellent source of carbohydrates (58–80%), protein (6–13%) with high profiles of essential amino acids, and lipids (6–16%) rich in oleic and stearic acids, according to this document (Siaka et al., 2014). Due to the fact that seed fat shares many physicochemical properties with commercial cocoa butter, it has garnered a lot of interest. Mango seed kernels have a 20% starch production and tapioca and marketable starch like properties. Mango skin contains between 51.2 and 78.4% of the recommended daily allowance of dietary fiber. Dietary fibers are indigestible carbohydrates found in plants and are regarded as a crucial part of a balanced diet for people.

1.3.3 PINEAPPLE

Originally from Central and South America, pineapples (*Ananas comosus*) are tropical fruits that have enticing sensory (mechanical properties, flavor, acidity/sweetness ratio, color) and nutritional (vitamin A, B, and C, minerals, fibers, and antioxidants) qualities. Bromalin is a group of enzymes found in pineapples that is frequently employed in food manufacturing. About 25,439,366 million tons of pineapples were produced globally in 2014. Thailand, Brazil, the Philippines, Costa Rica, and India are the top producers. In the same way, America (34.5%), Africa (16.4%), and Asia (48.2%) generate the bulk of the world's pineapples, respectively. During the processing, a lot of byproducts are created, including peel and pomace, which together make up around 30–35% of the edible pineapple fruit. An abundant source of total dietary fiber found in pineapple pomace (45.22%) is comparable to that found in mango peel. An agricultural waste with significant promise as a different source of starch is pineapple stem. Compared to commercial rice, maize, and cassava starches, the extracted starch shows distinctly distinctive characteristics.

1.3.4 BANANA

The banana is a fruit that is broadly farmed everywhere in the world, primarily in regions with tropical and subtropical climates. According to Mohapatra et al. (2010) and Tock et al. (2010), more than 130 nations cultivate bananas, with Brazil (Comim et al., 2010), Mexico (Rodriquez-Ambriz et al., 2008), India (Bankar et al., 2010b), Thailand (Wilaipon et al., 2009), Uganda, and Cameroon (Emaga et al., 2007) ranking among the top producers. Only 20 of the almost 300 species in the Musaceae family, which includes this fruit, are consumed (Venkateshwaran et al., 2010). Only 12% of a fruit's weight is edible, which leads to a lot of agro-industrial waste like peel that is typically utilized in industrial processes to create new items (chips, dried pulps, jams, wine, beer, and sauces). According to Babbar et al. (2011) and Bankar et al. (2010b), banana peels account for 30–40% of the fruit's overall weight. Due to their antioxidant properties, banana peels have a variety of uses, including as a feed for animals and poultry, a fertilizer, a source of ethanol and biogas, a way to extract banana oil, a way to filter heavy metals out of water, a way to produce proteins, a way to create biomass for use as energy, and a way to create nutraceuticals (Bankar et al., 2010b; González Montelongo et al., 2010).

High quantities of dietary fiber, primarily insoluble dietary fiber, can be found in the peel. The peel includes pectin, which is utilized extensively in industry due to its ability to create gels and serve as an emulsifier. Its fiber content is roughly 50% dry matter basis. Not only these, but a high percentage of minerals like Potassium, Zinc, Calcium, and so forth, present in the peels can be used. Dietary fibers have become more significant in current years due to their healthiness advantages, and as a result, the market is concentrating on the creation of new products with high fiber content. Dietary fibers are believed to promote the development of the healthy intestinal flora, which has been related to the prevention and treatment of a number of diseases, particularly those that impact the digestive system.

1.3.5 GRAPES

Fruits like grapes are healthy and benefit the economies of the nations that grow them. Juices, wine, raisins, table grapes, and other products are all made from these fruits. Because they contain antioxidant polyphenols like resveratrol, table grapes and wine have a variety of nutritional and physiological advantages for people. It is common to consume both raw and processed grape products, such as wine, jam, juice, jelly, dried grapes, vinegar, and grape seed oil. A total of 45.53% of the production, or 51,801.0 tons, is garbage. According to several investigations, the phenolic compounds in their peels have biological and pharmacological effects that are very interesting, including cardioprotective, neuroprotective, antibacterial, and antioxidant qualities. Another byproduct produced in the majority of wine industries is grape seed, which makes up little more than 5 to 6% of the weight of the grape bunch. Rich in tocopherols, which prevent their oxidation, and linoleic acid (60–70%), grape seed oil is distinguished by these compounds. Additionally, grape seed oil provides significant nutritional benefits for human health, including lowering cholesterol and lipid levels, avoiding the development of atheromatous lesions, and improving lipidaemia.

1.3.6 POMEGRANATE

This plant has many elongated leaves and a stem that is heavily branched, as well as bright red blooms that are packed with stamens (De la Cruz et al., 2015; Jurenka et al., 2008). Despite being a native of the Mediterranean, pomegranates have been widely cultivated throughout America, Asia, and Africa (Aguilar-Zárate et al., 2017). Pomegranate is a similarly significant fruit economically since it is either consumed fresh or used professionally to create juice, jam, and wine. As a result, up to 1,500,000 tons of pomegranates were produced worldwide. Additionally, it is useful for particular medical purposes. These characteristics are a result of the polyphenolic chemicals' high concentration and distinctive composition, particularly ETs. Furthermore, it has been proven that plant extracts from all plant parts, including the peel, tree bark, roots, and leaves, have medicinal properties and offer a range of health benefits (Jurenka et al., 2008).

1.3.7 CITRUS FRUITS

Citrus fruits are distinguished by their fleshy nature and high concentrations of citric acid, an organic tricarboxylic acid that gives them their distinctive acidic flavor (Rueda et al., 2007). Another distinctive element of this fruit family is vitamin C. Citrus, a popular antioxidant, is a rich source of this vitamin. There is a lot of fiber and minerals like calcium, magnesium, and potassium in addition to these important chemicals (Rezzadori et al., 2012). Citrus fruit production and consumption have seen rapid expansion since the middle of the 1980s. Fresh juice or other drinks are produced from one-third of citrus fruits.

1.3.8 TOMATO

The tomato's native land is South America's Andean area, specifically Peru, Ecuador, Bolivia, and Chile. However, Mexico is where it was domesticated. Since a sizeable portion of the economically active population is directly or indirectly tied to tomato agriculture, this crop has a very significant social impact. The majority of the leftovers from tomato processing are the seeds and peel. However, damaged entire fruits may also be thrown away. If these collected wastes are not properly treated, they could lead to environmental issues including foul odors, contaminated water, or the spread of pests. Additionally, this form of trash is produced at a specific time right after the end of the tomato harvest, thus a big amount of waste. It possesses a low caloric value of 17 kcal/100 g and is distinguished by having a high-water content (90–94%) and significant amount of soluble sugars (fructose, glucose, and sucrose). However, fewer amounts of amino acids, fiber, and natural acids (citric and malic), as well as a significant amount of vitamins (A and C), carotenoids, and mineral elements. For healthy growth and immune system support, vitamin A is a crucial component in the diet. Lycopene extracts can be used in a variety of applications due to its excellent antioxidant ability against singlet oxygen (O_2) and high capacity to scavenge free radicals.

1.3.9 UTILIZATION

Fruit and vegetable wastes are present throughout the supply chain and vary widely based on how they were processed. The amount of solid and liquid waste or byproducts generated during the processing of fruits and vegetables ranged from 10 to 60%. Waste material should be utilized in a way that maximizes benefits without resulting in financial loss or environmental issues. Therefore, waste products from these sectors can be processed either aerobically or anaerobically, producing biogas, animal feed, and value-added goods, as well as composting and landfilling.

1.3.9.1 BIOGAS GENERATION

Depending on how they were processed, fruit and vegetable wastes can be found throughout the supply chain and can vary considerably. Between 10% and 60% of the trash or byproducts generated during the processing of fruits and vegetables were both solid and liquid. Therefore, waste products from these sectors can be processed either aerobically or anaerobically, producing biogas by various reactions, composting, and landfilling.

1.3.9.2 ACTIVE PACKAGING

Recent research has shown that food waste byproducts can evolve into active packing, like pomace, peels, and seeds. Apple pomace, which contains phenolic acids and flavonoids, blueberry, and red grape skin pomace are a few examples of studies that utilized plant pomace. Anthocyanin is a natural, non-toxic, water-soluble pigment that acts as a good pH change indicator. Gutiérrez et al. produced a pH-sensitive film by combining unconventional trapezoidal arrowroot starch with grape waste (in the form of flour and extract) from the waste of the wine-making industry. Although chicken is a widely eaten meal in every country, its high protein and fat content makes it vulnerable to oxidation. Pectin from watermelon rind and kiwifruit peel extract were combined to develop a special active packaging film that may be used to package chicken thighs (Adilah et al., 2023).

1.3.9.3 FEED

One of the agricultural subsectors in emerging nations that is now growing at the highest rate is livestock. Recent research has demonstrated that processing fruit and vegetable leftovers may be utilized as valuable feed ingredients in farm animals' diets.

1.3.9.4 ALCOHOL PRODUCTION

Using dried active baker's yeast (*Saccharomyces cerevisiae*), ethanol synthesis from agricultural waste from *Carica papaya* (pawpaw) was studied. The mature pawpaw fruit that had been collected after papain tapping was the one that was deemed to be agricultural trash. With the aid of yeast, different grape skins and other fruit pomace are utilized to create ethanol (Akin-Osanaiye et al., 2008).

1.4 BYPRODUCTS AND WASTES OF MEAT, FISH AND POULTRY PROCESSING

Meat muscle, which is mostly attached to other muscle groups or an animal's skeletal system and is composed of fibers joined by connective tissue, is what we consume. Depending on the type and species, muscle contains 60% to 70% water, 10% to 20% protein, 2% to 22% fat, and 1% ash. When the bones are dissected cross-sectionally, it is simple to detect the muscle groups in bundles on bigger bones, such as the shanks of larger animals. Collagen fibers and a denser connective tissue called elastin, which together create a thin layer known as silver skin to divide the muscle groups or a tendon at the extremities of the muscle groups, surround the muscle groups. At a bone joint or close by, the tendon is joined to the bone. The building block of muscle fibers called a sarcostyle is composed of both thick and thin filaments that are piled next to one another in a recurrent pattern. One part of a bundle is a sarcomere, or small muscle. The thick filaments are composed of myosin, a contractile protein. Troponin and tropomyosin, two other proteins that help regulate muscle contraction, are also found in actin, the thin filaments. The lipids that are deposited all over particular regions of the animal prolong the meat's shelf life and improve its flavor and color when it is dry-aged. Intramuscular fat, also referred to as marbling, is a kind of fat that is present in the muscles of cows and appears as a pattern of wavy lines. In this cooking method, heavy collagen does not break down. Tendons at the ends of muscle groups and the protein elastin are two examples of heavy collagen. All meat animals have silver skin, tendons, and a thick, yellow section of collagen that runs down the top of the backbone from the base of the skull to the end of the rib cage. This section of collagen is frequently referred to as the "backstrap." The beef cuts that cook the fastest over a dry fire include the top sirloin, strip loin, and tenderloin from the cow's hind quarter. Since collagen dissolves into gelatin at temperatures exceeding 80°C (176°F), the blade, shoulder, and shank are tougher cuts from the front quarter of beef with more collagen connective tissue, necessitating a moist heat or mixed cooking method. Animal bones and connective tissue are used to make stocks that have body and thicken when refrigerated. Collagen dissolves in water. Bones, tendons, skin, gastrointestinal tract contents, blood, and internal organs are among the animal parts that are considered waste material.

1.4.1 TYPES OF MEAT BYPRODUCTS

The various meat byproducts produced and their contribution from different animal bodies are shown in Table 1.1, and the various byproducts are discussed in subsection.

TABLE 1.1 Meat Byproduct Produced and Their Contribution from Different Animal Body

Byproducts	Pigs (per 1000 g) (in g)	Cattle (per 1000 g) (in g)	Sheep (per 1000 g) (in g)
Hide and skin	60	60	150
Head	59	–	–
Viscera	100	160	110
Organs	70	160	100
Feet	20	20	20
Brain	1	1	2.6
Blood	30	180	24
Fatty tissue	30	40	30

Source: Liu et al. (2002).

1.4.1.1 CARCASS

Most animals are killed for their meat. Carcass makes up 63–77% of the overall weight of the body. Therefore, meat is the main product, while all other offal's are byproducts. There are edible and inedible components of these byproducts. Nearly 60% to 70% of the killed carcass is made up of byproducts, of which 20% are inedible and 40% are eatable (Ranganayaki et al., 1999). In sophisticated nations, banger casings are made of edible materials.

1.4.1.2 GASTRO-INTESTINAL TRACT AND OTHER ORGANS

Byproducts of the meat industry commonly make up more than one-eighth of the total protein in lean meat (Webster et al., 1982). There is a growing demand for protein hydrolysates despite the fact that it is theoretically and financially possible to utilize these byproducts in feeds and fertilizers after applying various technical processes, such as rendering, dry reduction, and so forth. These hydrolysis products can be used as nutritional supplements, functional ingredients, or taste enhancers in meals with low-quality proteins.

1.4.1.3 VISCERAL ORGANS

According to Ranganayaki et al. (1999) visceral mass, which includes the stomach and intestines, makes up almost 4% of the live weight of the animal that was killed. Although the majority of work has been done on the preparation of protein hydrolysates using plant, meat, and fish proteins as substrates, there has been very little research done on using bacterial and fungal proteases to solubilize the protein found in meat industry byproducts, such as that found in poultry. In order to improve the utilization of abattoir waste, this study describes a procedure for producing protein hydrolysates from the visceral mass of sheep. Also discussed are the properties of the protein hydrolysate made from sheep visceral mass.

1.4.1.4 BLOOD AND OTHER ORGANS

Tripe, blood, liver, lung, kidney, brains, and spleen are examples of byproducts with great nutritional value. Byproducts are used in medicine and pharmaceuticals. Among the most prominent applications for processed fish waste are animal feed, biodiesel/biogas, and dietetic products (chitosan).

1.4.2 UTILIZATION

India has a great wealth in livestock, which is expanding at a pace of 6% annually. The livestock industry, which contributes >40% of the entire agricultural sector and >12% of GDP, is growing significantly. This includes the poultry and fish industries. If the animal byproducts had also been used effectively, this contribution would have been significantly higher. Effective byproduct utilization directly affects the nation's economic and environmental damage. Meat, poultry, and fish processing wastes usually have the potential to be repurposed as raw materials or converted into useful items with higher value, in addition to eco-friendly and protection problems. When choosing to consume a beef byproduct, customs, culture, and religion are usually important considerations. Tripe, blood, liver, lung, kidney, brains, and spleen are examples of wastes with great nutritional importance. Animal feed, biodiesel/biogas, dietetic products (chitosan), natural colors (after extraction), and collagen in cosmetics are some of the most prominent uses for processed fish waste. This chapter has covered the information that is currently available on the use of waste products and byproducts from the processing of meat, poultry, and fish.

1.4.2.1 PROTEIN HYDROLYSATE

Protein hydrolysates are primarily used to regulate the diets of people who cannot digest full or intact proteins. Protein hydrolysates have been used most frequently to feed infants who have food hypersensitivity (Silvestre et al., 1997). Additionally, there are findings that enzymatically hydrolyzed proteins operate more effectively (Mahmood et al., 1992). According to research by Vijayalakshmi et al. (1986), protein hydrolysates rich in low molecular weight peptides, particularly di- and tri-peptides with as few free amino acids as feasible, have numerous dietary uses and have significant nutritional and therapeutic properties. The majority of the features of protein hydrolysates are influenced by the circumstances and sources of the protein (Mahmood et al., 1992; Silvestre et al., 1997). These can all be produced using meat byproducts.

1.4.2.2 NATURAL PIGMENTS AND COSMETICS

One of the best ways to obtain colors like myoglobin and other pigments from lean meat is by mixing blood and lean meat. Tendons and bone contain collagen, which can be harvested and used to improve the binding and emulsifying qualities of confections and beauty products.

1.4.2.3 FEED AND PET FOOD

Traditional feeds and pet meals have used raw or rendered animal byproducts as a component. Animal byproducts from the meat industry are rendered to provide high-quality fats and proteins on an annual basis of about 25 million tons in the US and 15 million tons in the EU (Hamilton et al., 2016). Animal byproducts are actually a good source of vitamins, minerals, trace elements, vital amino acids, fatty acids, B vitamins, and several fat-soluble vitamins.

1.4.2.4 ENERGY GENERATION

Biodiesel has recently been produced and is now gradually displacing diesel fuel because of its advantages, such as being biodegradable, non-toxic, and having a favorable combustion emission profile that results in reductions in carbon dioxide, carbon monoxide, particulate matter, and unburned hydrocarbons (Gerpen et al., 2005; Moreira et al., 2010). Additionally, using biodiesel does not require any engine adjustments.

1.4.2.5 MEDICAL AND PHARMACEUTICAL APPLICATION

Human burns or skin ulcers can be treated with pork skin. According to Toldrá et al. (2011), glands and organs are edible meat byproducts with high nutritional value that are consumed throughout the world. In fact, some of them are even consumed for medicinal purposes in nations like China, Japan, and India or are used as a source of specific pharmaceutical substances. Here, bile from the gall bladder and melatonin from the pineal gland are involved.

1.4.2.6 FERTILIZER APPLICATION

All nations produce significant amounts of meat and bone meal, and one intriguing method is thermochemical processing, which includes pyrolysis, combustion, and gasification. Pyrolysis and coal co-combustion receive the greatest analysis. The generated ashes exhibit a high phosphorus content that qualifies them for use as fertilizers, while the petrol emissions are compliant with international standards and contain combustibles that can be utilized to produce energy (Coutand et al., 2008).

1.4.2.7 CHEMICAL APPLICATIONS

The cosmetics sector uses rendered fats extensively in items like hand and body lotions, creams, and bath treatments. In the chemical industry, fatty acids are employed as plasticizers, softeners, lubricants, and catalysts in the polymerization of rubber and plastic. In the chemical industry, surfactants, paints, varnishes, adhesives, antifreeze, cleansers, and polishes all contain collagen, gelatin, and glycerin as active components (Pearl et al., 2004).

1.5 FISH

The head, tails, skin, intestines, fins, and frames are included in the solid byproducts of fish postprocessing. Between 30 and 50% of the flesh is routinely taken out of the rib cage and spine after filleting. Throughout whole fish filleting, more than 45% of the skin, 21–25% of the head, and 24–34% of the bones are wasted (Ghaly et al., 2013). According to Esteban et al. (2007) and Sutcliffe et al. (1977), these byproducts can be a significant source of proteins, amino acids, collagen, gelatin, oil, and a number of enzymes. Approximately 58% of the proteins, 19% of the lipids, and minerals come from fish faces. Additionally, fish waste contains a lot of monosaturated acids, including palmitic and oleic acids, which account for roughly 22% of the total.

1.5.1 TYPES AND UTILIZATION OF FISH BYPRODUCTS

1.5.1.1 MUSCLES

Enzymatic hydrolysis allows for the extraction of proteins and amino acids rather than their disposal as waste (Venugopal et al., 1996). Compared to other animal protein sources, proteins from fish provide a better balance of essential amino acids for the diet. Wet rendering is a particularly important process for extracting polyunsaturated fatty acids (PUFA) for use in the food and nutraceutical sectors. A fresh investigation on the energy industry's contribution of fish biodiesel production. On the other hand, fish protein hydrolysate requires an end product with a low lipid content and can therefore create lipid as a byproduct. Fish muscle proteins contain a number of peptides that exhibit a range of bioactivities, such as antihypertensive, antithrombotic, immune-modulating, and antioxidative characteristics (Choi et al., 2000). According to Je et al. (2004), the bioactive peptides extracted from fish muscle contain anticoagulant and antiplatelet activities.

1.5.1.2 SKIN AND HIVES

Gelatin and collagen, which are used in the food, cosmetic, and biomedical industries, are abundant in fish skin and bones. In contrast to sources from cattle, fish skin collagen and gelatin have no hazard of infuriated cow disease or bovine spongiform encephalopathy (BSE). Additionally, gelatin possesses a special glycine-proline-alanine amino acid sequence that gives it its antioxidative properties.

1.5.1.3 BONES

Researchers are also becoming more interested in using fish collagen and gelatin to modify mammalian collagen for use in food and pharmaceutical items. Because trash fish contains a higher proportion of bones, scales, and heads, its mineral content is typically higher. Calcium, phosphorus, and hydroxyapatite are among the 60–70% of minerals found in fish bones. It has also been extensively reported that fish skin scraps can be used to prepare protein hydrolysates. These include the Amur sturgeon Nikoo et al. (2015) which uses alkalize and flavor zyme, the Grass Carp (*Ctenopharyngodon idella*), which uses alkalize, the *Rachycentron canadum* (Yang et al., 2008) which uses bromelain, papain, pancreatin, and trypsin enzymes, and the *Lutjanus vitta* (Khantaphant et al., 2011).

1.5.1.4 INTERNAL ORGANS

Enzymes are abundant in the inner body part of fish. These enzymes, which are used to hydrolyze proteins, include pepsin, trypsin, chymotrypsin, and collagenase. They have improved catalytic characteristics, are more active at lesser temperatures and less sensitive to substrate concentrations, and are more stable over a wider pH range. In a wide pH range, fish myofibrillar proteins often have lower water solubility (Venugopal et al., 1994). Protein is gradually broken down by enzymes into smaller peptide units. For many dietary applications, fish protein hydrolysate's high solubility over a broad pH range is extremely helpful.

1.6 BYPRODUCTS AND WASTES OF CEREAL-BASED INDUSTRY

The exponential development in the requirement to meet the nutritional base of people in the period of anthropogenic waste creation has pushed the scientific community to examine rising topics and hot concerns about this global dilemma. In terms of global food security and nutrition, cereal grains provide a substantial contribution to the global food pool. Most cereals are a staple supply of various proteins, lipids, minerals, and vitamins, as well as an essential source of nutritional energy (Poole et al., 2021). Corn is the most farmed and utilized plant in the world, serving as a basic element in many gourmet cultures as well as being vital in animal feed, biofuel generation, and a variety of other industrial applications. Its output as a staple food is predicted to be above 1136.3 million metric tons (from September 2020 to August 2021), well exceeding wheat production (776.8 million metric tons) or rice production (504.4 million metric tons) (Zhang et al., 2021). According to the FAO's most recent updated worldwide prediction, overall cereal output in 2022 is expected to be 2774 million tons, 1.3% less than the previous year. Grain processing may be a sustainable option for transforming waste and byproducts into resources with value added for the food sector. The requirement for valorization is crucial for promoting the Circular Economy. This chapter's goal is to provide an overview of sustainable food industry applications resulting from the recovery and repurposing of underutilized grain processing waste and byproducts. Food waste must be recovered rather than disposed away, and this can only be done via the creation and widespread application of innovative solutions. Currently, only conventional methods are used on an industrial scale for producing animal feed, biofuels, or compost, which only partially utilizes waste from the processing of grain.

1.6.1 TYPES OF BYPRODUCTS

The yearly output of grain bran, a byproduct of grain milling, is estimated to be 150 million tons worldwide (Duţă et al., 2018) or 3–30% of the weight of the kernel in the case of dry milling. Additionally, hulls, husks, germ, broken grains, and powders (7–12%) are byproducts of the dry grinding of cereal grains. Cereal bran, which is the most popular of these byproducts for usage in food, is used mostly in bakery goods and was added with the express purpose of raising the nutritional fiber content by partially substituting the flour in bread, muffins, shortbread cookies, and cakes (Salazar-López et al., 2020).

1.6.1.1 BRAN

As a milling byproduct, maize bran (60–70 g/kg) is a cheap source of dietary fiber and inorganic dietary antioxidants (Hussain et al., 2021). When creating bread, some people see the high bran fiber content and the simultaneous presence of lipids and lipase in the germ as disadvantages. Utilizing fermented byproducts from the fermentation of lactic bacteria (*Lactobacillus plantarum* and *Weissella confusa*) as ingredients has improved the nutritional, textural, and sensory qualities of wheat bread (containing 25% fermented byproducts), resulting in a higher concentration of dietary fiber and proteins of 11 and 13% of the dry matter and a significantly higher increase in protein digestibility (up to 60%) (Hussain et al., 2021).

1.6.1.2 MAIZE BRAN

The food sector uses maize germs from both dry and wet milling operations to extract edible oil. Mechanical screw presses or a screw press and solvent extraction combination are used to collect edible oil from the maize germ fraction generated after dry milling (Papageorgiou et al., 2018). A better bread texture was shown in the case of bread fortification with maize germ protein hydrolysate (1%–4%) by lowering hardness and chewiness after storage (Karimi et al., 2021). Corn starch manufacturing byproducts that are high in fiber, protein, and antioxidants can also be added to food items as low-calorie, low-cost ingredients to partially replace fat or sugar (Version et al., 2020).

1.6.1.3 WHEAT BRAN

It is thought that adding wheat bran to flour may increase the nutritious content of baked goods. The fact that wheat bran contains a high quantity of insoluble dietary fibers, which will impact the quality of the bread goods and cause their color to darken after baking, is one disadvantage of this use of wheat bran. Their quality is to be coarse, and their volume to decrease. If too much wheat bran is introduced, it hinders the creation of the gluten network. However, by increasing the bran level up to 24%, the dough development time was extended. The benefit of adding wheat bran to flour is a reduction in the amount of electricity needed for mixing and the maximum power used. In terms of the rheology of the dough, the stickiness and extensibility reduced as the bran content rose. The addition of wheat bran to the flour increased the rate of fermentation and dough expansion during mixing.

1.6.1.4 WHEAT GERM

Wheat germ is another important byproduct of wheat milling. Due to its high fat content and susceptibility to rancidity and shortened shelf life, despite its high bioactive component concentration, it has been employed in food composition only infrequently (Abu-Ghannam et al., 2018). In a different research, wheat flour bread was fortified with 15% fermented milling byproducts (using *Lactobacillus plantarum* and *Lactobacillus rossiae*), using a dough made of wheat bran and germ, resulting in a product with 6.53% dietary fiber, or 5% more than in wheat flour bread (Pontonio et al., 2017). Fortified bread had a decreased glycemic index *in vitro* and, more significantly, *in vivo*, reaching 36.9%, much below the cutoff point needed for a food item to be classified as having a "low glycemic index" (Pontonio et al., 2017).

1.6.1.5 RICE BRAN AND GERM

The pericarp, seed coat, nucellus, aleurone layer, partial endosperm portions, and embryo make up the majority of the obtained rice bran from paddy rice after dehiscing and pearling. According to Hoed et al. (2006), rice husk accounts for 20% of byproducts, bran for 8%, and germ for 2% of paddy grains. Byproducts of rice milling are often utilized as animal feeds, either on their own or combined with other feeds. The byproducts of rice are now being sought after for use in culinary and medicinal applications. According to Cicero et al. (2001), rice bran and its byproducts may be helpful in preventing cardiovascular disorders. Because rice bran contains significant levels of micronutrients such oryzanols, tocopherols, tocotrienols, and phytosterols, it is a desirable substrate for their separation as nutraceuticals or functional food components (Hernandez et al., 2000).

1.7 BYPRODUCTS AND WASTES OF DAIRY INDUSTRY

The manufacture of many products (sterilized and pasteurized milk and processed dairy products) is the foundation of the dairy sector (Ahmad et al., 2019; Rivas et al., 2010). It is possible to distinguish between dairy companies' byproducts that may be eaten (such as whey, ice cream, cheese, butter, and milk residues) and dairy industry wastes that cannot be categorized as food, such as expired, contaminated, or ruined dairy goods. Although wastes and byproducts may both be used in microbial processes like anaerobic digestion (AD), it seems more feasible to use byproducts for the production of food goods like lactic acid or biomass. The production processes also result in the discharge of a number of byproducts from the processing of dairy products, including whey, spilt milk, curd chunks, and skim milk (Ahmad et al., 2019). A possible supply of lactose, protein, and fat is found in dairy industry byproducts. According to the Food and Agriculture Organization (FAO), 1.3 billion tons of the food produced for human use are misplaced or wasted. Reports state that a range of factors result in the loss of about 18% of milk output. Large volumes of whey (9–10 L for 1 kg of cheese production), in addition to this excessive amount of milk, are also discharged throughout the cheese-making process. The amount of whey, which has a greenish-yellow color and varies based on the milk source (cow, goat, sheep, buffalo, and other animals), as well as the cheese-making process (resulting in acid, sweet, and salty whey)

(Blaschek et al., 2007). Although secondary cheese whey is produced throughout the process, the original whey is still utilized to make ricotta and cottage cheeses. Whey may also be ultrafiltered to produce numerous byproducts, primarily for use as food ingredients (Atra et al., 2005). These byproducts include whey protein, lactose, and whey powder. As a result of its high lactose and other nutrient contents, whey powder outperforms whey in microbial processes (Dragone et al., 2011).

1.7.1 UTILIZATION

1.7.1.1 BIOMETHANE

The generation of biogas by AD can utilize dairy sector byproducts with high chemical oxygen demand (COD) as a suitable feedstock (Kozowski et al., 2019). According to Wainaina et al. (2019), the key phases of AD are the hydrolysis of complex materials into soluble materials, the formation of volatile fatty acids (VFAs) and hydrogen, and the conversion of VFA and hydrogen into biogas. The bacteria responsible for the hydrolysis are *Acetivibrio, Bacillus, Peptococcus,* and *Vibrio*. According to Wainaina et al. (2019), several methanogenic bacteria and archaea are also biogas producers in the AD system, in addition to *Butyribacterium, Propionibacterium,* and *Clostridium*. From the perspective of a circular economy, generated biogas may be used as a thermal energy source for steam production in the cheese-making process. The wastewater treatment unit can utilize it as a source of electrical power in any case (Comino et al., 2009). Additionally, the anaerobic digesters' solid and liquid fractions include beneficial elements that may be used as soil fertilizers (Kavacik et al., 2010). As a result, the AD strategy not only reduces energy consumption but also controls wastewater pollution caused by the dairy sector (Asunis et al., 2020; Hublin et al., 2012).

1.7.1.2 BIOHYDROGEN

For the generation of hydrogen, dairy waste, whey, whey powder, and milk powder have been utilized (Chandra et al., 2018; Karadag et al., 2014). Theoretically, a unit of lactose consumption may provide 8 mol of hydrogen; however, current technology can only create 3 mol of hydrogen

(Ryan et al., 2016; Sherif et al., 2014). It is challenging for bacteria that produce hydrogen to hydrolyze protein-rich dairy wastes (Okamoto et al., 2000).

1.7.1.3 ORGANIC ACIDS

The useful microbial compounds known as organic acids have acidic characteristics. According to Louasté et al. (2020) and Panday et al. (2016), the most prevalent organic acids are acetic, lactic, propionic, succinic, citric, oxalic, valeric, and capric acids. These acids, which are also known as carboxylic acids, are essential raw materials for several industries, and their demand has been gradually growing on a global scale. Finding accessible, affordable, and renewable resources like dairy wastes or high yield organic acid manufacturing grabs researcher's interest.

1.7.1.4 LACTIC ACID

A suitable medium for LAB culture may be made from waste dairy products like skim milk and whey. Numerous studies have demonstrated that when bacterial strains *(Lactobacillus casei, L. paracasei subsp. tolerans, L. rhamnosus, L. lactis, and Streptococcus thermophiles)* were cultured in whey media, the L (+) isomer of lactic acid was generated instead of the D (-) isomer of lactic acid (Ghasemi et al., 2017; Pescuma et al., 2015). The L (+) lactic acid possesses superior characteristics that make it acceptable for usage in the food and pharmaceutical sectors. Many types of milk, including those from cows, goats, sheep, and camels, naturally contain lactic acid bacteria that are utilized to make probiotics (Ao et al., 2012; Fguiri et al., 2016). Additionally, dairy products including yoghurt, ice cream, cheeses, sweet creams, and sweet karma's (a type of cream) have been shown to contain lactic acid bacteria (Góral et al., 2018; Vidojevic et al., 2014). Lactic acid has been created for this usage using cheese whey and old dairy products like milk.

1.7.1.5 CITRIC ACID

The citric acid industry will improve, which will help with waste disposal issues as well as lessen industry dependence on other manufacturers. As a result, the citric acid sector will reap both financial and ecological advantages

(Sawant et al., 2021). A combination of whey and other substrates can be used to regenerate the citric acid that is added as an emulsifier, acidifier, or antioxidant during the preparation of ice creams and cheese products (Soccol et al., 2006; Yalcin et al., 2009). As a consequence, the citric acid made by the dairy sector may be utilized again to make dairy products, which might lower overall costs in the cheese manufacturing unit.

1.7.1.6 ENZYME PRODUCTION

Enzyme production at an industrial scale began in the 1960s and expanded quickly. Microorganisms that have undergone genetic modification produce the majority of industrial enzymes. One of the less expensive sources for the manufacture of enzymes is byproducts from the dairy industry (Ryan et al., 2016). Additionally, creating enzymes from these substrates could be a long-term solution to the dairy industry pollution issue. Some bacteria, yeasts, and molds may grow well on lactose-containing wastes and produce a variety of enzymes, including β-galactosidase, α-amylase, protease, and manganese peroxidase (Romero et al., 2001).

1.7.1.6.1 Β-galactosidase

In-depth research has been done on the possible use of whey among dairy wastes for the synthesis of enzymes (Alves et al., 2019; Bosso et al., 2019). The whey's original pH and incubation temperature may affect how quickly β-galactosidase is produced (Alves et al., 2019). Another disadvantage of lactase synthesis is that it increases the cost of downstream processing due to its intracellular nature. To this end, several sorts of study were carried out to discover lactase purification techniques that were easy to use, effective, and economical (Bansal et al., 2008).

1.7.1.6.2 Α-amylase

The commercial enzyme α-amylase is an alternative one that may be made from dairy waste. Wheat bran, cheese whey, soybean cake, and hazelnut oil cake are examples of agro-dairy wastes that may be used to produce α-amylase by a semi-solid-state fermentation (Selen et al., 2017).

Anoxybacillus beppuensis may further manufacture a thermally stable amylase from agro-dairy wastes, which are a combination of sugarcane bagasse and whey (Jabeen et al., 2019). In addition, dairy wastes combined with orange peel byproducts from the fruit sector can be utilized to make β-amylase (Uygut et al., 2018).

1.7.1.6.3 Proteases

Proteases are hydrolytic enzymes that are widely employed in many different industries, such as the food, pharmaceutical, fabric, and detergent sectors. Their manufacturing from dairy waste is a viable method for resource recovery and biotechnology. When *Serratia marcescens* was cultivated on whey, it was examined if the different strains generated extracellular metalloprotease (Romero et al., 2001). *Aspergillus* and *Mucor* are two examples of fungi that may grow in whey and generate the protease enzyme (El Shora et al., 2008).

1.8 OTHER BYPRODUCT AND WASTE

1.8.1 BUTTERMILK

Recent years have seen an increase in milk production in emerging nations, mostly as a result of the rise in the number of animals raised for this purpose as a result of the global population growth. Due to their greater purchasing power and higher per capita consumption, industrialized nations currently account for the majority of the world's dairy consumption. Nearly 656 million tons of dairy products are produced globally each year. Whey is a byproduct of the dairy business that is produced in vast amounts in America and Africa but has no additional value. These nations' biggest source of pollution is whey, which they produce in approximately 1 million tons and which contains 5,000 tons of real protein and 50,000 tons of lactose. Although the availability of nutrient-rich substances that may be helpful, 47% of whey is discharged into the drainage system, which has a significant detrimental effect on soil and river pollution. The nutritional value of whey can vary greatly based on the characteristics of the milk used to make the cheese, which also depend on the kind of cheese made and the biotechnological method utilized. Thus, they may be divided into the two most popular types: acid and sweet. Sweet whey is produced during the cheese-making process using proteolytic enzymes or rennet, which cause milk caseins to destabilize and precipitate under specific temperature conditions of approximately 15–20°C and a pH that is slightly acidic. Its constituents include lactose (46–52 g/L), proteins (6–10 g/L), fats (5 g/L), calcium (0.4–0.6 g/L), and phosphorous (1.4–1.6 g/L). Mexico produced around 11,607,493 L of milk in 2016, producing an estimated 5,874,905 L of buttermilk as a byproduct (SIAP, 2017). Mexico ranked ninth in the world in terms of milk production. The key product in the dairy sector is cheese, which requires around 25% of the total global output to make. The primary byproduct of the dairy sector is the buttermilk required to make cheese, which maintains about 55% of the milk's natural components.

The agro-industrial sector is interested in items produced through biotechnological processing of whey, such as fermented drinks, organic acids (lactic and propionic acid), and microbial protein. These goods are produced using microbes that can absorb the lactose, one of the nutrients found in buttermilk, reducing the pollution risk while producing beneficial products (Quintero et al., 2001). The main advantages of whey proteins, which all have various biological functions, include the prevention of cancer (breast, colon, and prostate), an increase in glutathione levels, an increase in satiety, antiviral and antimicrobial activities, prebiotic effects, and immunomodulatory effects (Hernández et al., 2014).

The need to enhance milk's qualitative attributes is developing (Drapala et al., 2016). Bovine milk differs from human milk in several ways, including the amount of casein, lactose, mineral salts, and notably the proportion of proteins present (beta lactoglobulin is absent from breast milk). Bovine milk byproducts are a rich source of raw materials, and each one of them may be used in some form (Hernández et al., 2014). In addition to being utilized as a supplement in the manufacturing of bread (Visetnn et al., 2014) and confectionery items, some of the whey generated by the dairy sector is used in animal feed, yoghurt, fermented drinks (Parra Huertas et al., 2010) and other products.

1.9 CONCLUSION

The development of long-term solutions for handling byproducts and food waste is one of the primary problems that our society is now confronting.

For these solutions to have a beneficial effect on the economy, society, and environment, they must be able to optimize the biological potential of biomaterials. Given the nutritional problems that contemporary society is facing (hunger indices and the growing world population), using food waste for human sustenance should be of the utmost concern. In their formulation, wastes and byproducts produced in developing countries can be used for both nutritional and functional reasons, as well as a potent tool to eliminate hunger. Additionally, the diversity of the supply chains' added value might lead to job opportunities for locals, resulting in a further social benefit.

KEYWORDS

- **agro byproducts**
- **byproducts**
- **circular economy**
- **food sustainability**
- **food waste management**
- **grain processing**

REFERENCES

Abu-Ghannam, N., & Balboa, E. (2018). Biotechnological, food, and health care applications. In *Clinical Principles of Transfusion Medicine.* Elsevier. https://doi.org/10.1016/B978-0-08-102162-0.00009-5 (accessed on 11th August 2025).

Aguilar-Zárate, P., Wong-Paz, J. E., Michel, M., Buenrostro-Figueroa, J., Díaz, H. R., Ascacio, J. A., & Aguilar, C. N. (2017). Characterization of pomegranate-husk polyphenols and semi-preparative fractionation of punicalagin. *Phytochemical Analysis, 28*(5), 433–438.

Ahmad, T., Aadil, R. M., Ahmed, H., ur Rahman, U., Soares, B. C., Souza, S. L., & Cruz, A. G. (2019). Treatment and utilization of dairy industrial waste: A review. *Trends in Food Science & Technology, 88,* 361–372.

Ajila, C. M., Brar, S. K., Verma, M., Tyagi, R. D., Godbout, S., & Valéro, J. R. (2012). Bio-processing of agro-byproducts to animal feed. *Critical Reviews in Biotechnology, 32*(4), 382–400. https://doi.org/10.3109/07388551.2012.659172 (accessed on 11th August 2025).

Akin-Osanaiye, B. C., Nzelibe, H. C., & Agbaji, A. S. (2008). Ethanol production from *Carica papaya* (pawpaw) fruit waste. *Asian Journal of Biochemistry, 3*(3), 188–193.

Ao, X., Zhang, X., Shi, L., Zhao, K., Yu, J., Dong, L., & Cai, Y. (2012). Identification of lactic acid bacteria in traditional fermented yak milk and evaluation of their application in fermented milk products. *Journal of Dairy Science, 95*(3), 1073–1084.

Asunis, F., De Gioannis, G., Dessì, P., Isipato, M., Lens, P. N., Muntoni, A., & Spiga, D. (2020). The dairy biorefinery: Integrating treatment processes for cheese whey valorization. *Journal of Environmental Management, 276,* 111240.

Atra, R., Vatai, G., Bekassy-Molnar, E., & Balint, A. (2005). Investigation of ultra- and nanofiltration for utilization of whey protein and lactose. *Journal of Food Engineering, 67*(3), 325–332.

Ayala-Zavala, J. F., Vega-Vega, V., Rosas-Domínguez, C., Palafox-Carlos, H., Villa-Rodriguez, J. A., Siddiqui, M. W., Dávila-Aviña, J. E., & González-Aguilar, G. A. (2011). Agro-industrial potential of exotic fruit byproducts as a source of food additives. *Food Research International, 44*(7), 1866–1874. https://doi.org/10.1016/j.foodres.2011.02.021 (accessed on 11th August 2025).

Ayoo, C., & Bonti-Ankomah, S. (2019). Economic impacts of value addition to agricultural byproducts. In *Byproducts from Agriculture and Fisheries: Adding Value for Food, Feed, Pharma and Fuels* (pp. 685–698). https://doi.org/10.1002/9781119383956.ch31 (accessed on 11th August 2025).

Babbar, N., Oberoi, H. S., Uppal, D. S., & Patil, R. T. (2011). Total phenolic content and antioxidant capacity of extracts obtained from six important fruit residues. *Food Research International, 44*(1), 391–396.

Bankar, A., Joshi, B., Kumar, A. R., & Zinjarde, S. (2010). Banana peel extract mediated novel route for the synthesis of silver nanoparticles. *Colloids and Surfaces A: Physicochemical and Engineering Aspects, 368*(1–3), 58–63.

Bansal, S., Oberoi, H. S., Dhillon, G. S., & Patil, R. T. (2008). Production of β-galactosidase by *Kluyveromyces marxianus* MTCC 1388 using whey and effect of four different methods of enzyme extraction on β-galactosidase activity. *Indian Journal of Microbiology, 48,* 337–341.

Bhaskar, N., Modi, V. K., Govindaraju, K., Radha, C., & Lalitha, R. G. (2007). Utilization of meat industry byproducts: Protein hydrolysate from sheep visceral mass. *Bioresource Technology, 98*(2), 388–394.

Blaschek, K. M., Wendorff, W. L., & Rankin, S. A. (2007). Survey of salty and sweet whey composition from various cheese plants in Wisconsin. *Journal of Dairy Science, 90*(4), 2029–2034.

Bosso, A., Setti, A. C. I., Tomal, A. B., Guemra, S., Morioka, L. R. I., & Suguimoto, H. H. (2019). Substrate consumption and beta-galactosidase production by *Saccharomyces fragilis* IZ 275 grown in cheese whey as a function of cell growth rate. *Biocatalysis and Agricultural Biotechnology, 21,* 101335.

Castro-Muñoz, R., Barragán-Huerta, B. E., Fíla, V., Denis, P. C., & Ruby-Figueroa, R. (2018). Current role of membrane technology: From the treatment of agro-industrial byproducts up to the valorization of valuable compounds. *Waste and Biomass Valorization, 9*(4), 513–529. https://doi.org/10.1007/s12649-017-0003-1 (accessed on 11th August 2025).

Chandra, R., Castillo-Zacarias, C., Delgado, P., & Parra-Saldívar, R. (2018). A biorefinery approach for dairy wastewater treatment and product recovery towards establishing a biorefinery complexity index. *Journal of Cleaner Production, 183,* 1184–1196.

Cheeke, P. R. (1991). Cereal milling byproducts. In *Applied Animal Nutrition: Feeds and Feeding* (pp. 53–57). New York: Macmillan.

Choi, Y. R., Park, P. J., Choi, J. H., Moon, S. H., & Kim, S. K. (2000). Screening of biofunctional peptides from cod processing wastes. *Applied Biological Chemistry, 43*(3), 225–227.

Cicero, A. F. G., & Gaddi, A. (2001). Rice bran oil and γ-oryzanol in the treatment of hyperlipoproteinemia's and other conditions. *Pharmacological Research, 44*(4), 277–289.

Comim, S. R., Madella, K., Oliveira, J. V., & Ferreira, S. R. S. (2010). Supercritical fluid extraction from dried banana peel (*Musa* spp., genomic group AAB): Extraction yield, mathematical modeling, economic analysis and phase equilibria. *The Journal of Supercritical Fluids, 54*(1), 30–37.

Comino, E., Rosso, M., & Riggio, V. (2009). Development of a pilot scale anaerobic digester for biogas production from cow manure and whey mix. *Bioresource Technology, 100*(21), 5072–5078.

Coutand, M., Cyr, M., Deydier, E., Guilet, R., & Clastres, P. (2008). Characteristics of industrial and laboratory meat and bone meal ashes and their potential applications. *Journal of Hazardous Materials, 150*(3), 522–532.

de la Cruz, R., Ascacio, J. A., Buenrostro, J., Sepúlveda, L., Rodríguez, R., Prado-Barragán, A., & Aguilar, C. N. (2015). Optimization of ellagitannase production by *Aspergillus niger* GH1 by solid-state fermentation. *Preparative Biochemistry and Biotechnology, 45*(7), 617–631.

Devine, C., & Jensen, W. K. (2004). *Encyclopedia of Meat Sciences.* Academic Press.

Diarra, S. S. (2014). Potential of mango (*Mangifera indica* L.) seed kernel as a feed ingredient for poultry: A review. *World's Poultry Science Journal, 70*(2), 279–288.

Dou, Z., Ferguson, J. D., Galligan, D. T., Kelly, A. M., Finn, S. M., & Giegengack, R. (2016). Assessing U.S. food wastage and opportunities for reduction. *Global Food Security, 8,* 19–26. https://doi.org/10.1016/j.gfs.2016.02.001 (accessed on 11th August 2025).

Dragone, G., Mussatto, S. I., e Silva, J. B. A., & Teixeira, J. A. (2011). Optimal fermentation conditions for maximizing the ethanol production by *Kluyveromyces fragilis* from cheese whey powder. *Biomass and Bioenergy, 35*(5), 1977–1982.

Duţă, D. E., Culeţu, A., & Mohan, G. (2018). Reutilization of cereal processing byproducts in bread making. In *Sustainable Recovery and Reutilization of Cereal Processing Byproducts.* https://doi.org/10.1016/B978-0-08-102162-0.00010-1 (accessed on 11th August 2025).

El-Shora, H. M., & Metwally, M. A. A. (2008). Production, purification and characterization of proteases from whey by some fungi. *Annals of Microbiology, 58,* 495–502.

Emaga, T. H., Andrianaivo, R. H., Wathelet, B., Tchango, J. T., & Paquot, M. (2007). Effects of the stage of maturation and varieties on the chemical composition of banana and plantain peels. *Food Chemistry, 103*(2), 590–600.

Esteban, M. B., García, A. J., Ramos, P., & Márquez, M. C. (2007). Evaluation of fruit–vegetable and fish wastes as alternative feedstuffs in pig diets. *Waste Management, 27*(2), 193–200.

Fguiri, I., Ziadi, M., Atigui, M., Ayeb, N., Arroum, S., Assadi, M., & Khorchani, T. (2016). Isolation and characterization of lactic acid bacteria strains from raw camel milk for potential use in the production of fermented Tunisian dairy products. *International Journal of Dairy Technology, 69*(1), 103–113.

Food and Agriculture Organization of the United Nations. (2015). *Food and Agriculture Organization of the United Nations: Rome.*

Food and Agriculture Organization of the United Nations. (2019). *Moving Forward on Food Loss and Waste Reduction: The State of Food and Agriculture 2019.*

Ghaly, A. E., Ramakrishnan, V. V., Brooks, M. S., Budge, S. M., & Dave, D. (2013). Fish processing wastes as a potential source of proteins, amino acids and oils: A critical review. *Journal of Microbial & Biochemical Technology, 5*(4), 107–129.

Ghasemi, M., Ahmad, A., Jafary, T., Azad, A. K., Kakooei, S., Daud, W. R. W., & Sedighi, M. (2017). Assessment of immobilized cell reactor and microbial fuel cell for simultaneous cheese whey treatment and lactic acid/electricity production. *International Journal of Hydrogen Energy, 42*(14), 9107–9115.

Girotto, F., Alibardi, L., & Cossu, R. (2015). Food waste generation and industrial uses: A review. *Waste Management, 45,* 32–41. https://doi.org/10.1016/j.wasman.2015.06.008 (accessed on 11th August 2025).

González-Montelongo, R., Lobo, M. G., & González, M. (2010). Antioxidant activity in banana peel extracts: Testing extraction conditions and related bioactive compounds. *Food Chemistry, 119*(3), 1030–1039.

Gooch, M., Bucknell, D., LaPlain, D., Dent, B., Whitehead, P., Felfel, A., & Maguire, M. (2019). *The Avoidable Crisis of Food Waste: Technical Report.* Value Chain Management International and Second Harvest, Ontario, Canada.

Góral, M., Kozłowicz, K., Pankiewicz, U., & Góral, D. (2018). Magnesium enriched lactic acid bacteria as a carrier for probiotic ice cream production. *Food Chemistry, 239,* 1151–1159.

Gupta, S. K. (2012). *Technological Innovations in Major World Oil Crops, Volume 2: Perspectives* (Vol. 2). https://doi.org/10.1007/978-1-4614-0827-7 (accessed on 11th August 2025).

Gustavsson, J., Cederberg, C., Sonesson, U., Van Otterdijk, R., & Meybeck, A. (2011). *Global Food Losses and Food Waste.*

Hamilton, C. R. (2016). Real and perceived issues involving animal proteins. *Protein Sources for The Animal Feed Industry.*

Hernandez, N., & Gonzalez, F. (2000). Enzymatic treatment of rice bran to improve processing. *Cereal Chemistry, 77*(2), 177–180.

Hernández-Rojas, M., & Vélez-Ruiz, J. F. (2014). Suero de leche y su aplicación en la elaboración de alimentos funcionales. *Temas Selectos de Ingeniería de Alimentos, 8*(2), 13–22.

Hoed, V. van, Depaemelaere, G., Ayala, J. V., Santiwattana, P., & De Greyt, W. (2006). Influence of chemical refining on the major and minor components of rice bran oil. *Journal of the American Oil Chemists' Society, 83*(4), 315–321.

Hublin, A., Zokić, T. I., & Zelić, B. (2012). Optimization of biogas production from co-digestion of whey and cow manure. *Biotechnology and Bioprocess Engineering, 17,* 1284–1293.

Hussain, M., Saeed, F., Niaz, B., Afzaal, M., Ikram, A., Hussain, S., Mohamed, A. A., Alamri, M. S., & Anjum, F. M. (2021). Biochemical and nutritional profile of maize bran-enriched flour in relation to its end-use quality. *Food Science & Nutrition, 9*(6), 3336–3345. https://doi.org/10.1002/fsn3.2323 (accessed on 11th August 2025).

Jabeen, F., Hussain, A., Younis, T., Manzoor, M., & Samiullah, K. (2019). Isolation of thermophilic *Anoxybacillus beppuensis* JF84 and production of thermostable amylase utilizing agro–dairy wastes. *Environmental Progress & Sustainable Energy, 38*(2), 417–423.

Jayathilakan, K., Sultana, K., Radhakrishna, K., & Bawa, A. S. (2012). Utilization of byproducts and waste materials from meat, poultry and fish processing industries: A review. *Journal of Food Science and Technology, 49,* 278–293.

Je, J. Y., Park, P. J., Kwon, J. Y., & Kim, S. K. (2004). A novel angiotensin I converting enzyme inhibitory peptide from Alaska pollack (*Theragra chalcogramma*) frame protein hydrolysate. *Journal of Agricultural and Food Chemistry, 52*(26), 7842–7845.

Jurenka, J. (2008). Therapeutic applications of pomegranate (*Punica granatum* L.): A review. *Alternative Medicine Review, 13*(2), 128–144.

Karadag, D., Köroğlu, O. E., Ozkaya, B., Cakmakci, M., Heaven, S., & Banks, C. (2014). A review on fermentative hydrogen production from dairy industry wastewater. *Journal of Chemical Technology & Biotechnology, 89*(11), 1627–1636.

Karimi, A., Ahmadi, H., Amini, R., & Udenigwe, C. C. (2021). Effect of maize germ protein hydrolysate addition on digestion, *in vitro* antioxidant activity and quality characteristics of bread. *Journal of Cereal Science, 97,* 103148. https://doi.org/10.1016/j.jcs.2020.103148 (accessed on 11th August 2025).

Kavacik, B., & Topaloglu, B. (2010). Biogas production from co-digestion of a mixture of cheese whey and dairy manure. *Biomass and Bioenergy, 34*(9), 1321–1329.

Khantaphant, S., Benjakul, S., & Ghomi, M. R. (2011). The effects of pretreatments on antioxidative activities of protein hydrolysate from the muscle of brown stripe red snapper (*Lutjanus vitta*). *LWT–Food Science and Technology, 44*(4), 1139–1148.

Kozłowski, K., Pietrzykowski, M., Czekała, W., Dach, J., Kowalczyk-Juśko, A., Jóźwiakowski, K., & Brzoski, M. (2019). Energetic and economic analysis of biogas plant with using the dairy industry waste. *Energy, 183,* 1023–1031.

Kumar, S., Sangwan, P., V, D. R. M., & Bidra, S. (2013). Utilization of rice husk and their ash: A review. *Research Journal of Chemical and Environmental Sciences, 1*(5), 126–129.

Liu, D. (2002). Better utilization of byproducts from the meat industry. *FFTC Publication Database,* 1–15. http://www.agnet.org/library.php?func=view&id=20110706135001&type_id=4 (accessed on 11th August 2025).

Liu, D. C. (2002). *Better Utilization of Byproducts from the Meat Industry.* Taipei, Taiwan: Food and Fertilizer Technology Center.

Louasté, B., & Eloutassi, N. (2020). Succinic acid production from whey and lactose by *Actinobacillus succinogenes* 130Z in batch fermentation. *Biotechnology Reports, 27,* e00481.

Mahmoud, M. I., Malone, W. T., & Cordle, C. T. (1992). Enzymatic hydrolysis of casein: Effect of degree of hydrolysis on antigenicity and physical properties. *Journal of Food Science, 57*(5), 1223–1229.

Masibo, M., & He, Q. (2009). Mango bioactive compounds and related nutraceutical properties—A review. *Food Reviews International, 25*(4), 346–370.

Mohapatra, D., Mishra, S., & Sutar, N. (2010). Banana and its byproduct utilization: An overview. *Journal of Scientific & Industrial Research, 69*(5), 323–329.

Moreira, A. L., Dias, J. M., Almeida, M. F., & Alvim-Ferraz, M. C. (2010). Biodiesel production through transesterification of poultry fat at 30 °C. *Energy & Fuels, 24*(10), 5717–5721.

Nikoo, M., Benjakul, S., & Xu, X. (2015). Antioxidant and cryoprotective effects of Amur sturgeon skin gelatin hydrolysate in unwashed fish mince. *Food Chemistry, 181,* 295–303.

Okamoto, M., Miyahara, T., Mizuno, O., & Noike, T. (2000). Biological hydrogen potential of materials characteristic of the organic fraction of municipal solid wastes. *Water Science and Technology, 41*(3), 25–32.

Pandey, A., Negi, S., & Soccol, C. R. (2016). *Current Developments in Biotechnology and Bioengineering: Production, Isolation and Purification of Industrial Products.* Elsevier.

Papageorgiou, M., & Skendi, A. (2018). Introduction to cereal processing and byproducts. In *Sustainable Recovery and Reutilization of Cereal Processing Byproducts.* Elsevier. https://doi.org/10.1016/B978-0-08-102162-0.00001-0 (accessed on 11th August 2025).

Parra Huertas, R. A. (2010). Bacterias ácido lácticas: Papel funcional en los alimentos. *Biotecnología en el Sector Agropecuario y Agroindustrial, 8*(1), 93–105.

Pescuma, M., de Valdez, G. F., & Mozzi, F. (2015). Whey-derived valuable products obtained by microbial fermentation. *Applied Microbiology and Biotechnology, 99,* 6183–6196.

Pontonio, E., Lorusso, A., Gobbetti, M., & Rizzello, C. G. (2017). Use of fermented milling byproducts as functional ingredient to develop a low-glycemic index bread. *Journal of Cereal Science.* https://doi.org/10.1016/j.jcs.2017.08.022 (accessed on 11th August 2025).

Poole, N., Donovan, J., & Erenstein, O. (2021). Viewpoint: Agri-nutrition research: Revisiting the contribution of maize and wheat to human nutrition and health. *Food Policy, 100,* 101976. https://doi.org/10.1016/j.foodpol.2020.101976 (accessed on 11th August 2025).

Quintero, H., Marval, M. R., Páez, G., Ferrer, J., Mármol, Z., & Rincón, M. (2001). Producción continua de proteína microbiana (*K. fragilis*) a partir de suero de leche. *Revista Científica de la Facultad de Ciencias Veterinarias, 11*(2), 87–95.

Ranjhan, S. K. (1993). Agro-industrial byproducts as component of livestock rations. In *Animal Nutrition in The Tropics* (pp. 222–258). New Delhi, India: Vikas Publishing House PVT Ltd.

Res, J. P., Beeckman, T., De Rycke, R., Mane, R., Genetics, P., Interuniversity, F., & Ledeganckstraat, K. L. (2000). Histological study of seed coat development in *Arabidopsis thaliana. Journal of Plant Research, 113*(2), 139–148.

Rezzadori, K., Benedetti, S., & Amante, E. R. (2012). Proposals for the residue's recovery: Orange waste as raw material for new products. *Food and Bioproducts Processing, 90*(4), 606–614.

Rivas, J., Prazeres, A. R., Carvalho, F., & Beltran, F. (2010). Treatment of cheese whey wastewater: Combined coagulation–flocculation and aerobic biodegradation. *Journal of Agricultural and Food Chemistry, 58*(13), 7871–7877.

Robles-Sánchez, R. M., Rojas-Graü, M. A., Odriozola-Serrano, I., González-Aguilar, G., & Martin-Belloso, O. (2013). Influence of alginate-based edible coating as carrier of antibrowning agents on bioactive compounds and antioxidant activity in fresh-cut Kent mangoes. *LWT–Food Science and Technology, 50*(1), 240–246.

Rodríguez-Ambriz, S. L., Islas-Hernández, J. J., Agama-Acevedo, E., Tovar, J., & Bello-Pérez, L. A. (2008). Characterization of a fiber-rich powder prepared by liquefaction of unripe banana flour. *Food Chemistry, 107*(4), 1515–1521.

Romero, F. J., García, L. A., Salas, J. A., Díaz, M., & Quirós, L. M. (2001). Production, purification and partial characterization of two extracellular proteases from *Serratia marcescens* grown in whey. *Process Biochemistry, 36*(6), 507–515.

Rueda, X. Y., Mancilla, L. L., & Parada, D. P. (2007). Estudio del aceite esencial de la cáscara de la naranja dulce (*Citrus sinensis*, variedad Valenciana) cultivada en Labateca (Norte de Santander, Colombia). *Bistua: Revista de la Facultad de Ciencias Básicas, 5*(1), 3–8.

Ryan, M. P., & Walsh, G. (2016). The biotechnological potential of whey. *Reviews in Environmental Science and Bio/Technology, 15,* 479–498.

Salazar-López, N. J., Ovando-Martínez, M., & Domínguez-Avila, J. A. (2020). Cereal/grain by-products. In *Food Wastes and by-Products* (pp. 1–34). https://doi.org/10.1002/9781119534167.ch1 (accessed on 11th August 2025).

Sawant, O., Mahale, S., Ramchandran, V., Nagaraj, G., & Bankar, A. (2021). Fungal citric acid production using waste materials: A mini-review. *Journal of Microbiology, Biotechnology and Food Sciences, 2021,* 821–828.

Selen, V., & Saban Tanyildizi, M. (2017). Optimization of low-cost medium composition for the production of α-amylase by *Bacillus amyloliquefaciens* using a semi-solid substrate. *Current Physical Chemistry, 7*(4), 305–312.

Silvestre, M. P. C. (1997). Review of methods for the analysis of protein hydrolysates. *Food Chemistry, 60*(2), 263–271.

Soccol, C. R., Vandenberghe, L. P., Rodrigues, C., & Pandey, A. (2006). New perspectives for citric acid production and application. *Food Technology and Biotechnology, 44*(2), 141–149.

Solís-Fuentes, J. A., & del Carmen Durán-de-Bazúa, M. (2011). Mango (*Mangifera indica* L.) seed and its fats. In *Nuts and Seeds in Health and Disease Prevention* (pp. 741–748). Academic Press.

Spang, E. S., Moreno, L. C., Pace, S. A., Achmon, Y., Donis-Gonzalez, I., Gosliner, W. A., & Tomich, T. P. (2019). Food loss and waste: Measurement, drivers, and solutions. *Annual Review of Environment and Resources, 44,* 117–156.

Sutcliffe, P., & Disney, J. (1977). *Proceedings of the Conference on the Handling, Processing and Marketing of Tropical Fish.*

Tang, R., Liu, Y., Ma, R., Zhang, L., Li, Y., Li, G., & Yuan, J. (2023). Effect of moisture content, aeration rate, and C/N on maturity and gaseous emissions during kitchen waste rapid composting. *Journal of Environmental Management, 326,* 116662.

Terzic-Vidojevic, A., Mihajlovic, S., Uzelac, G., Veljovic, K., Tolinacki, M., Nikolic, M., & Kojic, M. (2014). Characterization of lactic acid bacteria isolated from artisanal Travnik young cheeses, sweet creams and sweet kajmaks over four seasons. *Food Microbiology, 39,* 27–38.

Tock, J. Y., Lai, C. L., Lee, K. T., Tan, K. T., & Bhatia, S. (2010). Banana biomass as potential renewable energy resource: A Malaysian case study. *Renewable and Sustainable Energy Reviews, 14*(2), 798–805.

Toldrá, F., & Reig, M. (2011). Innovations for healthier processed meats. *Trends in Food Science & Technology, 22*(9), 517–522.

Toldrá, F., Mora, L., & Reig, M. (2016). New insights into meat byproduct utilization. *Meat Science, 120,* 54–59.

Toriyama, K. (2005). *Rice is Life: Scientific Perspectives for the 21st Century.* International Rice Research Institute.

Torres-León, C., Rojas, R., Serna-Cock, L., Belmares-Cerda, R., & Aguilar, C. N. (2017). Extraction of antioxidants from mango seed kernel: Optimization assisted by microwave. *Food and Bioproducts Processing, 105,* 188–196.

Uygut, M. A., & Tanyildizi, M. Ş. (2018). Optimization of alpha-amylase production by *Bacillus amyloliquefaciens* grown on orange peels. *Iranian Journal of Science and Technology, Transactions A: Science, 42,* 443–449.

Van Gerpen, J. (2005). Biodiesel processing and production. *Fuel Processing Technology, 86*(10), 1097–1107.

Venkateshwaran, N., & Elayaperumal, A. (2010). Banana fiber reinforced polymer composites—A review. *Journal of Reinforced Plastics and Composites, 29*(15), 2387–2396.

Venugopal, V., & Shahidi, F. (1994). Thermostable water dispersions of myofibrillar proteins from Atlantic mackerel (*Scomber scombrus*). *Journal of Food Science, 59*(2), 265–268.

Venugopal, V., Chawla, S. P., & Nair, P. M. (1996). Spray dried protein powder from Threadfin bream: Preparation, properties and comparison with FPC type-B. *Journal of Muscle Foods, 7*(1), 55–71.

Version, A. (2020). Extruded snacks from industrial byproducts: A review.

Vijayalakshmi, M. A., Lemieux, L., & Amiot, J. (1986). High performance size exclusion liquid chromatography of small molecular weight peptides from protein hydrolysates using methanol as a mobile phase additive. *Journal of Liquid Chromatography & Related Technologies, 9*(16), 3559–3576.

Wainaina, S., Lukitawesa, Awasthi, M. K., & Taherzadeh, M. J. (2019). Bioengineering of anaerobic digestion for volatile fatty acids, hydrogen or methane production: A critical review. *Bioengineered, 10*(1), 437–458.

Wangkheirakpam, M. R., Mahanand, S. S., Majumdar, R. K., Sharma, S., Hidangmayum, D. D., & Netam, S. (2019). Fish waste utilization with reference to fish protein hydrolysate—A review. *Fishery Technology, 56,* 169–178.

Webster, J. D., Ledward, D. A., & Lawrie, R. A. (1982). Protein hydrolysates from meat industry byproducts. *Meat Science, 7*(2), 147–157.

Wilaipon, P. (2009). The effects of briquetting pressure on banana-peel briquette and the banana waste in Northern Thailand. *American Journal of Applied Sciences, 6*(1), 167–171.

Yalcin, S. K., Bozdemir, M. T., & Ozbas, Z. Y. (2009). Utilization of whey and grape must for citric acid production by two *Yarrowia lipolytica* strains. *Food Biotechnology, 23*(3), 266–283.

Zainal Arifin, M. A., Mohd Adzahan, N., Zainal Abedin, N. H., & Lasik-Kurdyś, M. (n.d.). Utilization of food waste and byproducts in the fabrication of active and intelligent packaging for seafood and meat products. *[Journal or source title missing]*.

Zhang, R., Ma, S., Li, L., Zhang, M., Tian, S., Wang, D., Liu, K., Liu, H., Zhu, W., & Wang, X. (2021). Comprehensive utilization of corn starch processing byproducts: A review. *Grain and Oil Science and Technology, 4*(3), 89–107. https://doi.org/10.1016/j.gaost.2021.08.003 (accessed on 11th August 2025).

CHAPTER 2

Pulses: Wastes, Byproducts, and Utilization

RADHA KUSHWAHA and DEVINDER KAUR*

Center of Food Technology, University of Allahabad, Prayagraj, Uttar Pradesh, India

*Corresponding author

ABSTRACT

In recent years, value has been placed on valorizing food waste and byproducts from different manufacturing processes, not only as a means of ensuring environmental sustainability but also to improve economic efficiency. Extracting valuable compounds from these byproducts could be one way to utilize them. As a result of their specialized functions, nutritive characteristics, or favorable effects on human or animal health, these compounds can be used as ingredients in food and pharmaceutical products. In particular, there is a continuous production of very large amounts of pulse residues. Plant proteins may help satisfy the steadily growing global protein demand through their use as a promising source. Pulses are widely grown in the world. About 25% of the whole is obtained as waste byproduct during the primary processing of the pulses. These byproducts include husk or hull, aleurone-rich husk, plumule, and germ fractions. A significant amount of primary and secondary plant metabolites can be extracted from these byproducts, which can be used to produce a variety of useful products based on their technological and nutritional benefits. Various studies showed that using these byproducts in various food products like biscuits, noodles, meat analogues, ice cream, sports

Utilization of Food Waste and Byproducts. Mudasir Ahmad Malik and Kshirod Kumar Dash (Eds.)

DOI: 10.1201/9781779641151-2

drinks, instant breakfasts, wine, and so forth, was found extensively profitable and healthy. This chapter provides a discussion on the utilization of pulse wastes and byproducts towards ensuring food and environmental sustainability with additional health benefits.

2.1 INTRODUCTION

Currently, byproducts and wastes, primarily those that originate from agriculture and food, are considered promising sources of bioactive compounds. Despite this, the majority of these byproducts, however, are not utilized and are instead disposed of or eaten by pests, which causes microbial decomposition and leachate to occur, resulting in hazardous environmental conditions. According to researchers (Arshadi et al., 2016; Mateos-Aparicio and Matias, 2019), as we go through the entire food chain, we encounter agro-food byproducts and wastes all the way from the field to the factory shelf to the home (Arshadi et al., 2016; Mateos-Aparicio and Matias, 2019). Byproducts and wastes vary from one to another; according to quality parameters such as the amount produced, the chemical composition of emerging wastes and byproducts will vary depending on the type of raw material being processed, the type of product being produced, and the production technique. Consequently, pulse processing produces wastes containing small amounts of fat, high levels of dietary fibers (DFs), protein, and secondary metabolites such as polyphenols. New functional foods and dietary supplements can be developed from these wastes. Besides providing economic benefits to the food industry, food waste and byproducts can be used to manufacture bioactive ingredients and added and novel food additives, while also removing the nutritional burden from the food industry. By extracting the necessary components from byproducts that have otherwise been discarded, products with high fiber and bran, high antioxidants, and high vitamins and minerals can now be formulated. Carbohydrates, fats, sugars, organic acids, pigments, phenolic compounds, vitamins and minerals, enzymes, and protein are among the primary and secondary metabolites found in byproducts. In addition to the byproduct type, extraction process, and biomaterial from which it is extracted, the composition of these substances changes (Luzardo-Ocampo et al., 2020).

A pulse is a leguminous plant that is eaten as a split or as a dhal. As the second most consumed food group after cereals, pulses serve as a rich source of nutrients for people all around the world (Sherasia et al., 2017).

In order to process pulses, they are cleaned, graded, husked, or removed from the outer seed coat, split, powdered into flour, or polished as whole dehusked pulses. This is a matter of determining the final product that must be produced. Dehusking can be accomplished by wet or dry processes, depending on whether the husk has been loosened prior to the dehusking. Identifying and assessing pulse byproducts to determine their health and functional potential will continue to be a successful trend in the future. Consequently, this chapter examines pulse byproducts, their potential industrial uses for their functionality, and for developing functional foods and extraction of bioactive compounds.

2.2 PULSE PROCESSING WASTE AND BYPRODUCTS

According to Ramakrishnaiah et al. (2004), in India, pulse milling has become the third largest industry in food grain processing. Pulses milled commercially have an efficiency of about 75%, with the remainder producing byproducts like husks, powder, broken grains, and shriveled grains that are pure and unprocessed (Figure 2.1). According to Ramakrishnaiah et al. (2004), pulse mills generate 2.5 million tons of byproduct each year, containing about 40% of valuable cotyledon material. In order to recover and increase the value of these byproducts, a technological platform has been developed (Figure 2.1).

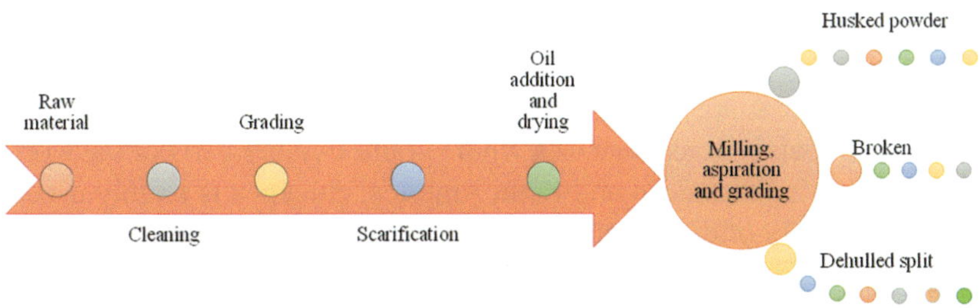

FIGURE 2.1 Pulse processing.

In commercial mills, abrasive dehusking results in dal recovery of only 70%, whereas it is possible to recover 85% of the dal during pulse milling. Abrasive dehusking is more efficient than other methods because the husks are separated from the rest of the seed without breaking the grain. Additionally, pulse milling produces fewer broken grains, resulting in higher dal quality. Pulse milling byproducts include broken grains (6%–13%), germs (7%–12%), and husks (4–14%). Various uses can be made of the byproducts of pulse milling (Mateos-Aparicio et al., 2010a). The husks are then used for various purposes, such as animal feed or fertilizer, resulting in a high-end complete utilization. The percentage of seed coat, cotyledon, and germ in different pulses is given below in Table 2.1.

TABLE 2.1 Percent of Seed Coat, Cotyledon, and Germ in Different Pulses

Pulses	Seed Coat (%)	Cotyledon (%)	Germ (%)
Pea	10.0	89.3	1.7
Pigeon pea	15.5	83.0	1.5
Cowpea	10.6	87.2	2.1
Common bean	11.0	84.0	5.0
Mung bean	12.1	85.6	2.3
Urad bean	13.0	84.0	4.0
Horse gram	12.0	83.0	5.0
Chickpea	12.0	84.0	4.0

Source: Dalgetty and Baik (2006), Girish et al. (2012a), and Salunkhe et al. (1985).

2.3 BENEFITS AND UTILIZATION OF PULSE WASTES

2.3.1 CHICKPEA

The way that chickpeas are used is different around the world due to cultural and regional preferences. In Asia and Africa, grain or its byproducts are utilized to make soups, stews, bread, and sauces. In other parts of the world, chickpea is mainly consumed as a whole grain. In other parts of the world, such as the Mediterranean and Latin America, chickpea is mainly used to prepare a variety of dishes, such as falafel, hummus, and nachos (Hawkins and Johnson, 2005; Jukanti et al., 2012). Commercially, chickpea is used for the production of flour (major) and oil (minor). Chickpea flour is known as gram flour or besan. As a result of the oil and flour production from chickpeas, byproducts are generated, the main byproduct being hulls, which can be used to create functional ingredients. In addition to being a good source of fiber, protein, and phenolics, chickpea byproducts are also rich in antioxidants. The dietary fiber present in chickpea hulls is mostly insoluble, which means it can help to improve digestive health. The protein content is also quite high, making it a good source of plant-based protein.

Additionally, phenolic compounds have antioxidant effects and can help to reduce the risk of chronic diseases (Niño-Medina et al., 2017). Several studies have reported the health benefits of these chickpea byproducts phenolic (Hawkins and Johnson, 2005; Jukanti et al., 2012; Niño-Medina et al., 2017). Chakraborty et al. (2022) formulated instant Idli mix and bran-rich paneer using milling byproducts of chickpea, wheat, and rice after fermentation, which was enriched with fiber, minerals, and antioxidants. Another study performed by Losano Richard et al. (2021) reported the chemical composition, thermal properties, and their utilization in the product (hummus) of chickpea export byproducts (discard and split). Compared to whole grains, split grains were higher in protein, lower in water hydration capacity (WHC), and lower in fiber, making them unsuitable for making hummus. Byproduct-based hummus has potential as an additive in the food industry because of its prebiotic effect. It was found that hummus made from whole grains and discarded grains had similar sensory characteristics. This indicates that the discarded grain can be used to create a product that still has the same sensory qualities as hummus made from whole grains but has an added prebiotic benefit. This could potentially be a more nutritious and cost-effective alternative for consumers. The properties of chickpea byproducts derived after milling that is, hull from two different varieties were reported by Costantini and their team (2021).

TABLE 2.2 Chemical Composition and Bioactive Components of Chickpea Hull

Composition	Black	White
Carbohydrates (%)	16.30	1.31
Total dietary fiber (%)	65.26	82.75
Protein (%)	11.26	7.85
Fat (%)	2.48	1.22
Ash (%)	4.70	6.87
Total phytates (mg/g)	3.82	1.83
Total phenolic compounds (mg/g)	0.81	0.63
Antioxidant activity (μmol/g Trolox[2])	4.91	3.34
Total anthocyanins (mg/kg)	225.64	64.77
Total carotenoids (mg/kg)	62.16	25.01
Sucrose (mg/g)	18.50	9.40

Source: Costantini et al. (2021).

In addition to dietary fiber (65–82%), chickpea hulls also contain antioxidants such as phenols and carotenoids. Food manufacturers can also extract considerable amounts of anthocyanins from chickpea hulls of the subtype Apulian black (225 mg/kg cyanidin 3-O-glucoside d.m.) in order to benefit from the value of anthocyanins (Table 2.2). Furthermore, they enhanced pasta fiber content and antioxidant activity by utilizing chickpea hulls (8% w/w) without altering the sensory attributes or acceptance of the food. Chickpea hulls are a good source of anthocyanins, which are associated with numerous health benefits due to their antioxidant properties. Anthocyanins are also linked to cardiovascular benefits, such as improved blood pressure control and anti-inflammatory effects. Chickpea hulls can also increase the dietary fiber content of food, which is important for maintaining healthy gut flora. Overall, chickpea hulls offer numerous health benefits when added to food products. Additionally, their use in food manufacturing increases the fiber content and antioxidant activity of the food without affecting the sensory attributes.

2.3.2 BLACK GRAM (VIGNA MUNGO)

Pulse processing removes the main functional components during milling, sieving, and deranging. The process works by using a combination of mechanical vibration and air pressure to break down the components of the grain. This helps to improve the texture, color, and flavor of the products. The black gram husk contained protein 18.21%, fiber 20.29%, NFE 54.60%, and 5.54% ash. The main fiber fractions were 45.54–51.74% NDF, 33.84–41.62% ADF, 25.32–28.34% cellulose, 9.99–11.88% hemicellulose, and 7.20–12.44% lignin. Husk is a good source of calcium (0.51%), phosphorus (0.26%), and magnesium (0.014%). The Cu, Zn, Co, and Mn levels were at 13.16 ppm, 37.47 ppm, 0.97 ppm, and 53.75 ppm, respectively (Arulnathan et al., 2013). Oligosaccharides are abundant in pulse byproducts. According to several publications (Palaniappan et al., 2017; Saman et al., 2017), this fraction has higher prebiotic potential than commercial prebiotic formulations. An agro-industrial byproduct such as wheat bran, rice bran, rice husk, barley byproducts, and finger millet seed coat can be used to produce prebiotic compounds such as XOS and AXOS (Palaniappan et al., 2017; Saman et al., 2017; Vazquez-Olivo et al., 2019). Oligosaccharides promote the growth of beneficial bacteria in the gut and can therefore improve gut health. Additionally, they are much less expensive than commercial prebiotic formulations, making them a more affordable option.

The number of prebiotics present in food is decreasing nowadays, so researchers are demonstrating that black gram husk can be used as a prebiotic. Consequently, these husks could serve as a potential prebiotic source because they encourage the growth of probiotic bacteria *in vitro*. Based on their DPPH scavenging activity, the husks have natural antioxidant properties (Table 2.3). According to Alam et al. (2013), free radicals may play a role in cancer, cardiovascular disease, neurodegenerative disorders, and aging, as well as atherosclerosis. Consuming sufficient antioxidants can enhance the body's ability to resist such free radicals. Therefore, it has become increasingly popular to consume food items with antioxidant activity on a daily basis. By consuming food items with antioxidant activity, our bodies are better able to combat the damaging effects of free radicals. This can help to reduce the risk of developing chronic diseases and the effects of aging. Furthermore, the husks of certain grains have been shown to have high antioxidant activity, making them an ideal choice for those looking to increase their antioxidant intake. According to Yoganandhan and Hameed (2000), black gram husks possess optimal growth and survival characteristics for brine shrimp, *Artemia* sp., in a laboratory for 25 days. By adding the husks to the laboratory tanks, the researchers were able to create a high-nutrient environment that was suitable for the brine shrimp. They found that the brine shrimp had a high survival rate and grew to a large size. In another study, performed by Das et al. (2008), the biosorption of heavy metals such as lead, cadmium, zinc, copper, and nickel using black gram husk was reported. It was reported by Kumar et al. (2012) that rhamnose, arabinose, xylose, mannose, galactose, and glucose were found in black gram peels when a solvent system of n-butanol, acetone, pyridine, and water in the ratio of 10:10:5:5, v/v/v/v, was used. Peroxidase enzyme was found to be abundant in black gram husks, a waste product of the milling industry. At 50°C, the enzyme displayed strong temperature stability and had a half-life of approximately five hours. The findings of the study indicate that the enzyme activity and stability of peroxidase in black gram husks can be used for various industrial applications, such as bleaching, dyeing, and paper production, and can also be used to produce biofuels and other value-added products from black gram peels.

TABLE 2.3 Composition of Black Gram Byproducts

Composition	Seed Coat/Husk	Germ	Plumule	Aleurone + seed coat (ALESC)
Protein (%)	11.32–12.41	29.29–42.16	12.25	23.00–31.38
Fiber (%)	33.72	19.48	23.30	22.40
Ash (%)	2.96–3.03	3.72–5.30	3.00	4.02–4.12
Total dietary fiber (%)	78.53–79.62	25.90–44.80	71.86	67.70
Insoluble dietary fiber (%)	69.23–77.04	21.63–39.60	65.93	59.76
Carotenoid (mg/100 g)	0.42	0.03	0.13	0.33
Anthocyanin (mg/100 g)	86.84	13.06	33.24	70.42
TPC (mg GAE/g)	134.66	12.81	78.83	72.83
IC_{50} (μg of GAE)	3.01	4.93	2.27	2.90
Calcium (mg/100 g)	66.21	75.95	–	73.09
Iron (mg/100 g)	8.38	4.38	–	10.26
Magnesium (mg/100 g)	138.36	130.02	–	157.54
Potassium (mg/100 g)	496.93	811.41	–	650.08
Zinc (mg/100 g)	1.48	2.36	–	3.48
Copper (mg/100 g)	0.89	0.77	–	0.95

Source: Girish et al. (2012a); Kamani and Meera (2021).

According to Girish et al. (2012a), the phenolic acid composition showed that the husk and aleurone layer were rich in gallic, protocatechuic acid, gentisic acid, syringic, and ferulic acid, while only plumule has vanillic acid. They also reported the black gram byproducts had α-glucosidase inhibitor activity. It is commonly known that Type 2 diabetes is caused by impaired insulin secretion, resulting in heightened levels of blood glucose postprandial following a meal. In terms of treating postprandial hyperglycemia, one of the most effective therapeutic approaches is to inhibit the carbohydrate hydrolyzing enzymes in order to retard the absorption of glucose. α-Glucosidase is known to be one of the enzymes that are involved in the release of glucose from starch for absorption into the intestinal tract. A significant benefit of inhibiting this enzyme is the reduction in blood glucose levels, making it one of the most important strategies used in the treatment of type-2 diabetes (Plus et al., 1977). By inhibiting the enzyme, the breakdown of starches is reduced, meaning that less glucose is released into the bloodstream and absorbed into the intestinal tract. This helps to keep blood glucose levels from rising too high, reducing the risk of complications associated with diabetes. A natural food source of α-glucosidase inhibitors is a promising strategy for managing postprandial hyperglycemia. Comparatively to black gram seeds, aleurone layers enhanced with seed coats, and plumule fractions, black gram seed coats showed better enzyme inhibition. Inhibition ranged from 58 to 70% for seed coat, aleurone layer enriched in the seed coat, and plumule fractions, at 2.5 μg level, while inhabitance ranged from 19 to 36% for the other fractions. The enzyme inhibitory activity of whole flour extracts was superior to cotyledon and germ extracts. The highest enzyme inhibitory activity was found in the seed coat fraction, indicating the potential of black gram seed coats as a natural source of enzyme inhibitors. These findings suggest that black gram seed coats could be used as a natural preservative in food products. A study by Girish et al. (2012b) reported that oxidative damage in DNA and erythrocytes was prevented by black gram husk water extract (BGH), a natural product derived from black gram husk. This is due to the presence of natural polyphenols in BGH, which have antioxidant properties that help protect the cells from oxidative damage. These polyphenols also reduce inflammation, which further helps protect the cells from damage. The total polyphenol content of the husk extract was 59.00 mg GAE/100 g. Several phenolic acids have been identified as possessing antioxidant properties, such as gallic, gentisic, protocatechuic, and ferulic acids. An IC_{50} value of 3.92 mg GAE was determined for the radical scavenging activity against DPPH. Also, α-glucosidase inhibition was observed in the husk extract, and its IC_{50} value was 2.78 mg GAE. Husk extract inhibited oxidative hemolysis caused by hydrogen peroxide in rat erythrocytes. IC_{50} values for hemolysis were 11.5 mg and 14 mg of GAE for BGH extract and BHA, respectively. Black gram husk extracts protected erythrocyte membranes from morphological changes caused by hydrogen peroxide. Various antioxidant properties of black gram husk extract make it an attractive candidate for the development of functional foods or nutraceuticals.

According to Kamani and Meera (2021), the black gram milling byproduct Rich in Germ and Aleurone (BRGA), a protein-rich byproduct, contains an essential amino acid composition comparable to that of whole seeds. Furthermore, husk fractions, low-protein byproducts, contained adequate amounts of essential and conditionally essential amino acids. The amino acid profiles of these byproducts are therefore beneficial for overcoming amino acid deficiencies in food. Unsaturated fatty acids (UFAs) are generally known to promote good health. A positive cholesterol-lowering effect has also been observed in black gram lipids (Devi and Kurup, 1972). In addition, the study found that BRGA contains only 20% saturated acids, while the majority of its unsaturated fats (56.59%) are PUFAs. Kamani and Meera (2021) also found that husk provided adequate MUFA and PUFA profiles. This indicates that BRGA contains healthy fats which can help reduce cholesterol levels. Furthermore, the presence of PUFAs, which are known to have anti-inflammatory properties, may provide additional health benefits.

2.3.3 PEA (PISUM SATIVUM)

Peas are valuable nutritional crops grown throughout the world and in Europe because they can withstand cold winters and are high in protein content (18–30%). The benefits of peas include providing a good source of vitamins, minerals, fiber, and antioxidants, helping to reduce cholesterol, and maintaining healthy blood sugar levels. Peas are also a great source of plant-based protein, making them a great option for vegetarians and vegans (Klupšaitė and Juodeikienė, 2015). A variety of peas is commonly marketed, namely green peas (*Pisum sativum* L.) and yellow peas (*Pisum sativum* L. var. macrocarpon), which are usually dried and dominate global production. Most pea plants are left to rot on land after harvest, including pods, stalks, leaves, and vines. This practice has several environmental impacts, such as releasing greenhouse gases, increasing soil erosion, and depleting soil nutrients. Therefore, alternative strategies, such as composting and using crop residues for animal feed, or in food value addition, should be considered. An earlier report published by Mateos-Aparicio et al. (2010) demonstrated that discarded pea pods contain dietary fiber (59%, mostly insoluble fiber) and monounsaturated and polyunsaturated fats (69%). In comparison with other agricultural wastes, pea pod polysaccharides have a lower IC_{50} (0.06 mg/mL) antiradical DPPH scavenging activity. This means that pea pods can be a great source of dietary fiber and fats, and also a powerful antioxidant. The polysaccharides in pea pods are more effective at scavenging free radicals than those in other agricultural wastes, making them a great source of antioxidants. Pea hull fibers have been extensively used in baking products for their high total and soluble fiber contents (80–88% and 2–10%, respectively). In the pea plant, pea pods are the byproducts of the pea processing industry. In addition to dietary fiber, these pods contain protein and minerals, and it was utilized for the development of instant pea soup powder (IPSP). Pea hull fibers have potential applications in a variety of bakery products due to their good binding properties. They can be used as a replacement for wheat flour, providing a healthier and more nutritious alternative. Additionally, pea hull fibers are a sustainable and cost-effective option.

Overall acceptability of 8.5 was found with formulations that contained pea pod powder at a level of 12.5%. In this study, dietary fiber levels at 13.25%, carotenoids at 6.65 mg/100 g, and chlorophyll at 1.5 mg/100 g through the development of powder soup with a calorific value of 345.1 kcal/100 g (Hanan et al., 2020). As a result, pea byproducts are a very high source of dietary fiber (472.6 g kg^{-1}), with the majority of it coming from insoluble fibers, and a number of polyphenols that have been extracted are shown to be antioxidants. This antioxidant showed 25.9 μmol Trolox equivalents per gram and a scavenging effect of 16.0 mg mL against 2-diphenyl-1-picrylhydrazyl radicals (Mateos-Aparicio et al., 2012). A diet supplemented with pea fiber improves several microbiota and metabolism variables in the gut (Hashem et al., 2017). Due to their high levels of polyphenolics and natural antioxidants, pea hulls exhibit antioxidant activity. Stanisavljevi et al. (2016) describe phenolic profiles and *in vitro* anticancer activities for seed coats from 10 different colored European pea varieties. Different varieties had different total phenolic content and antioxidant (DPPH) capacity, and the most abundant phenolics were flavonoids, especially catechins, epicatechins, epigallocatechins, epigallocatechin- and gallocatechin- gallates. All samples were equally abundant in quercetin, galangin, rutin, and kaempferol.

Byproducts of pea showed various health benefits as reported in earlier research, as the study published by Salmean and their team in 2015 supported this statement. They have reported that when pea hull fiber-supplemented products were administered to chronic kidney disease patients for 12 weeks, they significantly reduced p-Cresol while not affecting C-reactive protein, blood urea nitrogen (BUN), ammonia, creatinine, cystatin C, or estimated glomerular filtration rate (eGFR). The results of this study suggest that pea hull fiber supplementation may be an effective way to reduce p-Cresol levels in chronic kidney disease patients. These results suggest that pea hull fiber supplementation can benefit chronic kidney disease patients. It was also associated with modulation of gut microbiota due to fiber consumption since *p*-cresol is formed from the breakdown of tyrosine, a molecule that negatively impacts chronic kidney disease.

A study with 26 children (2–10 years old) suffering from constipation or abdominal pain in the previous 12 months supplemented them with pea hull fiber-rich snacks. The results concluded that fiber administration can be successfully administered to children with gastrointestinal problems, leading to significantly increased fiber consumption (17.1 grams/day) and reduced energy consumption. This led to a significant improvement in symptoms, such as reduced abdominal pain and improved stool consistency. The study showed that fiber-rich snacks can be a viable nutritional option for children with gastrointestinal issues (Flogan and Dahl, 2010). Guo et al. (2021) reported the D-gal model in rats was used to measure antioxidant activity *in vivo*. In plasma or urine samples, they identified 54 metabolites, while 42 ingredients were identified in yellow pea husk extracts. In the liver and kidneys, metabolites were mainly distributed, and the distribution of these metabolites may have a positive effect on SOD, MDA, T-AOC, and GSH-Px.

2.3.4 GREEN GRAM (MUNG BEAN)

Green gram is one of the most common foods in India. Approximately 18% of the grain is converted into byproducts during the processing of green gram into dhal. In the process of sieving and air classification, the byproduct is classified into four categories: the husk, the germ, the plumule rich fraction (PRF), and the aleurone rich fraction (ARF). As compared to PRF, husks had the highest concentrations of polyphenols, flavonoids, carotenoids, chlorophyll, and tannins. In the DPPH assay, the husk and PRF showed IC_{50} values of 12 and 23 μg; and 17 and 25 μg for ABTS + scavenging, respectively. For inhibition of α-amylases, husk and PRF generated IC_{50} values of 0.3 and 0.3 mg; for inhibition of α-glucosidase, 0.2 and 0.3 mg; and for oxidative DNA damage protection, 2 and 3 μg. The most prominent bioactive compounds in various fractions were gallic acid, sinapic acid, ferrulic acid, is vitexin, and vitexin. Husks and PRFs had the most concentrations of these compounds (Basha and Rao, 2017).

Sihag and Kawatra (2003) reported that Bengal gram and green gram seed coats reduced serum cholesterol levels in rats. A six-week feeding study was conducted on four groups of albino rats with isonitrogenous diets that contained 1% cholesterol and 10% dietary fiber from both pulse seed coats. Pulse seed coats significantly reduced serum levels of total lipids, DL-cholesterol, triglycerides, and cholesterol. The same study also found that pulse seed coats had no significant effect on HDL-cholesterol or phospholipid levels in serum. Bengal gram seed coats excreted more cholesterol and bile acid than green gram seed coats, according to fecal analyses. An array of anti-inflammatory effects can be observed in mung bean byproducts. Bioactive compounds from mung bean seed coat were evaluated for their immunomodulatory effects using *in vitro* tests. The flavonoid content of the mung bean seed coat and the expression of the IL-6 gene on RAW 264.7 macrophages stimulated by LPS were shown to be synergistic; IL-6 production was inhibited dose-dependently. Additionally, the polysaccharides in hulls appear to have antioxidant properties. They optimized water extraction from mung bean hulls. In the purified fractions, uranic acid fractions (19–28%), sugar fractions (69–88%), and protein fractions (15%) were present.

Mung bean husk and plumule-rich fractions contained higher amounts of polyphenols, anthocyanins, flavonoids, and carotenoids compared to whole bean flour. However, the tannin concentration was considerable (74 and 56 mg (+) - catechin equivalents/g, for the husk and plumule-rich fractions, respectively). They have the greatest potential for radical inhibition (DPPH and 2,2′azinobiphenylhydroxylradicals [ABTS]) and the greatest effect on amylase and glucosidase enzyme inhibition, which inhibit starch hydrolysis, raising postprandial blood glucose levels (Koch et al., 2015). Postprandial glucose levels may be regulated by the husk and plumule-rich fractions in the sample. In a study by Luo et al. (2016), chemical compounds from mung bean hulls were shown to have antimicrobial activity against *Pseudomonas aeruginosa*, *Staphylococcus aureus*, and *Bacillus subtilis*: vitexin (37 mg/g) and isovitexin (47 mg/g). Kang et al. (2016) also demonstrated anti-inflammatory effects in 3T3-L1 cells by reducing the level of IL-6 and MCP-1 as well as a significant reduction in fat accumulation in adipocytes in a dose-dependent manner over the course of the study. Vitexin and isovitexin, both functional compounds in the mung bean ethanol extracts, have been found to regulate lipogenesis and adipogenesis in the KK-Ay mouse model through anti-inflammatory mechanisms and MEK/ERK pathways. This indicates that mung bean ethanol extracts are effective in reducing inflammation and fat accumulation in adipocytes. It suggests that this extract has potential as an anti-obesity and anti-inflammatory agent.

2.3.5 BROAD BEANS

The seeds of broad beans (*V. faba*) or faba beans (*V. faba*) are typically marketed after they have been removed from the pods, thereby creating a byproduct that can provide an important source of fiber. Broad bean waste is mostly composed of the husk and pod of the bean (70%), which can be used to produce all kinds of components apart from the pods (Jafri et al., 2019). As a whole, broad bean pods are a rich source of potassium (2.3%), iron (0.65%), calcium (0.34%), manganese (0.40%), and zinc (0.34%) with significant concentrations; however, their bioavailability is low. There was a significant improvement in the texture profile of bread after the addition of broad bean pod fibers (Fendri et al., 2016). It has been found that broad beans have a significant amount of ash, which may contribute to the improvement of our daily intake of essential trace elements, which play a vital role in the human body's biological functions (metabolic reactions, acid-base balance, water balance, and energy production) (Mejri et al., 2018).

Several subspecies of broad beans have been analyzed and compared in terms of their phenolic contents and anti-inflammatory properties. A higher TEAC measured by ORAC could be caused by a higher phenolic content of the hull fractions of *Vicia faba* compared to the whole or cotyledon fractions (Boudjou et al., 2013). UFAs were found to be the most abundant fatty acid in the sample identified by Mejri et al. (2018). It was primarily composed of essential polyunsaturated linoleic acid (39.74%, w/w) and linolenic acid (24.99%, w/w). In addition to stearic acid (6.72%, w/w), palmitic acid has the largest proportion of saturated fatty acids (30.78% w/w). Other valuable nutrients, such as glucans and pectin, can also be extracted from *faba* bean pods. Pectin yield (17.2%, w/w) was highest at the first stage of maturation, while glucan yield (45 mg/g) was highest at later maturation (Fazio et al., 2020). A useful way to incorporate phenolic compounds into food is through the husks of *faba* bean plants, which contain approximately 110.0 mg of GAE per gram (Chaieb et al., 2011; Hashemi and Ebrahimzadeh, 2014; Karataş et al., 2017). Faba bean husks (FBH) have high antioxidant properties because of their high TPC levels. FBHs improved the cooking properties and delayed lipid and protein oxidation of enriched burger patties, prevented color changes, and decreased the microbiological load (Elsebaie et al., 2022). As a result of incorporating FBHs into Turkish noodles with 10% (w/w) enrichment, there were slight changes in texture, but sensory acceptability remained (Kaya et al., 2018). FBH are also capable of enriching bread with dietary fiber by incorporation into it. Despite a slight decrease in consumer acceptance, satiating effects increased (Daliani et al., 2019). This indicates that FBH are a viable option for enriching foods with dietary fiber. They can be incorporated into many foods with minimal changes in texture and flavor. Additionally, they can help to increase satiating effects, making consumers feel fuller for longer. According to Table 2.4, the residues of the *faba* bean plant are composed of the following compounds. High-fiber waste products of *faba* beans called FBH can be used as animal feed and as fiber supplements. By adding FBH to gluten-free bread, its fiber content has been improved (Daliani et al., 2019). Apparently, fava bean husk can be used to fortify dietary fiber in gluten-free bread without compromising quality.

TABLE 2.4 Composition of Faba Bean Plant Residues

Composition	Plant Residue
Protein (%)	4.3–18.8
Fiber (%)	26.4–46.7
Ash (%)	4.6–18.4
Calcium (g/Kg)	8.7–23.1
Magnesium (g/Kg)	1.7–4.8
Potassium (g/Kg)	13.2–25.2
Sodium (g/Kg)	0.3–1.6
Phosphorus (g/Kg)	1.0–5.3
Lignin (g/Kg)	39.0–324.0
Hemicellulose (g/Kg)	15.6–353.0
Cellulose (g/Kg)	168.0–560.0

Source: Frejnagel et al. (1997), Ivarsson et al. (2021), Karataş et al. (2017), Kibret et al. (2021), Minakowski et al. (1996), Ni et al. (2020), Nigus et al. (2015), and Saharan et al. (2001).

Cellulose microfibers (CMF) and nanocrystals (CNC) were isolated from broad bean pods by Kassab et al. (2020). This can be accomplished by recovering the underused residues from broad bean pods and reducing the amount of waste in the environment by recovering CMF and CNC.

2.4 CONCLUSION

Pulses are harvested and processed in a manner that produces several residual materials. Pulse byproduct utilization is an important area of research, as these materials offer potential health benefits and can be used to add texture and flavor to food products. Furthermore, the utilization of pulse byproducts could reduce waste and provide an economical alternative to other ingredients. There are a number of applications for pulse residues, whether they be food, feed, or non-feed. This chapter showed that these residues contain valuable components. It is therefore essential not to dispose of the material, but to further valorize it in order to maximize the resource's potential. As well as maintaining unaffected organoleptic properties, these wastes can also offer new textural properties for traditional foods. Yet, the procedures for valorizing food and non-food still need to be established, especially for food use. Husks and pods have a wide range of potential applications, but research is still needed to fully understand them. In particular, new methods for fractionation, extraction, and purification need to be developed in order to create ingredients with specific functionalities. This will allow for the use of these materials in a variety of food products. Further research should focus on creating sustainable and safe processes for valorizing these materials.

KEYWORDS

- **bioactive compounds**
- **byproducts**
- **extraction techniques**
- **functional food**
- **health-promoting**
- **nutraceutical**
- **utilization**

REFERENCES

Ajila, C. M., & Rao, U. P. (2009). Purification and characterization of black gram (*Vigna mungo*) husk peroxidase. *Journal of Molecular Catalysis B: Enzymatic, 60*, 36–44.

Alam, M. N., Bristi, N. J., & Rafiquzzaman, M. (2013). Review on *in vivo* and *in vitro* methods evaluation of antioxidant activity. *Saudi Pharmaceutical Journal, 21,* 143–152.

Arshadi, M., Attard, T. M., Lukasik, R. M., Brncic, M., da Costa Lopes, A. M., Finell, M., & Yuste-Córdoba, F. J. (2016). Pre-treatment and extraction techniques for recovery of added value compounds from wastes throughout the agri-food chain. *Green Chemistry, 18,* 6160–6204.

Arulnathan, N., Murugan, M., & Balakrishnan, V. (2013). Proximate principles, fiber fraction and mineral content of black gram (*Vigna mungo*). *International Journal of Livestock Research, 3,* 24–30.

Basha, S. A., & Rao, U. P. (2017). Bioactivities of fractions obtained from green gram (*Vigna radiata*) milled byproducts. *Food Bioscience, 19,* 134–141.

Belghith-Fendri, L., Chaari, F., Jeddou, K. B., Kallel, F., Bouaziz, F., Helbert, C. B., Mesrati, L. A., Chaabouni, S. E., & Ghribi-Aydi, D. (2018). Identification of polysaccharides extracted from pea pod byproducts and evaluation of their biological and functional properties. *International Journal of Biological Macromolecules, 116,* 947–954.

Boudjou, S., Oomah, B. D., Zaidi, F., & Hosseinian, F. (2013). Phenolics content and antioxidant and anti-inflammatory activities of legume fractions. *Food Chemistry, 138,* 1543–1550.

Chaieb, N., González, J. L., López-Mesas, M., Bouslama, M., & Valiente, M. (2011). Polyphenols content and antioxidant capacity of thirteen faba bean (*Vicia faba* L.) genotypes cultivated in Tunisia. *Food Research International, 44,* 970–977.

Chakraborty, M., Budhwar, S., & Kumar, S. (2022). Development of fermented products with enriched fiber and micronutrients by using underutilized cereal-legume milling byproducts as novel food ingredients. *International Journal of Gastronomy and Food Science, 27,* 100493.

Chockchaisawasdee, S., Mendoza, M. C., Beecroft, C. A., Kerr, A. C., Stathopoulos, C. E., & Fiore, A. (2023). Development of a gluten free bread enriched with faba bean husk as a fiber supplement. *LWT, 173,* 114362.

Costantini, M., Summo, C., Faccia, M., Caponio, F., & Pasqualone, A. (2021). Kabuli and Apulian black chickpea milling byproducts as innovative ingredients to provide high levels of dietary fibre and bioactive compounds in gluten-free fresh pasta. *Molecules, 26,* 4442.

Dalgetty, D. D., & Baik, B. K. (2006). Fortification of bread with hulls and cotyledon fibers isolated from peas, lentils, and chickpeas. *Cereal Chemistry, 83,* 269–274.

Daliani, D., Stott, S., Fiore, A., & Cottin, S. (2019). The effect of the enrichment of bread with fibres from fava bean husk byproducts on postprandial satiety response and palatability. *Proceedings of the Nutrition Society, 78*(OCE2).

Das, N., Karthika, P., Vimala, R., & Vinodhini, V. (2008). Use of natural products as biosorbent of heavy metals—An overview. *Natural Product Radiance, 7,* 133.

Elsebaie, E. M., Elmahdy, A., El-Gezawy, E. S., Badr, M. R., Asker, G. A., El-Gawish, A. M., & Essa, R. Y. (2022). Effects of faba bean hull nanoparticles on physical properties, protein and lipid oxidation, color degradation, and microbiological stability of burgers under refrigerated storage. *Antioxidants, 11,* 938.

Fazio, A., La Torre, C., Dalena, F., & Plastina, P. (2020). Screening of glucan and pectin contents in broad bean (*Vicia faba* L.) pods during maturation. *European Food Research and Technology, 246,* 333–347.

Fendri, L. B., Chaari, F., Maaloul, M., Kallel, F., Abdelkafi, L., Chaabouni, S. E., & Ghribi-Aydi, D. (2016). Wheat bread enrichment by pea and broad bean pods fibers: Effect on dough rheology and bread quality. *LWT, 73,* 584–591.

Flogan, C., & Dahl, W. (2010). Effects of fiber-fortified foods on children with constipation: Potential improved stool frequency and decreased energy intake. *ICAN: Infant, Child, & Adolescent Nutrition, 2,* 312–317.

Frejnagel, S., Zduńczyk, Z., & Krefft, B. (1997). The chemical composition and nutritive value of low- and high-tannin faba bean varieties. *Journal of Animal and Feed Sciences, 6,* 401–412.

Girish, T. K., Pratape, V. M., & Rao, U. P. (2012a). Nutrient distribution, phenolic acid composition, antioxidant and alpha-glucosidase inhibitory potentials of black gram (*Vigna mungo* L.) and its milled byproducts. *Food Research International, 46,* 370–377.

Girish, T. K., Vasudevaraju, P., & Rao, U. J. P. (2012b). Protection of DNA and erythrocytes from free radical induced oxidative damage by black gram (*Vigna mungo* L.) husk extract. *Food and Chemical Toxicology, 50*(5), 1690–1696.

Guo, F., Tsao, R., Wang, X., Jiang, L., Sun, Y., & Xiong, H. (2021). Phenolics of yellow pea (*Pisum sativum* L.) hulls, their plasma and urinary metabolites, organ distribution, and *in vivo* antioxidant activities. *Journal of Agricultural and Food Chemistry, 69*(17), 5013–5025.

Hanan, E., Rudra, S. G., Sagar, V. R., & Sharma, V. (2020). Utilization of pea pod powder for formulation of instant pea soup powder. *Journal of Food Processing and Preservation, 44,* e14888.

Hashemi, Z., & Ebrahimzadeh, M. A. (2014). Evaluation of three methods for the extraction of antioxidants from *Vicia faba* L. bean and hulls. *Latin American Applied Research, 44,* 203–208.

Hashemi, Z., Fouhse, J., Im, H. S., Chan, C. B., & Willing, B. P. (2017). Dietary pea fiber supplementation improves glycemia and induces changes in the composition of gut microbiota, serum short chain fatty acid profile and expression of mucins in glucose intolerant rats. *Nutrients, 9,* 1236.

Hawkins, A., & Johnson, S. K. (2005). *In vitro* carbohydrate digestibility of whole-chickpea and chickpea bread products. *International Journal of Food Sciences and Nutrition, 56,* 147–155.

Ivarsson, E., Grudén, M., Södergren, J., & Hultberg, M. (2021). Use of faba bean (*Vicia faba* L.) hulls as substrate for *Pleurotus ostreatus*—Potential for combined mushroom and feed production. *Journal of Cleaner Production, 313,* 127969.

Jafri, H., Khan, M. S. A., & Ahmad, I. (2019). *In vitro* efficacy of eugenol in inhibiting single and mixed-biofilms of drug-resistant strains of *Candida albicans* and *Streptococcus mutans*. *Phytomedicine, 54,* 206–213.

Jukanti, A. K., Gaur, P. M., Gowda, C. L. L., & Chibbar, R. N. (2012). Nutritional quality and health benefits of chickpea (*Cicer arietinum* L.): A review. *British Journal of Nutrition, 108,* S11–S26.

Kamani, M. H., & Meera, M. S. (2021). Assessment of black gram milling byproduct as a potential source of nutrients. *Journal of Food Science and Technology, 58,* 3844–3852.

Kang, I., Choi, S., Ha, T. J., Choi, M., Wi, H. R., Lee, B. W., & Lee, M. (2015). Effects of mung bean (*Vigna radiata* L.) ethanol extracts decrease proinflammatory cytokine-induced lipogenesis in the KK-Ay diabetic mouse model. *Journal of Medicinal Food, 18,* 841–849.

Karataş, S. Ç., Günay, D., & Sayar, S. (2017). *In vitro* evaluation of whole faba bean and its seed coat as a potential source of functional food components. *Food Chemistry, 230,* 182–188.

Kassab, Z., Abdellaoui, Y., Salim, M. H., & El Achaby, M. (2020). Cellulosic materials from pea (*Pisum sativum*) and broad beans (*Vicia faba*) pods agro-industrial residues. *Materials Letters, 280,* 128539.

Kaya, E., Yılmaz Tuncel, N., & Tuncel, N. B. (2018). Utilization of lentil, pea, and faba bean hulls in Turkish noodle production. *Journal of Food Science and Technology, 55,* 1734–1745.

Kibret, W., Asmare, B., & Mekuriaw, Y. (2021). Effects of faba bean hull, noug seed (*Guizotia abyssinica*) cake, wheat bran and their mixtures supplementation on biological performance and economic benefits of Farta sheep fed natural pasture hay in Ethiopia. *Scientific Papers: Animal Science & Biotechnologies, 54.*

Klupšaitė, D., & Juodeikienė, G. (2015). Legume: Composition, protein extraction and functional properties—A review. *Chemical Technology, 66,* 5–12.

Koch, W., Baj, T., Kukula-Koch, W., & Marzec, Z. (2015). Dietary intake of specific phenolic compounds and their effect on the antioxidant activity of daily food rations. *Open Chemistry, 13,* 869–876.

Kumar, C. C., Chandraju, S., Mythily, R., Ahmad, T., & Gowda, N. M. (2012). Extraction of sugars from black gram peels by reversed-phase liquid chromatography systems and identification by TLC and mass analysis. *Advances in Analytical Chemistry, 2,* 32–36.

Losano Richard, P., Steffolani, M. E., Allende, M. J., Carreras, J., & León, A. E. (2021). By-products of the classification of chickpea as an alternative in the production of hummus. *International Journal of Food Science & Technology, 56,* 1759–1765.

Luo, J., Cai, W., Wu, T., & Xu, B. (2016). Phytochemical distribution in hull and cotyledon of adzuki bean (*Vigna angularis* L.) and mung bean (*Vigna radiata* L.), and their contribution to antioxidant, anti-inflammatory and anti-diabetic activities. *Food Chemistry, 201,* 350–360.

Luzardo-Ocampo, I., Cuellar-Nuñez, M. L., Oomah, B. D., & Loarca-Piña, G. (2020). Pulse by-products. In R. Campos-Vega, B. D. Oomah, & H. A. Vergara-Castañeda (Eds.), *Food Wastes and By-Products: Nutraceutical and Health Potential* (pp. 59–92). Wiley.

Mateos-Aparicio, I., & Matias, A. (2019). Food industry processing byproducts in foods. In C. M. Galanakis (Ed.), *The Role of Alternative and Innovative Food Ingredients and Products in Consumer Wellness* (pp. 239–281). Academic Press.

Mateos-Aparicio, I., Redondo-Cuenca, A., & Villanueva-Suárez, M. J. (2012). Broad bean and pea by-products as sources of fiber-rich ingredients: Potential antioxidant activity measured *in vitro*. *Journal of the Science of Food and Agriculture, 92,* 697–703.

Mateos-Aparicio, I., Redondo-Cuenca, A., Villanueva-Suárez, M. J., Zapata-Revilla, M. A., & Tenorio-Sanz, M. D. (2010a). Pea pod, broad bean pod and okara, potential sources of functional compounds. *LWT–Food Science and Technology, 43,* 1467–1470.

Mejri, F., Selmi, S., Martins, A., Baati, T., Chaabane, H., Njim, L., & Hosni, K. (2018). Broad bean (*Vicia faba* L.) pods: A rich source of bioactive ingredients with antimicrobial, antioxidant, enzyme inhibitory, anti-diabetic and health-promoting properties. *Food & Function, 9,* 2051–2069.

Minakowski, D., Sajko, H. S., & Falkowska, A. (1996). Nutritive value of faba bean hulls for ruminants. *Journal of Animal and Feed Sciences, 5,* 225–233.

Ni, Q., Ranawana, V., Hayes, H. E., Hayward, N. J., Stead, D., & Raikos, V. (2020). Addition of broad bean hull to wheat flour for the development of high-fiber bread: Effects on physical and nutritional properties. *Foods, 9,* 1192.

Nigus, A., Yusuf, M., & Animut, G. (2015). Determination of digestibility of faba bean (*Vicia faba* L.) hull and wheat bran in Afar sheep. *Journal of Biology, Agriculture and Healthcare, 5,* 186–191.

Nino-Medina, G., Muy-Rangel, D., & Urias-Orona, V. (2017). Chickpea (*Cicer arietinum*) and soybean (*Glycine max*) hulls: Byproducts with potential use as a source of high value-added food products. *Waste and Biomass Valorization, 8,* 1199–1203.

Palaniappan, A., Balasubramaniam, V. G., & Antony, U. (2017). Prebiotic potential of xylooligosaccharides derived from finger millet seed coat. *Food Biotechnology, 31,* 264–280.

Puls, W., Keup, U., Krause, H. P., Thomas, G., & Hoffmeister, F. (1977). Glucosidase inhibition: A new approach to the treatment of diabetes, obesity, and hyperlipoproteinaemia. *Naturwissenschaften, 64,* 536–537.

Ramakrishnaiah, N., Pratape, V. M., Sashikala, V. B., & Narasimha, H. V. (2004). Value addition to byproducts from dhal milling industry in India. *Journal of Food Science and Technology, 41,* 492–496.

Saharan, N. K., & Saroj, B. K. (2001). Processing of newly released rice bean and faba bean cultivars: Changes in total and available calcium, iron and phosphorus. *International Journal of Food Sciences and Nutrition, 52,* 413–418.

Salmean, Y. A., Segal, M. S., Palii, S. P., & Dahl, W. J. (2015). Fiber supplementation lowers plasma p-cresol in chronic kidney disease patients. *Journal of Renal Nutrition, 25,* 316–320.

Salunkhe, D. K., Kadam, S. S., & Chawan, J. K. (1985). *Post-Harvest Biotechnology of Food Legumes* (pp. 29–52). CRC Press.

Saman, P., Tuohy, K. M., Vázquez, J. A., Gibson, G., & Pandiella, S. S. (2017). *In vitro* evaluation of prebiotic properties derived from rice bran obtained by debranning technology. *International Journal of Food Sciences and Nutrition, 68,* 421–428.

Sherasia, P. L., Garg, M. R., & Bhanderi, B. M. (2017). Pulses and their by-products as animal feed (1st ed., pp. 158–159). FAO.

Sihag, N., & Kawatra, A. (2003). Hypolipidemic effect of pulse seed coats in rats. *Plant Foods for Human Nutrition, 58,* 1–10.

Stanisavljević, N. S., Ilić, M. D., Matić, I. Z., Jovanović, Ž. S., Čupić, T., Dabić, D. Č., & Tešić, Ž. L. (2016). Identification of phenolic compounds from seed coats of differently colored European varieties of pea (*Pisum sativum* L.) and characterization of their antioxidant and *in vitro* anticancer activities. *Nutrition and Cancer, 68,* 988–1000.

Vazquez-Olivo, G., Gutiérrez-Grijalva, E. P., & Heredia, J. B. (2019). Prebiotic compounds from agro-industrial by-products. *Journal of Food Biochemistry, 43,* e12711.

Yoganandhan, K., & Hameed, A. S. (2000). Evaluation of red gram (*Cajanus cajan*) and black gram (*Vigna mungo*) husks as food for brine shrimp, *Artemia* sp., culture. *Journal of Applied Aquaculture, 10,* 79–85.

CHAPTER 3

Oilseeds: Wastes, Byproducts, and Utilization

MARIYA NAYEEM[1] and AASIMA RAFIQ[2,*]

[1]*Division of Food Science and Technology, Sher-e-Kashmir University of Agricultural Sciences and Technology, Srinagar, Jammu and Kashmir, India*

[2]*Krishi Vigyan Kendra, Srinagar, Sher-e-Kashmir University of Agricultural Sciences and Technology, Jammu and Kashmir, India*

*Corresponding author

ABSTRACT

Oilseeds are one of the major agricultural commodities. Oilseeds are plants that produce oil-rich seeds or fruits, which are commonly used to extract edible oils. During the oil extraction process, various byproducts and wastes are generated. These byproducts and wastes can have different levels of usefulness and can be utilized in various ways to minimize waste and enhance sustainability. A major portion left after oil extraction is seed cake or meal, which is rich in protein, fiber, and other nutrients and can be used as animal feed, particularly in livestock and poultry farming. Hulls and shells, a byproduct of the production of oilseeds, have antioxidant properties, are rich in fiber, and can be utilized as animal feed or used as a source of biomass for energy production. Also, the seed hulls and fines were used to manufacture soap, waxes, and candles. Overall, oilseed waste is rich in bioactive compounds and is thus regarded as a valuable vehicle for food fortification, primarily to prevent malnutrition and improve human health. Furthermore, numerous products such as edible

Utilization of Food Waste and Byproducts. Mudasir Ahmad Malik and Kshirod Kumar Dash (Eds.)

DOI: 10.1201/9781779641151-3

films, protein concentrates, isolates, and emulsifiers may be obtained by properly processing these wastes and byproducts.

3.1 INTRODUCTION

In developing countries, the production of food results in significant amounts of waste and byproducts, which has a negative environmental and economic impact. The production of waste and byproducts has an effect on the realms of society, economy, and the environment. These impact the environment through increasing greenhouse gas emissions (Girotto et al., 2015). They are not used; thus they end up in municipal landfills where they produce leachate and cause major environmental issues through microbial breakdown. In other instances, the byproducts are burned to get rid of parasites and fungi. The expenditures associated with processing solid waste in landfills have a negative economic impact. In the greater framework of global food security, social consequences are co-dependent both ethically as well as morally, given that 805 million people globally experience hunger (FAO, 2014). We need more heterogeneous nutritional elements in order to address the dietary issues facing modern civilization. The existence of adequate amounts of lipids, micronutrients, carbohydrates, proteins, bioactive, and dietary fiber makes food wastes and byproducts of the utmost relevance. Food fortification is a key tactic in the fight against malnutrition, and in Europe and the USA, significant efforts have been made to make use of residues and byproducts (Girotto et al., 2015; Mirabella et al., 2014). Poorer countries with more severe hunger and malnutrition issues have a bigger potential for exploitation because they produce an abundance of byproducts. Oilseed crops are among the most significant food crops cultivated globally. Before the refining operations, the oilseeds must be cleansed for the manufacture of high-quality oil. There should also be efficient refining steps, such as mechanical expression, solvent extraction, degumming, neutralization, bleaching, and deodorization. These procedures lead to the formation of high quantities of byproducts and high-quality oil. Getting rid of these byproducts creates adverse environmental effects in addition to a multitude of other issues. Agro-industrial byproducts are widely available, particularly the leftover oilseed cakes and meals following oil extraction. According to OECD/FAO (2016), global production will increase to 386,000,000 t by 2025 from the current 317,000,000 t.

3.2 OILSEED TYPES AND THEIR NUTRITIONAL PROFILE

According to their extent of production, oilseeds are divided into major and minor oilseeds. Rapeseed, soybean, groundnut, sesame, sunflower, safflower, castor, linseed, and Niger are major oilseeds, while palm and coconut oil are minor oilseeds. There are small amounts of other oils, like cottonseed oil and rice bran oil. Protein, fat, fiber, vitamins, carbs, minerals, and vitamins are abundant in oilseeds (Ajila et al., 2012). India also leads the world in the production of important oilseed crops like castor, safflower, sesame, and nigella. India ranks first for the production of groundnuts, second for the production of rapeseed, and fifth for the production of soybeans (Smeu et al., 2022). Because of the impact of policy alternatives like minimum support prices and market intervention policies, India's oilseed and edible oil production has been exposed to the global market. But due to environmental policy and priority concerns in India, the productivity of oilseeds creates variance in the trend of the annual production of oilseeds. About 40% oil is found in rapeseed, which also supplies 18% of the world's demand for vegetable oil, while the two main crops, soybean and groundnut, contain 60% oil. Oilseeds are an excellent source of energy and also serve as a rich source of protein. Oil-intact seeds and meal are both made from this protein content. The pharma, food, and soap industries use the oil produced by the minor oilseed crops, which account for 5–10% of total oil production. Polyunsaturated fatty acids (PUFA), monounsaturated fatty acids (MUFA), and other fatty acids like oleic and stearic acid are all present in soybeans, which contain 18–22% fat (Young et al., 2010). About 50% of the fat in peanut oil is monounsaturated, and 14% is saturated. According to Smeu et al. (2022), sunflower seeds contain 194% PUFAs. The removal of harmful cholesterol by unsaturated fatty acids (UFAs) lowers the risk of obesity, atherosclerosis, and heart disease. Fat-soluble vitamins are transported via fatty acids, which also have antioxidant properties that reduce oxidative stress in the body. Oilseed fats are used as dietary ingredients. Groundnut is the most abundant source of arginine and has 20–26% protein content, while soybean, a major crop, only has 20 to 25% protein (Shahidi, 2009). Sesame oilseed contains 20% protein with albumin 8.6% and glutelin 6.9%, while rapeseed contains 34.50% protein, 39.88% albumins, and 46.25% glutelin's. Albumin makes about 20–42% of the protein in linseed, an oilseed, whereas globulins make up 26–58%. Additionally, oilseed provides water-soluble vitamins

B1, B2, B3, and C as well as fat-soluble vitamins A, E, and K, but not vitamin B12 (Shahidi, 2009; Smeu et al., 2022). The vitamins are crucial for the digestive system, nervous system, skin, and immune system. Ca, Mg, Zn, Mn, P, K, Fe, and Cu are minerals found in oilseeds that act as cofactors for enzymes. Additionally, oilseeds contain phytochemicals that are good for human health, such as polyphenols, carotenoids, phytosterols, and tocopherols. For the making of foods such extruded food, bakery goods, and confections, oilseeds are employed as functional components.

3.3 OILSEED WASTE

Pre-treatment, separation, and refinement are the three stages of seed processing. The oil is separated from the remainder of the seed material by expelling (commonly referred to as pressing) or extraction after physical-mechanical pre-treatment to remove husks and debris. Oilcake is a byproduct of expelling that is left behind. Expelling is typically followed by the extraction of any remaining oil from an oilcake containing seeds with a high oil content (sunflower, rapeseed, or canola), whereas extraction alone is performed with seeds with minimal oil content, such as soybeans.

3.4 VALORISATION OF OILSEEDS AS WASTE

3.4.1 HULLS

Historically, seed hulls and fines were utilized as low-grade animal bedding or feed. Acid waste water thereby obtained was frequently used to manufacture soap, and waxes were used to make candles. Hulls and shells, a byproduct of the production of oilseeds, have antioxidant properties. Due to the phenolic chemicals they contain, they are used as animal feed. Hulls include phenolic chemicals, flavonoids, catechins, and epicatechins, which are used as feed and fuel. Due to the fact that they are extracted from aqueous ethanolic extraction following saponification, sunflower oilseeds contain a lot of caffeic acid (Ribeiro et al., 2022). The hulls and fines may be used as a component of structural materials. In fact, recent research suggests that it may be possible to extract levoglucosan and/or furfural from bio-oils composed of pyrolyzed sunflower seed hulls (Casoni et al., 2019).

3.4.2 BLEACHING CLAY

There are financial and environmental incentives for reusing bleaching clay because it is an expensive natural resource that is generally dumped after use (Boukerroui et al., 2018). The spent earth can, at the very least, be used as fuel in boilers (Pandey et al., 2003). Solvents, acids, or alkalis have been used successfully to regenerate clay, but these techniques can be expensive to operate and produce ancillary liquid waste streams that could be harmful (Oladosu et al., 2017). A portion of the oil from bleaching clays can be recovered using more recent technologies such sub-critical water extraction and sub-critical fluid extraction, but no studies so far have looked at the possibility of using the clay again (Oladosu et al., 2017). There is a critical need for investigation into bleaching clay regeneration utilizing environmentally friendly chemical and physical procedures with minimal waste, and greener bleaching techniques, such as the use of boiler ash as a substitute bleaching chemical, as outlined by Lau et al. (2019).

3.4.3 WASTE WATER

Large amounts of water are used in extraction and refining, and the waste that they produce is challenging to recycle. Given the high organic concentration of edible seed oil waste, biological treatment may be a desirable method for organic elimination, particularly when oils are emulsified (Chipasa and Medrzyca, 2006). However, numerous studies have demonstrated that lipids persist in the treated effluent due to their simultaneous biodegradation and biosynthesis as well as their release during endogenous microbial decay (Chipasa and Medrzyca, 2008).

3.4.4 WASTEWATER SLUDGE

An indirect type of solid waste that is typically disposed of in landfills is wastewater sludge from the ESO industry (Welz et al., 2017). The sludge has a 76% organic content and a nutritional level that made it suitable for compost as a low-cost product (Pandey et al., 2003). Additionally, it has been demonstrated that a significant portion of this clarifier sludge contains oil that is suitable for the development of second-generation biofuel (Muanruksa and Kaewkannetra, 2020). While most next-generation biofuels cannot currently be produced commercially, it is only a matter of time before this will change as nations transition to circular economies and industrial methods advance (Corea et al., 2019). Additionally, ways should be explored to reuse other types of solid waste as well, like deodorizer distillate and spent nickel catalyst (Pandey et al., 2003). For instance, saponified deodorizer distillates contain significant amounts of important nutraceuticals, such as tocopherols and sterols (Sherazi et al., 2016).

3.5 OILSEEDS AS BYPRODUCTS

Oilseeds are a byproduct that is extracted from the seed parts and are widely used in industry to produce various oil byproducts. Proteins and water make up the majority of food items (De Castro et al., 2017). These two elements work together to provide the food items their useful qualities. According to Bejar et al. (2011), these interactions have an impact on the qualities of food products, such as swelling capacity, oil holding capacity, and oil absorption capacity. These characteristics, which are regarded as functional characteristics, have an impact on how generally food impacts health. These are crucial for many other critical industrial processes, such as food preservation, storage, and texture (Neethirajan and Jayas, 2011). Sunflower, peanut, pumpkin, and sesame seeds are among the different oilseeds utilized to make byproducts. Following are descriptions of various oilseeds and their functional groups:

3.5.1 SUNFLOWER OILSEED

Sunflower is part of the Asteraceae family and gets its name from its similarity to the sun. One of the often-produced crops and seeds is sunflower seed. It has two different kinds of seeds: seeds used for making confections and seeds used to make oil. These seeds are high in dietary fiber, UFAs, antioxidants, flavonoids, amino acids, proteins, and vitamins (Aslaksen et al., 2007). Calcium, iron, magnesium, manganese, zinc, and many other minerals are among the additional minerals. There are linoleic and oleic fatty acids among them. Apigenin, quercetin, and luteolin are some of the flavonoids. Cysteine, phenylalanine, leucine, glutamic, tyrosine, and methionine are among the amino acids, which account for 20% of the total contents (Petraru et al., 2021). Sunflower oilseeds are a great source of components with functional qualities and serve as a great functional food. It contains the PUFA linolenic acid. It is an essential precursor to omega 3, which is required for the body's antithrombotic and immune-modulating functions. It contains numerous byproducts and precursors for numerous more compounds in addition to the basic constituents. It contains proteins and other phenolic byproducts, for instance (Filho and Egea, 2021). It also contains ALA, a precursor to eicosatetraenoic acid (EPA) and docosahexaenoic acid (DHA). Due to its various qualities, including anti-inflammatory, hypotriglyceride, and hypotensive effects, this substance is essential in controlling how the human body functions. Last but not least, it contains byproducts from the vitamin and tocopherol categories, which are highly valuable biological and nutritional components and good functional ingredients for human body.

3.5.2 GROUNDNUT OILSEED

The very nutritious groundnut oilseeds, also known as peanut oilseeds and arachis oil, are noted for their exceptional composition. Vitamins, UFAs, and PUFA are all abundant in it. It contains a variety of byproducts, including proteins, fatty acids, phenolic compounds, and vitamins. More than 32 proteins make up the constituents of peanut oil, many of which are byproducts. It contains a variety of vitamins, including pantothenic acid, thiamine, riboflavin, and niacin, which serve as useful elements in various foods (Arya et al., 2016). Additionally, several phytochemicals are derived from peanuts during the oil extraction process. These phytochemicals have anti-tumor, anti-cancer, and antiproliferative properties. They include resveratrol, campestral, and β-sitosterol. Breast, colon, and prostate cancer are all prevented by the presence of these functional groups. Numerous triglycerides, which have atherogenic potential and aid in the prevention of cardiovascular illnesses, are formed as byproducts from peanut oilseeds (Akhtar et al., 2014).

3.5.3 PUMPKIN SEED OIL

Pumpkin oilseeds contain a wide variety of components with a variety of beneficial attributes. Numerous PUFAs, MUFA, vitamins, and other mineral components are present in it. Due to its advantages in promoting hair development, nourishing the skin, and enhancing health, it is well-liked in the market. Urinary tract infections are frequently treated with it. During the extraction process, several alcohols are produced, and phenols are byproducts with potent antioxidant capabilities (Saavedra et al., 2015).

3.5.4 SESAME SEED OIL

A popular oil and oilseed due to its anti-inflammatory, antioxidant, and antiproliferative properties is sesame. It is abundant in vitamins, proteins, carbs, and dietary fiber. It contains several vitamins as well as linoleic, oleic, stearic, and palmitic acids. Numerous byproducts are also created during the extraction process in addition to the basic products. Phytosterols, sesamol, lignans, and sesamin are some of these byproducts (Asghar et al., 2014). The ability of phytosterols to lower the risk of cardiovascular disease and stop intestinal cholesterol buildup is one of its many useful features. According to Jayaraj et al. (2020), sesamol is a great antioxidant that aids in the prevention of numerous chronic diseases. The other byproducts contribute to a decrease in stress, blood sugar, and inflammation.

3.5.5 RAPESEED OIL

Due to its mild flavor, smooth texture, and high smoke point, rapeseed oil is utilized in cooking. It is used for cooking methods like baking, grilling, and stir-frying. Salad dressings, sauces, and marinades all make use of it. Rapeseed oil's fatty acid content has positive effects on human health. The rapeseed oil contains bioactive components and other bio-peptides that are used to produce a variety of byproducts that customers can use and give value to the food industry (Abdul-Hamid and Luan, 2000; Jayaraj et al., 2020).

3.6 OILSEEDS UTILIZATION AS BYPRODUCTS AND WASTES

Proteins, lipids, minerals, and vitamins make up oilseed byproducts, which serve as useful additives and add nutrition to food. There are numerous oilseed byproducts accessible in the markets, the most popular of which are discussed in subsections.

3.6.1 COOKING OIL

As a useful culinary item, cooking oil or vegetable oils are an excellent alternative to animal fat. The vegetable oils are utilized as health-added products for customers and contain monosaturated fatty acids (MUFA) and PUFA. The choice of cooking oil must be made carefully because it directly affects the consumer's health (Young et al., 2010). Cooking oils that are healthy and include the necessary fatty acids 3 and 6 include canola, mustard, rice bran, and groundnut oils. These fatty acids are not produced by the body; instead, they are obtained from the outside through eating food. The same 6-fatty acids are present in cottonseed, sunflower, and groundnut. Because it has good cooking qualities and a pleasant flavor, palm oil is suitable for use as vegetable oil. Due to the presence of PUFA and MUFA, mustard oil is also suitable for frying and cooking (Kammoun et al., 2011).

3.6.2 SHORTENINGS

Shortenings, which are mostly utilized in the manufacture of baked goods, are fats made from animal sources like butter and lard. The production of shortenings uses a variety of vegetable oils, including cottonseed, palm, soybean, and sunflower oil. When making shortening, mixtures of two or more oils are best measured. Shortenings work well in baked goods because they provide the food moisture and softness. The shortenings are used to make a range of commodities, including deep-fried goods, baked goods, snacks, and confections (Aluyor and Ori-Jesu, 2008).

3.6.3 MARGARINE/SALAD OILS

Margarines are mostly oil. For the creation of margarine, vegetable oils such palm, soybean, sunflower, cottonseed, and corn oil are used. Margarine can be used in place of butter to make bread, biscuits, cakes, and frosting. As margarine made from vegetable oils contains plant sterols, which lower blood cholesterol in our bodies, it is used as a functional food (Aluyor and Ori-Jesu, 2008). Plastic margarine was made by combining sunflower oil with additional fats and oleic acid. Margarine is made from vegetable oils and has 20–40% fat content. The use of groundnut oil in food salads and its suitability for human consumption has widely been studied (Ribeiro et al., 2022).

3.6.4 OIL CAKES/MEALS

Oilseeds are used to make byproducts like oil cakes and meals. Oilcake and meals are produced using mechanical and solvent extraction techniques. Soybean cake, cottonseed cake, rapeseed, groundnut cake, sunflower seed cake, safflower seed cake, flaxseed cake, and so forth, are some of the oilseed's cakes that are available on the market. Meals are used to enhance the oilcake in the diets of undernourished people, and oilseed cakes and isolates are employed as protein concentrates, isolates, and defatted oilcake. They are employed in the production of extruded and baked products, and morning cereals (Aluyor and Ori-Jesu, 2008).

3.7 BIOACTIVE COMPOUNDS OF OILSEED BYPRODUCTS

Active substances that improve health are known as bioactive components, which give food its disease-preventing qualities. These include: flavonoids, carotenoids, phytosterols, phenols, and PUFA. These ingredients are used to inhibit bacterial toxins, reduce cholesterol absorption, and function as antioxidants. They guard against cancer, heart disease, and other illnesses. Oilseeds contain phenolic, flavonoids, glucosinolates, lignan, tannins, and indoles, among other bioactive components. One form of abundant supply of phytosterols is through oil seeds (Roche et al., 2016). Phytosterols have anti-inflammatory, anti-bacterial, anti-carcinogenic, and anti-oxidative properties. They serve as an emulsifier and aid in extending the shelf life of vegetable oils. Phenolics are hydroxyl groups with aromatic hydrocarbon groups and are the most common class of secondary metabolites and function as plant defense substances (Akbarirad et al., 2016). They possess anti-tumor, anti-oxidative, and anti-inflammatory effects. Carotenoids are regarded as exceptional pigments that play the job of giving green, red, and yellow colors to the plant and have anti-oxidant effects against cancer, cardiovascular disorders, and skin conditions (Fadavi et al., 2006). Tocopherols (antioxidants that are lipophilic) are phenolic compounds found in plants that provide oilseeds their antioxidant and nutritive characteristics. These tocopherols have anti-inflammatory and anti-hypertensive properties in addition to inhibiting cholesterol levels. Human health is improved by the use of polyphenolic substances. Since they protect against free radicals and other oxidative stressors and give antioxidant benefits, phenolic compounds are referred to as secondary components. Vegetable oils are shielded from oxidation by them. They function as antioxidants and control metabolic processes linked to inflammatory and malignant conditions. Groundnut and soybean seeds are rich in flavonoids, which have anti-inflammatory and antioxidant characteristics (Ashaolu, 2020). Anti-microbial, anti-tumor, anti-inflammatory, anti-oxidant, anti-radical, and anti-bacterial properties of these active ingredients offer health advantages and successfully capture market share. The lignans like sesamin, sesamolin, sesaminol, and tocopherol isomers can be found in sesame seeds and oil which operate as antioxidants and have oxidative rancidity resistance.

3.8 FUNCTIONAL COMPOUNDS FROM OILSEED BYPRODUCTS

The byproducts of oilseeds are rich in bioactive and functional chemicals that are extracted because of their high nutritional value. When different oilseeds are extracted for their oil, many sorts of chemicals are obtained, including phenols, lipids, polyestrous, and polyphenols. These useful components from oilseed byproducts serve to hydrate skin, enhance bodily processes, lower the risk of tumor development, and enhance general health. These goods are often separated into saponifiable and unsaponifiable fractions of useful oilseed components.

3.8.1 SAPONIFIABLE FRACTION OF FUNCTIONAL COMPOUNDS FROM OILSEED BYPRODUCTS

It is the majority of the product that was obtained by the extraction procedure. About 90–98% of the oil output is made up of it (Díaz et al., 2019). Triglycerides are a significant group of chemicals from this category because of their double bonds and lengthy, hefty chains. These substances also only contain a single link, making them both saturated and UFAs. Both cis and trans configurations exist in their unsaturated form. UFAs are substances that are frequently acquired. Following that, numerous molecules from the omega nomenclature, such as omega-3, omega-6, and omega-9, are present (Cox et al., 1990). The majority of oilseeds are saponifiable, such as coconut (Marina et al., 2009), almonds (Barku et al., 2012), groundnut, and caster. Mineral oils, on the other hand, cannot be saponified. FA (fatty acids) is abundant in oil seeds, but saponification reduces their quantity and transforms them into useful substances. For instance, according to Odia et al. (2015) and Kao and Chen (2006), palm oil has a palmitic acid content of approximately 44%. The human metabolic system also requires this saponifiable fraction, which contains byproducts including wax, glycerides, triglycerides, and phospholipids.

3.8.2 UNSAPONIFIABLE FRACTION OF FUNCTIONAL COMPOUNDS FROM OILSEED BYPRODUCTS

Only 1 to 2% is unsaponifiable, and they are those that solubilize in organic solvents but are either insoluble or slightly soluble in water. Triterpenes, phenolic compounds, β-sitosterol, and other colors are a few examples of these components (Bulotta et al., 2014). One of the key elements in this category is phytosterols. These phytosterols have a side chain with about 10 carbons at C-17 and are tetracyclic cyclopenta[α-phenanthrenes]. Phytostanols are their saturated form; nonetheless, they are rare. Phytosterol content varies and is typically between 1000 and 5000 mg/kg (Fernandes and Cabral, 2007). Alongside the phytosterols, tocopherols, with a chroman ring and hydrophobic side chains, are another element of the unsaturated portion. Like tocotrienols, tocopherols also have functions similar to those of Vitamin E (Munné-Bosch and Alegre, 2002). Owing to their capacity to oxidize unsaturated FA, tocopherols and tocotrienols are in demand. Corn, peanuts, and cottonseed all have large amounts of these functional groups, although grape seeds and palm seeds only contain little amounts of them. These enhance the nutritional value of the oilseeds by being able to oxidize the fatty acids. It has been discovered to be beneficial against several malignancies (Sylvester et al., 2014). This group includes phenolic substances, which are crucial for the foods' anti-proliferative, anti-hypertensive, antidiabetic, and antioxidant activities. Phenolic chemicals have been shown to enhance the immune system and inhibit cell division (Hollman, 2001). The efficiency of these functional classes of oilseeds has been demonstrated in numerous researches. It might be argued that this class has numerous functional and bioactive groups that are advantageous to human health (Nirmala et al., 2014; Santos-Buelga et al., 2019).

3.9 CONCLUSION

The development of sustainable strategies for handling byproducts and edible waste is one of the primary concerns facing our civilization. These solutions must be able to maximize the biological potential of byproducts in order to have a positive impact on the economy, society, and environment. Given the nutritional problems that contemporary society is facing (hunger indices and the growing world population), using food waste for human nutrition should be prioritized. Wastes and byproducts produced in developing countries can be used in the composition of foods for excellent nutritive and functional benefits, as well as an effective tool to combat hunger. Additionally, the extra value brought about by the diversification of the producing chains may result in the creation of employment possibilities for locals, adding yet another social advantage.

KEYWORDS

- **bioactive compounds**
- **fatty acids**
- **human health**
- **oilseed waste**
- **phytochemicals**
- **seed meal**

REFERENCES

Abdul-Hamid, A., & Luan, Y. S. (2000). Functional properties of dietary fiber prepared from defatted rice bran. *Food Chemistry, 68*(1), 15–19.

Ajila, C. M., Brar, S. K., Verma, M., Tyagi, R. D., Godbout, S., & Valéro, J. R. (2012). Bio-processing of agro-byproducts to animal feed. *Critical Reviews in Biotechnology, 32*(4), 382–400.

Akbarirad, H., Ardabili, A. G., Kazemeini, S. M., & Khaneghah, A. M. (2016). An overview on some of important sources of natural antioxidants. *International Food Research Journal, 23*(3).

Akhtar, S., Khalid, N., Ahmed, I., Shahzad, A., & Suleria, H. A. R. (2014). Physicochemical characteristics, functional properties, and nutritional benefits of peanut oil: A review. *Critical Reviews in Food Science and Nutrition, 54*(12), 1562–1575.

Aluyor, E. O., & Ori-Jesu, M. (2008). The use of antioxidants in vegetable oils–A review. *African Journal of Biotechnology, 7*(25).

Arya, S. S., Salve, A. R., & Chauhan, S. (2016). Peanuts as functional food: A review. *Journal of Food Science and Technology, 53,* 31–41.

Asghar, A., Majeed, M. N., & Akhtar, M. N. (2014). A review on the utilization of sesame as functional food. *American Journal of Food and Nutrition, 4*(1), 21–34.

Ashaolu, T. J. (2020). Immune boosting functional foods and their mechanisms: A critical evaluation of probiotics and prebiotics. *Biomedicine & Pharmacotherapy, 130,* 110625.

Aslaksen, M. A., Kraugerud, O. F., Penn, M., Svihus, B., Denstadli, V., Jørgensen, H. Y., Hillestad, M., Krogdahl, Å., & Storebakken, T. (2007). Screening of nutrient digestibility's and intestinal pathologies in Atlantic salmon, *Salmo salar*, fed diets with legumes, oilseeds, or cereals. *Aquaculture, 272*(1–4), 541–555.

Barku, V. Y., Nyarko, H. D., & Dordunu, P. (2012). Studies on the physicochemical characteristics, microbial load and storage stability of oil from Indian almond nut (*Terminalia catappa* L.). *Food Science and Quality Management, 8,* 9–17.

Bejar, K. A., Ghanem, N., Mihoubi, D., Kechaou, N., & Mihoubi, B. N. (2011). Effect of infrared drying on drying kinetics, color, total phenols and water and oil holding capacities of orange (*Citrus sinensis*) peel and leaves. *International Journal of Food Engineering, 7*(5).

Boukerroui, A., Belhocine, L., & Ferroudj, S. (2018). Regeneration and reuse waste from an edible oil refinery. *Environmental Science and Pollution Research, 25,* 18278–18285.

Bulotta, S., Celano, M., Lepore, S. M., Montalcini, T., Pujia, A., & Russo, D. (2014). Beneficial effects of the olive oil phenolic components oleuropein and hydroxytyrosol: Focus on protection against cardiovascular and metabolic diseases. *Journal of Translational Medicine, 12,* 1–9.

Casoni, A. I., Gutierrez, V. S., & Volpe, M. A. (2019). Conversion of sunflower seed hulls, waste from edible oil production, into valuable products. *Journal of Environmental Chemical Engineering, 7*(1), 102893.

Chipasa, K. B., & Mędrzycka, K. (2006). Behavior of lipids in biological wastewater treatment processes. *Journal of Industrial Microbiology and Biotechnology, 33*(8), 635–645.

Chipasa, K. B., & Mędrzycka, K. (2008). Characterization of the fate of lipids in activated sludge. *Journal of Environmental Sciences, 20*(5), 536–542.

Correa, D. F., Beyer, H. L., Fargione, J. E., Hill, J. D., Possingham, H. P., Thomas-Hall, S. R., & Schenk, P. M. (2019). Towards the implementation of sustainable biofuel production systems. *Renewable and Sustainable Energy Reviews, 107,* 250–263.

Cox, R. A., García-Palmieri, M. R., Walker, H. K., Hall, W. D., & Hurst, J. W. (1990). Cholesterol, triglycerides, and associated lipoproteins. In *Clinical Methods: The History, Physical, and Laboratory Examinations* (3rd ed., Chapter 31). Boston: Butterworths.

De Oliveira, F. J. G., & Egea, M. B. (2021). Sunflower seed byproduct and its fractions for food application: An attempt to improve the sustainability of the oil process. *Journal of Food Science, 86*(5), 1497–1510.

Díaz, G., Pega, J., Primrose, D., Sancho, A. M., & Nanni, M. (2019). Effect of light exposure on functional compounds of monovarietal extra virgin olive oils and oil mixes during early storage as evaluated by fluorescence spectra. *Food Analytical Methods, 12,* 2709–2718.

Fadavi, A., Barzegar, M., & Azizi, M. H. (2006). Determination of fatty acids and total lipid content in oilseed of 25 pomegranate varieties grown in Iran. *Journal of Food Composition and Analysis, 19*(6–7), 676–680.

Fernandes, P., & Cabral, J. M. S. (2007). Phytosterols: Applications and recovery methods. *Bioresource Technology, 98*(12), 2335–2350.

Food and Agriculture Organization of the United Nations. (2014). *Panorama De La Seguridad Alimentaria Y Nutricional En América Latina y el Caribe 2014.*

Food and Agriculture Organization of the United Nations. (2016). *The State of World Fisheries and Aquaculture: Contributing to Food Security and Nutrition for all.* Rome: FAO.

Girotto, F., Alibardi, L., & Cossu, R. (2015). Food waste generation and industrial uses: A review. *Waste Management, 45,* 32–41.

Hollman, P. C. H. (2001). Evidence for health benefits of plant phenols: Local or systemic effects? *Journal of the Science of Food and Agriculture, 81*(9), 842–852.

Jayaraj, P., Narasimhulu, C. A., Rajagopalan, S., Parthasarathy, S., & Desikan, R. (2020). Sesamol: A powerful functional food ingredient from sesame oil for cardioprotection. *Food & Function, 11*(2), 1198–1210.

Kao, T. H., & Chen, B. H. (2006). Functional components in soybean cake and their effects on antioxidant activity. *Journal of Agricultural and Food Chemistry, 54*(20), 7544–7555.

Lau, S. Y., Phuan, S. L., Danquah, M. K., & Acquah, C. (2019). Sustainable palm oil refining using pelletized and surface-modified oil palm boiler ash (OPBA) biosorbent. *Journal of Cleaner Production, 230,* 527–535.

Marina, A. M., Che Man, Y. B., Nazimah, S. A. H., & Amin, I. (2009). Chemical properties of virgin coconut oil. *Journal of the American Oil Chemists' Society, 86,* 301–307.

Mirabella, N., Castellani, V., & Sala, S. (2014). Current options for the valorization of food manufacturing waste: A review. *Journal of Cleaner Production, 65,* 28–41.

Muanruksa, P., & Kaewkannetra, P. (2020). Combination of fatty acids extraction and enzymatic esterification for biodiesel production using sludge palm oil as a low-cost substrate. *Renewable Energy, 146,* 901–906.

Munné-Bosch, S., & Alegre, L. (2002). The function of tocopherols and tocotrienols in plants. *Critical Reviews in Plant Sciences, 21*(1), 31–57.

Neethirajan, S., & Jayas, D. S. (2011). Nanotechnology for the food and bioprocessing industries. *Food and Bioprocess Technology, 4,* 39–47.

Nirmala, C., Bisht, M. S., & Laishram, M. (2014). Bioactive compounds in bamboo shoots: Health benefits and prospects for developing functional foods. *International Journal of Food Science & Technology, 49*(6), 1425–1431.

Odia, O. J., Ofori, S., & Maduka, O. (2015). Palm oil and the heart: A review. *World Journal of Cardiology, 7*(3), 144.

Oladosu, W. A., Manan, Z. A., & Alwi, S. W. (2017). Recovery of vegetable oil from spent bleaching earth: State-of-the-art and prospect for process intensification. *Chemical Engineering Transactions, 56,* 133–138.

Pandey, R. A., Sanyal, P. B., Chattopadhyay, N., & Kaul, S. N. (2003). Treatment and reuse of wastes of a vegetable oil refinery. *Resources, Conservation and Recycling, 37*(2), 101–117.

Petraru, A., Ursachi, F., & Amariei, S. (2021). Nutritional characteristics assessment of sunflower seeds, oil and cake: Perspective of using sunflower oilcakes as a functional ingredient. *Plants, 10*(11), 2487.

Ribeiro, A. R., Silva, S. S., & Reis, R. L. (2022). Challenges and opportunities on vegetable oils derived systems for biomedical applications. *Materials Science and Engineering,* 112720.

Roche, J., Alignan, M., Bouniols, A., Cerny, M., Mouloungui, Z., & Merah, O. (2010). Sterol concentration and distribution in sunflower seeds (*Helianthus annuus* L.) during seed development. *Food Chemistry, 119*(4), 1451–1456.

Saavedra, M. J., Aires, A., Dias, C., Almeida, J. A., De Vasconcelos, M. C. B. M., Santos, P., & Rosa, E. A. (2015). Evaluation of the potential of squash pumpkin byproducts (seeds and shell) as sources of antioxidant and bioactive compounds. *Journal of Food Science and Technology, 52,* 1008–1015.

Santos-Buelga, C., González-Paramás, A. M., Oludemi, T., Ayuda-Durán, B., & González-Manzano, S. (2019). Plant phenolics as functional food ingredients. *Advances in Food and Nutrition Research, 90,* 183–257.

Shahidi, F. (2009). Nutraceuticals and functional foods: Whole versus processed foods. *Trends in Food Science & Technology, 20*(9), 376–387.

Sherazi, S. T. H., & Mahesar, S. A. (2016). Vegetable oil deodorizer distillate: A rich source of the natural bioactive components. *Journal of Oleo Science, 65*(12), 957–966.

Smeu, I., Dobre, A. A., Cucu, E. M., Mustățea, G., Belc, N., & Ungureanu, E. L. (2022). Byproducts from the vegetable oil industry: The challenges of safety and sustainability. *Sustainability, 14*(4), 2039.

Sylvester, P. W., Akl, M. R., Malaviya, A., Parajuli, P., Ananthula, S., Tiwari, R. V., & Ayoub, N. M. (2014). Potential role of tocotrienols in the treatment and prevention of breast cancer. *Biofactors, 40*(1), 49–58.

Welz, P. J., Le Roes-Hill, M., & Swartz, C. (2017). *NATSURV 6 Water and Wastewater Management in the Edible Oil Industry* (2nd ed.). WRC Report (TT702/16).

Young, G., Nippgen, F., Titterbrandt, S., & Cooney, M. J. (2010). Lipid extraction from biomass using co-solvent mixtures of ionic liquids and polar covalent molecules. *Separation and Purification Technology, 72*(1), 118–121.

CHAPTER 4

Vegetables: Wastes, Byproducts, and Utilization

RIA AHUJA[1,*] and KUNAL BHARDWAJ[2]

[1]*Department of Foods and Nutrition, The M.S. University of Baroda, Vadodara, Gujarat, India*

[2]*Public Health Foundation of India, Saidulajab, Sainik Farm, New Delhi, India*

*Corresponding author

ABSTRACT

Food waste is generated in abundant amounts in agricultural lands during production, harvest, sorting, or as industrial waste during product development, or in households during the preparation of meals. The waste is produced throughout the supply chain from harvesting to its final consumption, which needs to be utilized in the best possible way, without having invested huge amounts in it and without creating environmental hazards, supporting the concept of a circular economy. The FAO has estimated the global volume of food wastage at 1.6 billion tons of food waste, which also accounts for 1.3 billion tons of "wastage of edible portions." Most of the food wastage ends up in landfills and gets completely rotten, contributing to greenhouse gas emissions from these dumping grounds. Since vegetable wastes are perishable in nature, actions to utilize these wastes need to be quick. Recent research has documented the generation of renewable energy by bioconversion as one of the proficient means of utilizing these vegetable wastes, in addition to strategies like farm animal nutrition, production of nutraceuticals deriving fibers from various vegetables, or derivation of bioactive compounds like antioxidants, minerals,

Utilization of Food Waste and Byproducts. Mudasir Ahmad Malik and Kshirod Kumar Dash (Eds.)

DOI: 10.1201/9781779641151-4

polyphenolic compounds, and so forth. The use of modern techniques like pulsed electric fields, hydrostatic pressure, and radiation are widely being used to put the waste to its best use, though every advantage comes with certain drawbacks, which need to be understood before putting it into use.

India is indubitably the largest global agricultural powerhouse. A growing population will exacerbate further food insecurity. Diet habits are changing with the growing population, which leads to alteration in the production and processing of foods to meet the increased demands. The processing of food generates 25–30% waste of the absolute product. The FAO has estimated the global volume of food wastage at 1.6 billion tons of food waste, which also accounts for 1.3 billion tons of "wastage of edible portions." According to MOHUA, it has been estimated that India produces about 42.0 million tons of municipal solid waste (MSW) annually. The per capita waste generation in major Indian cities ranges from 0.2 kgs to 0.6 kgs. Out of the total municipal waste collected, on average, 94% is dumped on land, and 5% is composted, which results in environmental hazards. The waste is produced throughout the supply chain from harvesting to its final consumption, which needs to be utilized in the best possible way. The Sustainable Development Goals (SDGs) also accentuate food security and sustainability, which emphasizes designing a better way to ensure sustainable consumption and production patterns. Vegetable waste majorly contains potential bioactive compounds, which can be further processed and utilized in different industries such as the health industry (pharmaceutical), food industry, and textile industry. This chapter describes the nature of the waste generated from vegetables, modern techniques of processing, the bioactive components, and the potential techniques to minimize and utilize its byproducts.

4.1 INTRODUCTION

India's population is growing at a steady pace day by day. Currently, it has a population of 1.27 billion people with an increased demand for food. While the GDP of India has declined from 1951 to 2011, achieving food security with the growing population is quite a challenge for India. As per the recent Global Hunger Index Report of 2022, India ranks at 107th out of 121 countries on stunting, wasting, and under 5 mortalities (Mogge and MacSorley, 2022). Agriculture is incontrovertibly the largest livelihood provider in India, especially in rural areas. It contributes to increasing the overall Gross Domestic Product (GDP) of the country. Sustainable agriculture, as it relates to food security, rural employment, and environmentally friendly technologies like soil conservation and sustainable resource management, are all parts of holistic rural development.

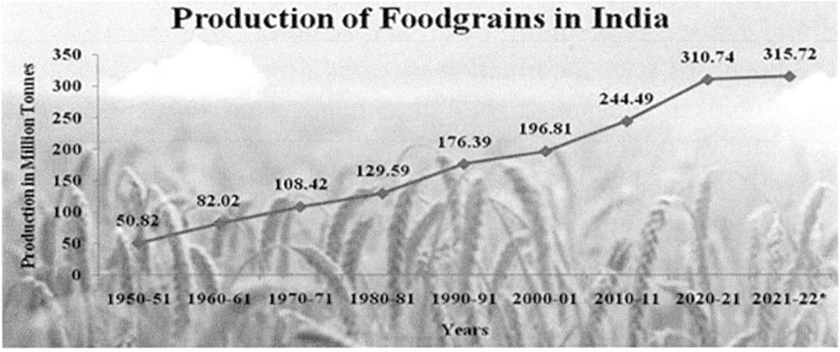

FIGURE 4.1 Production of food grains in India.

Source: Data for 2021–22 is from 4th Advance estimates of food grain production.

The Directorate General of Commercial Intelligence and Statistics (DGCI&S) has released data showing that agricultural exports to India increased by 19.92% in 2021–2022 to reach $50.21 billion (Bureau, 2022) (Figure 4.1). However, India still has many emerging concerns and challenges. Agriculture's percentage part of the total GVA of the economy has a long-term trend of roughly 18% as the Indian economy has expanded and diversified (Figure 4.2).

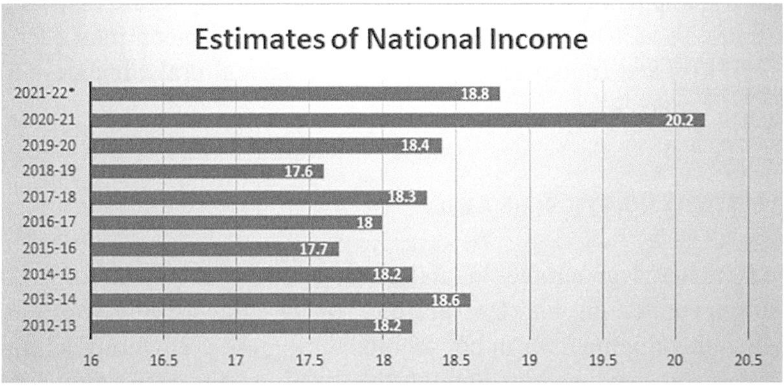

FIGURE 4.2 Estimates of national income.

Source: Based on data from DAFW As per 1st Advance Estimates of National Income (2021–2022).

The exponential growth is expected to be seen in the Indian food ingredients market, currently estimated at $1584.08 million, which is projected to grow at a CAGR of 6.7% till 2029. Future prospects for the supply and demand of food are anticipated to vary as a result of factors including population increase, rising per capita income, urbanization, shifting tastes and preferences, economic growth, and so forth. Due to its 32% share of the country's total food market in terms of production, consumption, exports, and anticipated growth, the Indian food processing industry placed fifth overall (TH, 2022).

India continues to have a fourth of the world's hungry population despite reaching food sufficiency in production. The Global Nutrition Report 2022 claims that since the Covid-19 outbreak, the number of individuals who experience hunger has increased by 150 million, from 618 million in 2019 to 768 million in 2021, and the number of people who cannot afford a healthy diet has increased by 112 million, to 3.1 billion in 2020 alone. In 2021, 2.3 billion individuals over the world (29.3% of the total population), an increase from 25.4% prior to the pandemic, experienced moderate to severe food insecurity. According to reports, India has not made any progress in meeting the objective for under-five child wasting (17.3%), which is among the highest in the world and higher than the average for the Asia area (8.9%) (Nutrition, 2022). SDG-2's "Zero Hunger" goal aims to eradicate hunger while also advancing sustainable agriculture, achieving food security, and improving nutrition. This can be done by promoting sustainable agriculture while using cutting-edge technology and equitable distribution methods. To assure a rise in food production and a concomitant decrease in food loss and food waste, the numerous agricultural advances might be mainstreamed.

4.2 FOOD WASTE SCENARIO

Food waste is generated in abundant amounts in agricultural lands during production, harvest, sorting, or as industrial waste during product development, or in households during the preparation of meals. The waste is produced throughout the supply chain from harvesting to its final consumption, which needs to be utilized in the best possible way, without having invested huge amounts in it and without creating environmental hazards, supporting the concept of the circular economy. FAO has estimated the global volume of food wastage at 1.6 billion tons of food waste, which also accounts for 1.3 billion tons of "wastage of edible portions." To combat food security, steps to reduce wastage have to be adopted. It has been estimated that a 50% reduction in waste would create 1314 trillion Kcal per year, which represents an additional 22% of calories that could be fed to the population estimate by 2050. Apart from food security, waste management would also contribute towards climate change and reduction in greenhouse emissions. Most of the food waste ends up in landfills and gets completely rotten, contributing to greenhouse gas emissions from these dumping grounds. Since the waste vegetables are perishable, actions to utilize this waste need to be quick. Recent researchers have documented the generation of renewable energy by bioconversion as one of the proficient means of utilizing vegetable wastes, in addition to this strategy like farm animal nutrition, production of nutraceuticals deriving fibers from various vegetables, or derivation of bioactive compounds like antioxidants, minerals, polyphenolic compounds, and so forth. The use of modern techniques like pulsed electric fields, hydrostatic pressure, and radiation is widely being used to put the waste to its best use, though every advantage comes with certain drawbacks, which need to be understood before putting it into use.

According to MOHUA, India is projected to produce roughly 42.0 million tons of municipal solid trash per year (GOI, 2022). Incorrect waste management or incineration will result in pollution of the air and water, further endangering human health. Methane emissions from burning waste, which are a major source of greenhouse gases and smog, have worsened the air quality and climate in India and contributed to the country's rising respiratory disease rates. In addition, pungent odors and water leachates have contributed to the rise in public health problems. According to the Central Pollution Control Board's GOI 2021 projection, 119 grams of solid waste are generated daily per person (Central Pollution Control Board, 2020). The entire amount of solid waste produced in the nation is 160038.9 TPD, of which 152749.5 TPD is collected, treated, and landfilled, accounting for 50% of the total amount, while 29427.2 TPD, or 31.7%, is still unaccounted for (Ahuja, 2022). To combat this issue of burning Parali, due to the lack of facilities to decompose it, IARI institute is working on the development of a Pusa decomposer that can decompose the paralic in 25 days. Further development of Sampoorna, which is a liquid biofertilizer, will help in replenishing the fertility of the land by increasing phosphorus, potassium, and nitrogen in the soil (Singh, 2021).

Globally, 675 million tons of vegetables and fruits are produced, of which 1.3 billion are wasted. India contributes 86 million tons of production and 5.6 million of wastage annually. The reason behind excessive dumping is the lack of control over food processing and agri-economy industries. There has been a rough estimate that India produces approx. About 400 tons of waste, taking into consideration vegetables, flowers, household peel waste, and so forth, which is 40% of the 1000 tons of solid waste generated daily. Since the proposed disposal of waste material is becoming an issue due to lack of its utilization and lack of strategy use, efforts should be taken to best utilize the waste and improve the polluted cities (Omre et al., 2018).

In 2019, 17% of the food available to consumers was wasted (11% in households, 5% in food service, and 2% in retail), leaving 690 million people hungry (Figure 4.3), according to the United Nations Environment Program's (UNEP) "Food Waste Index Report 2021." The report also stated that around 931 million tons of food waste were produced in 2019, with households accounting for 61% of this total, food services for 26%, and retail for 13% (UNEP, 2021). SDG 12.3 (SDG 12.3) encapsulates a commitment to reduce food loss throughout supply chains and to cut back by half on retail and consumer food waste (UN, 2022).

Food waste has equally become an issue across all income countries, such as high, upper-middle, and lower-middle income countries. A developed country like Austria has recorded very little food waste, that is, 39 kg/capita/year (Figure 4.4). Whereas, a country like Nigeria reported a waste of 189 kg/per capita/year, which is much higher than other countries. In India, the waste calculated was 50 kg/capita/year. However, compared to nations with higher GDPs, lower-GDP countries did show a broader range of variation in their estimations of food waste. The cause might be a greater range of food waste in low-income nations or a flaw in the studies that measure food waste, such as a small sample size (UNEP, 2021).

Food waste undermines the sustainability of the food systems. This also results in the wastage of all the resources (water, land, energy, labor,

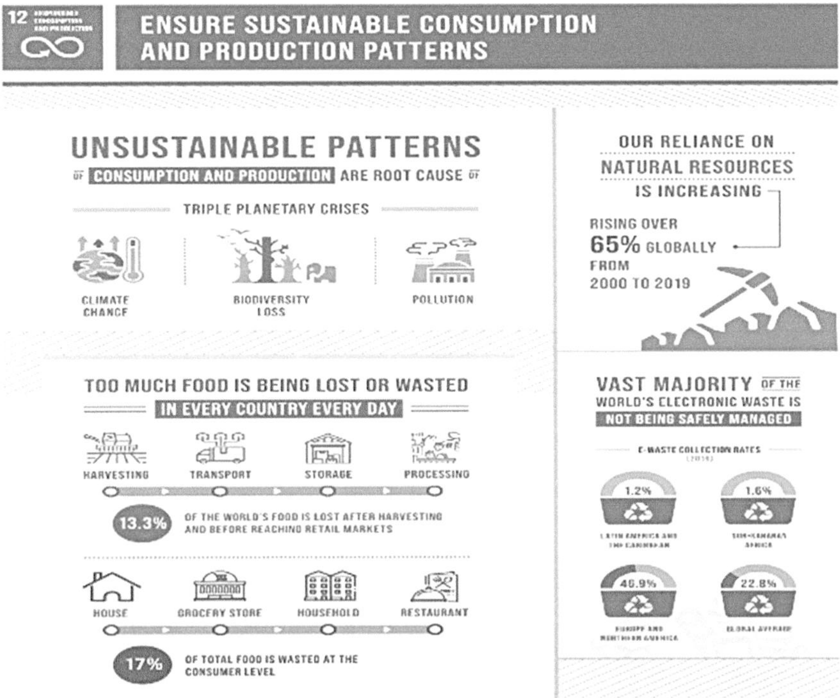

FIGURE 4.3 Consumption and production pattern.

Source: The Sustainable Development Goals Report (2022).

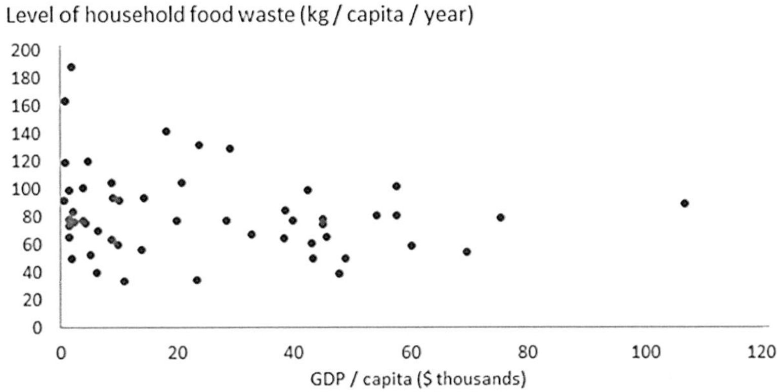

FIGURE 4.4 Food waste generated per year.

Source: Scatter plot of food waste (household) against the GDP of the country.

and capital) that were used to produce the food (Figure 4.5). The disposal of food loss and waste contributes to land pollution and hazardous greenhouse gas emissions, which contribute to climate change. This can also negatively impact food security and food availability, which leads to an increase in the cost of food (IFCO Systems, 2020).

Food loss needs to be primarily addressed at the national and regional levels. Solutions will differ and be strongly influenced by local conditions, production, transportation systems, and customer behavior. There have already been a lot of creative approaches to reducing agricultural and food waste put into use. The most effective ones demonstrate that concentrating solely on global averages is insufficient. Yet, effective practices usually involve comprehensive procedures to assure food safety across the supply chain (UN, 2023).

FIGURE 4.5 Waste generation per year.

Source: Food and Agriculture Organization of the United Nations.

4.3 DEFINING BYPRODUCT/WASTE

Byproducts are referred to as the secondary produce left after deriving the primary product from any given source. This product is often discarded unused; however, it contains rich nutritional and extra-nutritional bio-actives. As far as fruits and vegetables are concerned, it contains a significant number of pharmaceutical components such as pectin, polyphenols, enzymes, and antioxidants. This chapter will mainly focus on the utilization of these byproducts in the best possible ways, along with the new technologies and innovations for extractions.

Food wastes are typically solids and liquids (wastewater) left over after food is processed from agriculturally sourced raw ingredients. The wastewater comes from cleaning procedures or appears as too much-contaminated process water. Less than 5% of its mass is typically made up of dry stuff. Moreover, it could include organic or inorganic disinfectants or cleaners.

The water content of solid food wastes of an organic origin is unusually high (usually around 80% by mass). Because of the operations that were skipped, they are frequently distinguished by consistency in quality and purity. These materials are wastes because they are eliminated from the manufacturing process as unwanted elements.

The industry-standard phrase "byproduct" emphasizes that these are primarily hidden useful compounds, frequently with a market value. So, if there were appropriate technical methods to make items whose worth exceeded the cost of reprocessing, the wastes may be termed valuable byproducts. In this situation, residues are not wastes but rather new resources that can be used to supplement natural resources already available. Food processing waste has the potential to be recycled, reprocessed, and eventually used again, as opposed to being released into the environment where it has negative environmental effects.

4.4 FORMS OF FOOD BYPRODUCTS AND WASTES

In the food supply chains (FSC), five system boundaries have been identified as being primarily responsible for the development of waste.

i. Crop sorting, on-farm temporary storage, curing/pre-treatments, mechanical damage and/or spill losses, and so forth, are all examples of losses in agricultural output and agricultural waste brought on by environmental conditions.
ii. Post-harvest handling and storage, which includes damages brought on by spills and degradation during handling, storage, and transit between the farm and the point of distribution.
iii. Processing losses include those brought on by spills and degradation during commercial or domestic processing, such as making juice, canning food, and baking bread. Losses may occur during the washing, peeling, slicing, and boiling of crops or when crops are sorted out if they are not fit for processing.

iv. Distribution: Taking into account waste and losses in the market system at places like wet markets, supermarkets, and wholesale markets.
v. Consumption: This includes domestic consumption losses and waste.

An economical technology is anaerobic digestion (AD) of sorted organic wastes from MSW, particularly food waste (De Baere, 2000). Fermentation of methane is a difficult process. According to Veeeken et al. (2000), the overall process of AD entails several metabolic stages, including enzymatic hydrolysis, acidogenesis, acetogenesis, and methanogenesis. Each metabolic stage is supported by several microorganisms. The rate-limiting stage for FVW among the four stages is hydrolysis (Vavilin et al., 1996).

The microorganisms that produce acid break down macromolecules like cellulose, proteins, starches, and carbs into organic acids (steps 1 and 2). Organic acids are transformed into acetate in the third step, and then methanogens take over in the fourth step to turn the acetate into CH_4 and CO_2. Steps 1 and 2 take up the majority of the time because they require more time.

Faster production of volatile fatty acids (VFAs) occurs in the presence of substrates high in carbohydrates, such as FVW (Mata-Alvarez et al., 2000), which results in acidity, low pH, and process inhibition. Hence, greater substrate concentration for FVW causes pH to decrease, which results in less biogas being produced.

According to Uzodinma et al. (2011), the typical composition of biogas is 45%–70% methane, 30%–45% carbon dioxide, 0.5%–1.0% hydrogen sulfide, 15% water vapor, and minor amounts of other gases (hydrogen, ammonia, nitrogen, etc.). The source and mixture of biodegradable biomass affect a specific biofuel's composition (Anunputtikul and Rodtong, 2004). The nature and biochemical makeup of biogenic waste have a significant impact on its potential for gas production (Sagagi et al., 2010).

4.5 AGRICULTURE BYPRODUCTS

Due to the growing agricultural productivity, an estimated 998 million tons of agricultural waste are created annually. Cereal makes up 19% of worldwide food waste, while roots and tubers provide 20%, and vegetable and fruit byproducts account for 44%. Manure output can reach up to 5.27 kg per day per 1000 kg of live weight on a wet weight basis, whereas organic wastes generate up to 80% of the total solid wastes in any form. The incorrect use of seasonal overproduction of fruits and vegetables is another major contributor to agriculture byproducts (Lau et al., 2021).

Extrusion is a versatile method of handling raw materials and is an energy-efficient process, which allows wastes from agriculture to be utilized into value-added goods produced. This is so that extrusion, a high temperature, high shear, and quick processing technology, can combine the simultaneous application of unit activities including mixing, kneading, cooking, and shaping. The most popular raw material for extrusion processing is cereal flour (Faraj et al., 2004). Because of the quick processing durations, one benefit of extrusion processing over traditional thermal processing processes is the reduction of nutritional losses (Sevatson and Huber, 2000). Furthermore, trypsin inhibitors, phytates, and other antinutritional substances, as well as enzymes like lipases and lipoxidases, are destroyed during the high-temperature processing extrusion stage (Dey et al., 2021).

4.6 VEGETABLE BYPRODUCT

One-third of the total vegetable waste is generated during the process of manufacturing at different stages. However, some of the parts are disposed of due to unfavorable taste or texture, which still holds the scope to be utilized as a functional food or in another form.

For instance, vegetable scraps like hulls, bagasse, and seeds are typically thrown away in the production process. The stems and leaves of several vegetables, like broccoli, cauliflower, and pumpkin, are discarded rather than eaten. One of the vegetable byproducts is the seed coat, which is the outer coating that covers the seeds or beans and serves to protect them from external harm. Except for very thin seed coats, like peanut seed coats, most seed coats are not eaten because of their texture and flavor.

The Indian Agricultural Institute (IARI) plays a key role in increasing the production of agricultural products; however, since there is 40% wastage due to inappropriate storage facilities, the institute took initiatives to preserve the products too, with the launch of a refrigerator—Pusa Farm Sun Fridge—which holds a capacity to preserve tons of fruits and vegetables.

The solar-powered refrigerator can store two tons of fruits, vegetables, and flowers. The ACs are run by solar energy during the day, and cold water is circulated on the roof at night to keep the temperature between 4°C and 12°C so that the goods can be maintained more effectively. One refrigerator costs about Rs. 5–7 lakh to produce.

4.7 WASTES-DERIVED VALUE-ADDED PRODUCTS

In an organized sector globally, enormous amounts are being dumped despite having nutritional properties. Materials like peels, pomace, and seeds can be used wisely to protect the environment and advance the one-health philosophy.

4.7.1 PRODUCTION OF OILS

The areas of fruit cultivation and processing generate enormous volumes of trash and byproducts that have a negative influence on the environment and create major economic losses. We can attain SDGs by reducing greenhouse gas emissions and our carbon footprint by appropriately utilizing these fruit leftovers. Dietary fiber, flavonoids, phenolic compounds, antioxidants, polysaccharides, and a number of other beneficial minerals and phytochemicals are just a few of these byproducts that are good for your health. They also include a wide range of additional bioactive compounds. These bioactive materials can be separated and used as high-value products in a variety of industrial contexts. It is possible to make functional foods, food additives, and nutraceutical products using the extracted bioactive components. The application of new fruit waste valorization techniques in the food industry is thoroughly examined in this chapter (Nirmal et al., 2023).

The yield of citrus peels is 0.5 to 3.0 kg oil/tons of fruit. The oil derived from such peels have its role in alcoholic beverages, soft drinks, perfumes, confectionaries, soaps, household products like dishwashers, and aromatic flavors, and so forth.

Also, these wastes are renewable and viable resources that might be valorized in accordance with the "circular economy" concept to produce goods with a high market value. The efficient valorization of fruit waste requires waste recycling and resource recovery. Figure 4.1 depicts a schematic depiction of how fruit waste is utilized.

4.7.2 EDIBLE OILS

Oil obtained from oilseeds and nuts through an extraction process is referred to as vegetable edible oil. It can be extracted via a pressing method from a variety of oil seeds, including mustard, coconut, soybean, peanut, rapeseed, cottonseed, sesame, and others. For pan frying, deep fat frying, and salad dressing, plant-based derived oils are typically utilized.

The fatty acid and triglyceride profiles of the fat in mango seed kernels makes it a prospective source of edible oil. Guava (*Psidium guajava* L., Myrtaceae) seeds contain 5–13% oil rich in important fatty acids but are typically discarded during the processing of juice and pulp. According to Malacrida and Jorge (2012), the oil from the seeds of the passion fruit is high in unsaturated fatty acids (UFAs) (87.6%), particularly linoleic (73.1%), and oleic (13.8%) acids. The oil has the ability to scavenge free radicals.

4.7.3 MANUFACTURING OF EDIBLE OILS

At first, the sole steps in the oil extraction process were cleaning, crushing, heating, and pressing. However, starting around 1900, solvent extraction was used to either recover the remaining oil from the pressed cake or to fully replace the pressing process (e.g., for soybean oil). At the same time, the oil purification procedure evolved from straightforward decanting and filtering to include caustic neutralization, active clay bleaching, and high-temperature, steam-powered deodorization.

Oil is first mechanically (using an expeller press) and then chemically (using a hexane extraction method) removed from the seed in a conventional edible oil production facility. Less than 1% of the oil is retained in the dish after employing both techniques (PMG, 2021).

4.7.4 NUTRACEUTICAL PRODUCTS USING EXTRACTION PROCESS

Mango produces the most waste out of all the different fruits and vegetables, at 60%, followed by citrus (50%), passion fruit (45%), peas (40%), pineapple (33%), and pomegranate (40%). Other significant waste sources include the tomato, potato, apple, and grape (Bayram et al., 2021).

The tomato is one of the most extensively grown vegetable crops and is consumed by people all over the world. Millions of tomatoes are processed to make goods like ketchup and sauce, which generates a lot of tomato peel, pulp, and seed waste that is nearly comparable to 40% of the raw material. Different value-added products can be produced from tomato peel. In fresh fruit, tomato lycopene concentrations range from 30 to 200 mg/kg, while on a dry basis, they range from 430 to 2950 mg/kg. More than 85% of all carotenoids are made up of lycopene, which tomato peel has around five times more of than tomato pulp.

In the human diet, carotenoids have a significant role as immune system controllers, anticarcinogens, and preventers of cardiovascular disease. Moreover, they can function as antioxidants and vitamin A precursors (Singh and Goyal, 2008). Carotenoids have also been employed in pharmaceuticals and cosmetics since they are linked to human health. Lycopene is one of the most significant carotenoids ($C_{40}H_{56}$) (Machmudah et al., 2012).

It might be helpful in treating chronic illnesses like cancer, coronary heart disease, and others. Due to the lycopene and fiber included in this tomato manufacturing byproduct, tomato peel can be added to meat products to produce a healthier end product. Carotenoids can also be found in abundance in carrot pomace.

Banana bracts, which are the undersides of the calyx, and beetroot root pulp anthocyanin pigments were evaluated for their possible use as natural food colorants. This is another example of the valorization of peel 11–23 mg of Xanthins are present in the beetroot root pomace per gramme of dry extract. Betacaine, which are categorized into two main families by color—red betacyanin's and yellow betaxanthins—are nitrogenous pigments that are water-soluble and may be found in the peel of beetroot roots.

High flavonoid content citrus peels have anticancer properties. Citrus peels contain cancer-preventive properties. Prostate cancer tumor sizes were considerably reduced by citrus peels, which also had potent anti-inflammatory, antiproliferative, and antiangiogenic properties as well as improved apoptosis-inducing actions. Citrus peel's abundant polymethyl flavone content may likewise successfully prevent colonic carcinogenesis brought on by azoxymethane.

Comparing apple pomace to other fruit byproducts like citrus peel, apple pomace has a higher level of nutritional fiber. There are studies in the literature that demonstrate the use of apple pomace as a functional ingredient and source of extra fiber in the creation of bread, cakes, and biscuits. Moreover, apple pomace pectin has been employed as a substitute source of the gelatinizing ingredient typically required to make jam or jelly. Several functional foods, including bread, muffins, and jam with little calories, have been created using tomato pomace, which is rich in dietary fiber and minerals (Lau et al., 2021).

4.8 WASTE AS PACKAGING MATERIAL

According to recent studies, the use of natural materials derived from waste and byproducts in the food packaging industry as the next environmentally conscious stage of waste valorization may be an effective way to reduce the amount of food waste and byproducts. For instance, food goods can be made safer and more resistant to spoiling processes by including antioxidant chemicals made from food waste into the packing material.

In 2018, there was an approximate USD 974 billion global demand for packaging. By 2024, it is anticipated that the value of smart packaging will surpass that of intelligent and active packaging systems, reaching USD 26.7 billion. In essence, the primary goal of food packaging is to extend the shelf life of food goods (Bayram et al., 2021).

According to an FAO study, 1.3 billion tons of food are wasted or lost annually. A large amount of waste and loss may occur at just a few stages of the food processing process, including pre-harvest, harvest, on-farm post-harvest, transport, storage and distribution, processing and packaging, retail, and public household consumption. Globally, the cost of food loss and waste was projected to be USD 936 billion. The agri-food sector contributes 40–50% of the total discard from various plant sources, including peels, pulps, skins, pomaces, shells, roots, stems, stones, leaves, and seeds, generating a substantial amount of trash and byproducts.

There is growing interest in using alternative raw materials from agricultural and food processing wastes in food packaging applications because of factors such as the present global plastics consumption and the recovery of biopolymers to use them in the creation of biodegradable materials.

The polymeric hydrocarbons, starch, is made up of glucose polymers: 20–25% amylose and 75–80% amylopectin. Because it is readily available, abundant, biodegradable, inexpensive, and has strong mechanical qualities, starch can be used as a filler, coating, or film. It can also be processed using tools often employed in traditional polymer processing, such as extrusion and injection molding.

Pectin can be changed by demethylation in the presence of calcium ions to create edible films, and it is commonly found in agricultural waste products. Pectin films can only be used for food packing because of how little moisture resistance they have.

However, the whey protein isolate had no discernible impact on the food's color or texture. With high gloss coatings, edible films can also be used to enhance the aesthetic or visual appeal of food, particularly fruits and vegetables. Zein and dextrin are more susceptible to humidity and are therefore more likely to get de-adhered and cracked, whereas whey protein isolates, shellac, dextrin, and zein are excellent materials for a gloss coating.

Plasticizers are added to the protein solution after heating, which increases the intermolecular distance between the proteins and increases the film's flexibility and gas permeability. During the dehydration process, the prepared solution is applied as a food packaging by dipping or spraying; in this case, it is referred to as a coating. Alternatively, the prepared solution may be dried as solid sheets prior to packaging; in this case, it is referred to as a film. The temperature is then raised further, a plasticizer is added to boost the coating's flexibility and water permeability, and finally the solution is dried to allow for film formation. However, due to the great rigidity of such lipid-based coatings, they are frequently mixed with protein solutions.

4.9 CONCLUSION

Byproducts from the production, post-harvest, processing, distribution, or consumption of fruits and vegetables are known as byproducts. While a significant amount of the byproducts is being thrown away, some of them have been employed as building materials, animal feed, or fire fuel. Agriculture byproducts from the food industry come in a wide variety and have proven to be highly nutritious (Bayram et al., 2021). The cost of production can be reduced by using these byproducts as functional food and ingredients. Also, it is one of the methods for addressing the growing issue of food security and has the potential to promote food sustainability. Findings from the literature, which are extremely important, demonstrated that the nutrition from functional foods may be increased and enhanced by using wastes from the kitchen, byproducts, particularly vegetable and fruit (Falguera et al., 2011; McHugh et al., 1994).

KEYWORDS

- **black gram**
- **byproducts**
- **edible coatings**
- **glucosidase inhibitor**
- **husk aleurone layer**
- **phenolic acids**

REFERENCES

Ahuja, R. (2022). Waste management! A part and parcel for India. *Times of India.*

Anunputtikul, W., & Rodtong, S. (2004). Laboratory scale experiments for biogas production from *Euphorbiaceae* tubers. In *Proceedings of the Joint International Conference on Sustainable Energy and Environment (SEE).*

Bayram, B., Ozkan, G., Kostka, T., & Capanoglu, E. (2021). Valorization and application of fruit and vegetable wastes and byproducts for food packaging materials.

Bureau, P. I. (2033). Achieving Aatmanirbharta in agriculture. *Press Information Bureau (PIB).*

Central Pollution Control Board. (2021). *Annual Report 2020–2021* (pp. 2–2).

De Baere, L. (2000). Anaerobic digestion of solid waste: State-of-the-art. *Water Science and Technology: A Journal of the International Association on Water Pollution Research, 41*(3), 283–290.

Dey, D., Richter, J. K., Ek, P., Gu, B. J., & Ganjyal, G. M. (2020). Utilization of food processing byproducts in extrusion processing: A review. *Frontiers in Sustainable Food Systems, 4.*

Falguera, V., Quintero, J. P., Jiménez, A., Muñoz, J. A., & Ibarz, A. (2011). Edible films and coatings: Structures, active functions and trends in their use. *Trends in Food Science & Technology, 22*(6), 292–303.

Faraj, A., Vasanthan, T., & Hoover, R. (2004). The effect of extrusion cooking on resistant starch formation in waxy and regular barley flours. *Food Research International, 37*(5), 517–525.

Global Nutrition Report. (2022). The burden of malnutrition at a glance.

Government of India. (2022). *Guidelines for Preparation of Detailed Project Report and Selection of Technologies for Processing and Final Disposal of Municipal Solid Waste.*

IFCO Systems. (2020). Food waste by country: Who's the biggest waster?

Lau, K. Q., Sabran, M. R., & Shafie, S. R. (2021). Utilization of vegetable and fruit byproducts as functional ingredient and food. *Frontiers in Nutrition, 8.*

Machmudah, S., Zakaria, Winardi, S., Sasaki, M., Goto, M., Kusumoto, N., & Hayakawa, K. (2012). Lycopene extraction from tomato peel byproduct containing tomato seed using supercritical carbon dioxide. *Journal of Food Engineering, 108*(2), 290–296.

Majumder, P., & Annegowda, H. V. (2021). Fruit and vegetable byproducts: Novel ingredients for a sustainable society. In *Valorization of Agri-Food Wastes and Byproducts* (pp. 133–156). Elsevier.

Malacrida, C. R., & Jorge, N. (2012). Yellow passion fruit seed oil (*Passiflora edulis f. flavicarpa*): Physical and chemical characteristics. *Journal of the American Oil Chemists' Society, 55*(2), 127–134.

Mata-Alvarez, J., Macé, S., & Llabrés, P. (2000). Anaerobic digestion of organic solid wastes: An overview of research achievements and perspectives. *Bioresource Technology, 74*(1), 3–16.

McHugh, T. H., Aujard, J. F., & Krochta, J. M. (1994). Plasticized whey protein edible films: Water vapor permeability properties. *Journal of Food Science, 59*(2), 416–419.

Mogge, M., & MacSorley, D. (2004). *Global Hunger Index.*

Ninčević Grassino, A., Djaković, S., Bosiljkov, T., Halambek, J., Zorić, Z., Dragović-Uzelac, V., Petrović, M., & Rimac Brnčić, S. (2020). Valorization of tomato peel waste as a sustainable source for pectin, polyphenols and fatty acids recovery using sequential extraction. *Waste and Biomass Valorization, 11*(9), 4593–4611.

Nirmal, N. P., Khanashyam, A. C., Mundanat, A. S., Shah, K., Babu, K. S., Thorakkattu, P., Al-Asmari, F., & Pandiselvam, R. (2023). Valorization of fruit waste for bioactive compounds and their applications in the food industry. *Foods, 12*(3), 556.

Omre, P., Singh, S., & S., S. (2018). Waste utilization of fruits and vegetables: A review. *South Asian Journal of Food Technology and Environment, 4*(1), 605–615.

PMG. (2021). *Processing of Edible Vegetable Oils.*

Sagagi, B., Garba, B., & Usman, N. (2010). Studies on biogas production from fruits and vegetable waste. *Bayero Journal of Pure and Applied Sciences, 2*(1).

Sevatson, E., & Huber, G. R. (2000). Extruders in the food industry. In *Extruders in Food Applications* (1st ed., p. 38).

Singh, D. A. (2021). 40% vegetables, fruits get wasted in India: IARI director. *National Herald.*

Singh, P., & Goyal, G. K. (2008). Dietary lycopene: Its properties and anticarcinogenic effects. *Comprehensive Reviews in Food Science and Food Safety, 7*(3), 255–270.

The Hindu. (2022). Demand for food ingredients in India may rise 6.7% till 2029.

United Nations Environment Programmed. (2021). *Food waste index report.*

United Nations. (2022). *Sustainable Development Goal 12.* Department of Economic and Social Affairs Sustainable Development.

United Nations. (2023). Stop food loss and waste, for the people, for the planet.

Uzodinma, E., Ofoefule, A., & Enwere, N. (2021). Optimization of biogas fuel production from maize (*Zea mays*) bract waste: Comparative study of biogas production from blending maize bract with biogenic wastes. *American Journal of Food and Nutrition, 1*(1), 1–6.

Vavilin, V. A., Rytov, S. V., & Lokshina, L. Y. (1996). A description of hydrolysis kinetics in anaerobic degradation of particulate organic matter. *Bioresource Technology, 56*(2, 3), 229–237.

CHAPTER 5

Fruits: Wastes, Byproducts, and Utilization

RANJANA VERMA,[1,*] NILAKSHI CHAUHAN,[1] PREETI CHOUDHARY,[2] ABHA ANAND,[3] and FARHAN M. BHAT[1]

[1]*Department of Food Science, Nutrition and Technology, CSK Himachal Pradesh Krishi Vishvavidyalaya, Palampur, Himachal Pradesh, India*

[2]*Department of Food Science and Technology, Dr. Yashwant Singh Parmar University of Horticulture and Forestry, Neri, Hamirpur, Himachal Pradesh, India*

[3]*Dr. B. R. Ambedkar State Institute of Medical Sciences, Mohali, Punjab, India*

*Corresponding author

ABSTRACT

Fruit wastes, often discarded as byproducts of food processing and consumption, are rich sources of bioactive components with potential health benefits. This chapter explores the diverse array of bioactive components present in fruit wastes and their associated therapeutic potential on human health. These compounds, including polyphenols, phenolic compounds, flavonoids and dietary fiber, vitamins, minerals, and essential oils, have garnered attention due to their antimicrobial, anti-inflammatory, antioxidant, and other bioactive properties. Polyphenols, a group of bioactive components found abundantly in fruit peels and seeds, exhibit strong antioxidant capabilities that counteract oxidative stress implicated in chronic diseases such as cardiovascular disorders, cancer,

Utilization of Food Waste and Byproducts. Mudasir Ahmad Malik and Kshirod Kumar Dash (Eds.)

DOI: 10.1201/9781779641151-5

and neurodegenerative conditions. Moreover, dietary fiber from fruit pulps and peels supports digestive health, regulates blood sugar levels, aids in weight management, and contributes to the prevention of heart disease. The presence of vitamins (e.g., vitamin A and vitamin C) and minerals (e.g., potassium and magnesium) in fruit wastes underscores their importance in maintaining immune function, supporting bone health, and promoting overall well-being. Furthermore, the anti-inflammatory properties of specific fruit waste compounds hold promising potential in managing inflammatory diseases, while their antimicrobial and antiviral effects could contribute to natural preservatives or infection-fighting agents. Some bioactive compounds derived from fruit wastes may even exhibit hypoglycemic effects, making them potential allies in blood sugar management. Promising research also suggests that certain bioactive compounds in fruit wastes support cardiovascular health by improving blood pressure, lipid profiles, and blood vessel function. In the realm of neurological health, fruit waste compounds are being explored for their neuroprotective potential. Additionally, they may prevent cancer by inhibiting the growth of cancer cells and promoting programmed cell death. While these findings highlight the therapeutic potential of fruit waste bioactive compounds, further investigation is essential to unravel their mechanisms of action, potential adverse effects, and optimal utilization in therapeutic applications. Proper extraction techniques and formulation strategies are crucial to harness and amplify the bioactivity of these compounds, transforming fruit wastes into valuable resources for enhancing human health and well-being.

5.1 INTRODUCTION

In the last few years, the demand for fruit production has spiked worldwide due to the large increase in population and demographic transition to consume healthy diets. In this context, there has been a continuous international increase in acres of land accomplished for fruit production. The data from the National Horticulture Board estimated that in 2011, a total of 640 million tons of fruits were harvested, while 990 million tons were harvested in 2020 (FAO, 2021). During the last 25 years, the processing of fruits has increased considerably, which results in the generation of high amounts of fruit waste materials like peels, pomace, rags, seeds, stones, and kernels (Joshi, 2020). Total wastes, which may account for 20–80%

of the whole fruit, are produced in large amounts in the form of liquid and solid waste by the fruit processing industry, and if not managed properly, cause environmental pollution problems. Extremely diverse wastes are obtained from the food processing industry due to the broad range of processes and the heterogeneity of the fruit product (William, 2005). The management challenges of this large volume of fruit wastes is currently a concern for waste management, when seeking to address the number of malnourished people and the depletion of natural resources. Hence, valorization and minimization of waste became a worldwide discussion to enhance the feasibility of the fruit industry.

5.2 BIOACTIVE COMPONENTS OF FRUITS WASTES – NUTRACEUTICAL POTENTIAL FOR HUMAN HEALTH

The waste generated from various fruits, like peels, pomace, seeds, stems, leaves, and stones, has remarkable amounts of bioactive compounds (Marago et al., 2015; Monagas et al., 2005). The major byproducts obtained during the processing of various fruits are peels, which are a good source of various bioactive compounds and possess various beneficial health effects. Fruit wastes are thrown as agro-waste into the environment, but they can be further utilized as a source of antimicrobics and have other potential byproduct value. Therefore, the use of fruit wastes as byproducts for the further production of food additives and high nutritional value supplements has gained increasing interest and become a new aspect of concern, as these are economically attractive due to maximum recovery and high-value products production. The byproducts of fruit wastes are becoming popular due to being an important source of sugars, phenolics, organic acids, minerals, and dietary fiber, which show a wide range of nutraceutical potential properties, including antimutagenic, antitumoral, antibacterial, antiviral, and cardioprotective activities. However, the utilization of fruit waste byproducts is limited because of a poor understanding of their nutritional, therapeutic, and economic value, and the economic inability to reuse and salvage such materials results in the unnecessary waste and exhaustion of natural resources (Joshi, 2020). Actually, solid waste is dumped in local landfills, eaten, and scattered by stray animals. Some important fruits and the extent of waste produced by processing industries are given in Table 5.1. About 90% of the processing units in

India are small-scale units, and these units have no organized methods of waste disposal. Even in certain places, this waste is taken away by farmers for disposal as cattle feed or field manure. Solid waste from fruit industries can be utilized further for the preparation of different products.

TABLE 5.1 Quantities of Various Fruit Processing Waste in India

Fruit	Nature of Wastes	Production (Tons) NHB, 2020	Approximate Weight (g)	Percent Wastes
Apple	Peel, pomace, and seeds	2734.0	750–950	20–30
Citrus fruits	Peel, rag, and seed	1397.6	30–150	50–60
Berries	Peel, pomace, and seeds	5235.0	4–12	3–58
Stone fruits	Peel, stone/kernel, and pomace	2191.0	60–80	40–60
Melons	Peel and seed	4117	1800–3600	12–46
Tropical fruits	Peel, pomace, and seeds	9752.0	300–650	15–20
Pineapple	Core and crown	1799	1000–1500	33
Banana	Peel	3150.4	100–120	12–15

5.2.1 BIOACTIVE COMPONENTS IN FRUIT WASTES

Fruit wastes are promising, rich in biologically active compounds, and rich sources of bioactive components, which have beneficial effects on human health. These compounds can include vitamins, minerals, dietary fibers, antioxidants, phenolic compounds, flavonoids, carotenoids, and other phytochemicals, which are contributed from the trimmings, peels, stems, seeds, pomace, and shells after juice extraction. Several scientific researchers have manifested the presence of a wide range of biologically active compounds in various fruit wastes. The mainly components present in it are lipids, carbohydrates (pectin, cellulose, hemicellulose), and proteins, while phenolics, glycosides, alkaloids, gums, mucilage, and volatile oils act as a secondary metabolite. In general, seeds of fruit waste are considered a good source of bioactive lipids, while dietary fibers are rich in fruit peels (Ran et al., 2019). The main bioactive components present in fruit wastes are given in Table 5.2. It is important to note that the composition of bioactive components in fruit byproducts can vary based on the fruit type, its ripeness, and the methods of processing used. Researchers are actively exploring ways to utilize extraction of fruit wastes and production of valuable bioactive components, which may contribute to both reducing food waste and promoting human health. The various bioactive components which are present in fruit wastes in different ratios and proportions are discussed in sub section.

TABLE 5.2 Major bioactive compounds present in fruit byproducts

Fruit	Bioactive Compounds		
	Peel/ Core/ Crown	**Seed/ Stone/Kernel**	**Pomace**
Apple	Anthocyanins, dihydrochalcones, chlorogenic acid, ferulic acid	Phlorizin, hyperin, and chlorogenic acid, amygdalin	Chlorogenic acid, catechins, epicatechins, phlorizin
Drupe fruits	*β*-Carotene, catechin, chlorogenic acid, cyanidin-3-galactoside, cyanidin-3-glucoside	Gallic acid, vanillic acid, benzoic acid, phlorizin, quercitrin,	Neochlorogenic acid, cyanidin-3-glucosyl-rutinoside, *β*-Carotene
Citrus fruits	Naringin, tangerine hesperidin, nobiletin,	Limonin, nomilin, obconic acid, ichangin	β-Carotene, hesperidin, narirutin, and naringin
Tropical fruits	Ethyl gallate and penta-*O*-galloyl-glucoside, gallotannins, gallic acid, catechin, epicatechin, ferulic acid, punicalagin	Gallotannins, gallates, gallic acid, ellagic acid	Quinic acid, caffeic acid, catechin, chlorogenic acid, *p*-coumaric acid, protocatechuic acid
Berries	Gallic acid, gallotanins, anthocyanins epicatechin, caftaric acid	Procyanidins, epicatechin, gallic acid, vanillic acid, syringic acid	Rutin, naringenin, anthocyanins, catechin, epicatechin,
Melons	Gallic acid, catechin, ellagic acid, kaempferol, cinnamic acid, ferulic acid, chlorogenic acid, rutin	Gallic acid, caffeic acid, rosmarinic acid, protocatechuic acid, ellagic acids	-
Pineapple	Bromelain	-	-

Source: Adapted from Kandemir et al. (2022).

5.2.1.1 DIETARY FIBER

Fruit wastes, such as peels and pulp, are often high in dietary fiber. The tendency to recover dietary fiber compounds is increasing from fruit processing industrial wastes. Research has found that fruit wastes are mainly composed of non-starch polysaccharides, including cellulose, hemicellulose, pectin, gums, and lignin (Banerjee et al., 2017). An experiment resulted in the finding that the peel of an apple has more dietary fiber than its pulp, in which 15% soluble and 36% insoluble dietary fiber was found in apple pomace (Sudha et al., 2007). The blackcurrants, gooseberries, rowanberries, red currants, and chokeberries pomace powder have a good fiber content, which was observed to be more than 550 g/kg (Reissner et al., 2019). The total amount of dietary fiber reported in grape pomace is approximately 78% of the total dietary fiber reported, in soluble (9.5%) and insoluble (68%) dietary fiber form (Valiente et al., 1995). The peach ranged from 20–24% dietary fiber proportion in its pulp and peels as insoluble dietary fiber. Mango peels are the source of neutral sugars with glucose, arabinose, and galactose, as soluble and insoluble total dietary fiber, and their content ranges from 40–72% (Ajila and Prasada, 2013). Citrus fruits yield huge amounts of wastes like peel, pomace, rags, and seeds, including membrane residues, in which orange, grapefruit, lemon, and mandarin are the most widely grown crops. Citrus byproducts are excellent sources of dietary fiber with the presence of both types of dietary fiber, soluble and insoluble. Many studies reported that due to the high fiber content of citrus wastes, some functional and intriguing properties are shown, like the capacity to water-holding, swelling, oil-holding, emulsion, and foam. Citrus byproducts act as a substitute in meat products and raise the dietary fiber content in meat products. In this context, an impact assessment study showed that the content of cholesterol and saturated fatty acids was lowered by lemon fiber in the diet (Soncu et al., 2015). Similarly, other studies have shown that a low-fat diet with citrus fiber reduced the saturated fatty acids levels and also shows better water-holding properties (Song et al., 2016). Many health disorders are directly linked with dietary fiber and play a significant role in lowering it. Due to the water-holding as well as intestinal tuning properties of dietary fiber (Cellulose, hemicellulose, lignin, pectin, and gums), major effects were shown in lowering cholesterol and maintenance of blood glucose. Moreover, dietary fiber has shown widespread utilization in bakery products of the food industry due to improving freshness, prolonging shelf life, and more water binding as well as digestion (Maurya et al., 2015).

5.2.1.2 PHENOLIC COMPOUNDS

The phenolic compounds commonly found in fruits wastes (peels and seeds) have properties like antioxidant, anti-cancer, and anti-inflammatory.

Many experimental studies have isolated various phenolic compounds from byproducts of fruits such as hydroxycinnamic acids (HCA), hydroxybenzoic, lignans, and stilbenes. The fruit wastes, especially apple, citrus, pomegranate, and grape, showed the highest phenolic content, as compared with other fruit wastes. Apple pomace mainly contains the skin of the fruit, which is a good source of polyphenol. The phenolic compounds in the byproducts of the apple vary in different concentrations and compositions, like apple peel having a higher concentration of phenolic compounds as compared to apple flesh. Epicatechin, phloretin-2′-xyloglucoside, caffeic acid, phlorizin, and quercetin derivatives are the major phenolic compounds reported in apple pomace (Vieira et al., 2011). Citrus peels are also rich in phenolics compounds such as p-coumaric acid, caffeic acid, ferulic acid, tangerine, and naringin. Different cultivars of citrus fruits extract such as hybrid mandarin, navel orange clementine mandarin, common orange, satsuma mandarin, and pigmented orange were examined, and results showed that the highest vitamin C was reported in pigmented orange at 46 mg/ 100 g in fresh weight, navel orange had high hesperidin, containing 73.8 mg/100 g in fresh weight, and narirutin was reported greatest in satsuma at 27.6 mg/100 g fresh weight (Cano et al., 2008). An experiment reported that peels of citrus fruits like orange (179 mg/100 g fresh weight), grapefruit (155 mg/100 g fresh weight), and lemons (190 mg/100 g fresh weight) contained higher amounts of polyphenols as compared with fruit pulp (Kim and Kim, 2016). Catechins, glycosylated flavanols, and proanthocyanidins polyphenols are found in grape pomace in large quantities and act as free radical scavengers. Due to which, citrus byproducts are gaining more interest in the production of functional food ingredients (Theagarajan et al., 2019). Anthocyanins are the main phenolic compounds found in different berries. Scientific studies on red berries, blueberries, raspberries, red currants, and cranberries reported 11 different anthocyanins found mainly malvinidin-3-O-galactoside, malvinidin-3-Oarabinoside, delphinidin-3-O-galactoside, petunidin-3-O-galactoside, and residues of cyanidin derivatives in its pomace (Jara-Palacios et al., 2019). Researchers have shown that apricot kernels have been found to have more phenolic content than the flesh (Soong and Barlow, 2004). Peach fruit is rich in phenolic compounds, and polyphenols of peach are mainly found in pulp and peel. The primary phenolics compounds in peach fruit are chlorogenic acid, proanthocyanidin B1, catechin, neochlorogenic acid, epicatechin, rutin, different glycosides of quercetin, and anthocyanins (Eleren et al., 2018). Melon Seeds have been investigated for total phenolic content, and caffeic acid, quercetin-3-rutinoside, vanillic acid derivatives, ellagitannins, derivatives of syringic, and ellagic acid, a total of nine phenolics were identified, which are important for preventing cardiovascular diseases and neurodegenerative diseases (Milala et al., 2013). Quinin acid, catechin, chlorogenic acid, caffeic acid p-coumaric acid, derivatives of protocatechuic, kaempferol quercetin are the major phenolic compounds recorded in kiwi pomace. The interest in phenolic compounds is mainly increasing due to their capacity to regulate the generation of free radicals *in vivo*, scavenge free radicals, and ensure the prevention of oxidation reactions. The wastes constitute about 50% of the total fruit in many fruits, such as apples, peaches, banana, pears, and pomegranate, and are found to contain the highest amount of phenolics compounds than its edible part.

5.2.1.3 FLAVONOIDS COMPOUND

Flavonoids are a type of polyphenolic compound which have anti-inflammatory and antioxidant properties. They are present in various fruit wastes and linked to various health benefits. Peels of apple contain flavanols, anthocyanins, flavan-3-ols, phenolic acids, and dihydrochalcone having associated health-promoting effects. Researchers reported that apple pomace contains more than half of polyphenols quercetin glycosides (Boyer and Liu, 2004). Triterpenoids, secondary metabolites constituted from isopentenyl pyrophosphate oligomers biomolecules, are majorly found in apple pomace and present in six isoprene units (C_5H_8) (Castro et al., 2021). The researchers showed that the skin of apple cultivars has five times the highest amount of total flavonoid compounds, such as quercetin-3-O-galactoside, quercetin-3-O-rhamnoside, and quercetin-3-Oglucoside (Sun-Waterhouse et al., 2013). Several scientific studies reported that citrus byproducts, especially seeds, are a rich source of limonoids including limonin, ubiquinone, nobilin, and so forth, which are readily oxidized triterpenoids and divided either as glycosides or aglycones. Limonoid glucoside, a tasteless and nontoxic substance produced from citrus processing industry wastes, generates approximately up to 15,000 tons of limonoids, which presents good pharmacological properties such as antitumor, anticarcinogenic, antifungal, and antibacterial activities (Montoya et al., 2019). The experimental studies showed that grape pomace accounts for 4 to 5% (Theagarajan et al., 2019). Studies declared that apricot kernels have apigenin 7-O-glucoside flavones, quercitrin and quercetin flavanones, cyanidin-3-4″-acetylrutinoside anthocyanins, and salicylic acid and ferulic acid phenolic acids, which possess antioxidant activity. The flesh of the fruit contains more phenolic compounds than other byproducts (Senica et al., 2017). The plum, peach, and nectarine juice industries yield a large quantity of pomace byproduct (skin and stone), which contains almost all the flavanols and caffeic acid derivatives (neochlorogenic acid, chlorogenic acid, and cryptochlorogenic acid) as well as anthocyanins and flavanols. The major polyphenols reported in pomace are anthocyanins, proanthocyanidins, HCA, and quercetin glycosides (Milala et al., 2013). An experimental study reported that ethyl gallate and penta-O-galloyl-glucoside are the two major compounds found in mango peel, having hydroxyl radical potent, singlet oxygen scavenging activities, and superoxide anion activities. Same as mango peels, mango seed kernels are also byproducts of mango processing industries, and its extracts have very efficacious antioxidant with the highest metal-chelating, radical scavenging potent, and tyrosinase inhibitory potent. Gallotannins compound found mango byproducts as reported in many studies and having high antioxidant and good antiproliferative activity, a potential source of anticancer agents. The studies illustrated that mango seed kernels are richer in total phenolics than the mango peel (Luo et al., 2014). In kiwi seed, caffeic acid, catechin, quercetin, p-hydroxybenzoic acid, punicalagin, p-coumaric acid, and penicillin are the main phenolics compounds reported in different cultivars of kiwi. The ellagitannins compound reported in the peel of pomegranate while punicalagin is a specific pomegranate compound found in its all byproducts (seeds, peels, and juice). Indeed, characteristics of pomegranate peel biological mostly linked with tannins and flavonoids content. Owing to high polyphenol constituents, the fruits have the strongest antioxidant activity (Wang et al., 2018).

5.2.1.4 VITAMINS AND MINERALS

Fruit wastes may contain significant amounts of micronutrients, including vitamins (vitamin C and A) and minerals (potassium and magnesium). Vitamins play an important role in the metabolism of fatty acids, proteins, and carbohydrates and assist in the proper functioning of the body by synthesizing various necessary compounds. Fruit waste has fat-soluble vitamins which exhibit antioxidant properties and thus help to reduce the risk of various lifestyle and degenerative diseases like cardiovascular disease, neurological diseases, neuropsychiatric disorders, and even cancer (Verhaar et al., 2002). Fruits, as well as their discarded parts, also contain considerable amounts of several water-soluble vitamins such as B1, B2, B3, and vitamin C (Tedesco et al., 2021). A scientific study conducted on watermelon seeds revealed that minerals like phosphorus, potassium, sodium, and magnesium and vitamins A and C were present in considerable concentrations, and vitamins like thiamine, riboflavin, niacin, pyridoxine, cobalamin, D, E, and K in minor concentrations (Wani et al., 2011). The peel of mango has been reported to be rich in calcium, potassium, magnesium, sodium, iron, manganese, zinc, copper, vitamins C and E (Lebaka et al., 2021). Similarly, the banana peel also contained minerals like potassium, copper, calcium, iron, and phosphorus and vitamins like A, C, and E (Muhammad et al., 2021). In the previous study on the mineral content in the peel of pineapple, orange, banana, papaya, and watermelon, it was found that calcium content ranged from 8.30–162.03 mg/100 g, indicating that all the peels are good sources of calcium, with orange peel having the maximum calcium content. Therefore, orange peel could be used as a good source of calcium, thereby helping in the transmission of nerve impulses, regulation of muscle contraction, and strengthening of bone and teeth formation. Iron and magnesium content in these peels ranged from 15.15–45.58 and 1.48–65.98 mg/100 g, respectively. Therefore, these fruit peels could serve as a good alternative source of magnesium. The results of this study indicated that the peel of these fruits, which is regarded as waste, can be utilized in the formulation of pharmaceutical products as they serve as a source of micro and macronutrients. The sodium content was found to be below the detection limit in pineapple peel, and it could serve as a good additive in dietary supplements for people suffering from hypertension (Gbenga-Fabusiwa et al., 2022).

5.2.1.5 ESSENTIAL OILS

Some fruit peels and seeds contain essential oils that can have antimicrobial and aromatic properties. Apple seeds were found to be high in oil content, with oleic and linoleic acids as the main fatty acids, which are unsaturated fatty acids (UFAs), making apple seed oil nutritionally favorable with positive effects on lowering low-density lipoprotein (LDL) cholesterol, thereby preventing the risks of cardiovascular diseases (Yu et al., 2007). A significant amount of seed oil could be extracted from citrus fruits, which comprised saturated, unsaturated, and omega fatty acids (Zayed et al., 2021). Being a rich source of fatty acids, apricot kernels comprise 40–50% UFAs, mainly oleic (60–70%) and linoleic acid (25–30%) (Femenia et al., 1995). Similarly, the seeds of sour cherries were found to be rich in oleic and linoleic acids and essential fatty acids like carotenoids, sterols, and tocopherols. Whereas, sour cherries were reported to have oil content from 17 to 16% in different studies (17–36%) (Yilmaz et al., 2018). The pomace of sour cherry was found to have a high amount of phenolics, which have positive effects on humans due to their antioxidant, antimicrobial, and anti-inflammatory properties (Yılmaz et al., 2019). The seed was found to be rich in UFAs, mainly oleic acid and linoleic acid (Yilmaz et al., 2018). Kiwi seeds were observed to have palmitic, oleic, and linoleic acids as the major fatty acids (Duman et al., 2011). The pomegranate seeds were found to be rich in oils and included a unique fatty acid, such as linoleic acid and conjugated linoleic acid (Viladomiu et al., 2013). The plum seeds were found to have a high amount of oil and included several bioactive compounds, such as tocopherols, proteins, lipids, and phenolics (Savic et al., 2016). A good potential for extracting phytochemicals for valorization in different industries was observed in melon seeds. In addition to phenolics, melon seed oil also had similar properties to vegetable oils, such as soybean and sunflower oils (Foster et al., 2009).

5.3 POTENTIAL OF BIOACTIVE COMPOUNDS AS NUTRACEUTICALS FOR HUMAN HEALTH

Exploratory investigations on the bioactivities of their constituent substances are the only way to use portions of fruit byproducts. According to Zeb (2020), fruit wastes are a rich source of high-value components like bioactive compounds, which are thought to have an impact on human health due to their biological features such as anti-inflammatory, antioxidant, antibacterial, and antimutagenic properties. The synergistic action of these various chemicals is what gives the various byproducts of fruits their physiological function. Investigating the antioxidant and antibacterial qualities of fruit wastes such as seeds, meat, peels, and pulp can help create novel functional meals that will improve human health and wellbeing.

The management of food waste within the same industry ensures circulation of the leftover food biomass and facilitates, since fruits byproducts are mostly a result of the food business. According to the available preliminary scientific evidence, fruit wastes are a valuable source of phyto constituents for use in pharmaceutical products. According to Tremocoldi et al. (2018), avocado peel has anti-inflammatory activity, and seed extracts were shown to reduce the production of nitric oxide and the release of tumor necrosis factor in lipopolysaccharide-stimulated RAW 264.7 macrophages.

Avocado seed extract was found to have gastroprotective properties against a stomach ulcer that indomethacin caused in mice (Park et al., 2014). Polyphenols from orange flesh and peel have the ability to inhibit the activity of inducible nitric oxide synthase and cyclooxygenase-2 in murine macrophage cell line, as well as to protect human leukocytes against oxidative DNA damage and HepG2 cells against peroxyl radical-induced oxidation *in vitro* (Chen et al., 2016). A significant byproduct of winemaking, black raspberry seeds include ellagic acid, ellagitannins, and its metabolites (urolithin A and urolithin B), which have been shown to have anti-cancer properties against HT-29 colon cancer cells. The cell cycle was halted by the chemicals under investigation, which also caused intrinsic and extrinsic apoptotic pathways (Cho et al., 2015).

Pomegranate peel extracts both reduced osteoporosis-related bone loss in mouse models and promoted osteoblastic development *in vitro* (Spilmont et al., 2015). Aromatic glycosides isolated from berry seed waste significantly reduced H_2O_2-induced damage to HepG2 cells and inhibited α-glucosidase at concentrations comparable to the positive control acarbose (Liu et al., 2018). The anti-inflammatory effects of grape seed extract in LPS-induced RAW264.7 macrophages were attributed to flavonoids (Harbeoui et al., 2019), however catechins contained in grape seeds indicated capacity to inhibit angiotensin-I-converting enzyme *in vitro*. According to Biaek et al. (2017) pomegranate seed oil had a favorable impact on the fatty acid profile and reduced the activity of desaturases in the rat liver. Fruit wastes have a high biological potential that the pharmaceutical business can use. In fact, treating fruit waste for potential therapeutic applications necessitates stringent extract standardization or isolated chemical purity determination. Furthermore, preliminary *in vitro* investigations must be validated not only in animal models but also in clinical trials before fruit waste compounds may be used in the form of medications. Reusing food byproducts is an easy technique to produce active chemicals sustainably, and this concept has gained popularity recently. Figure 5.1 depicts a thorough analysis of the bioactive substances found in fruit wastes and their advantages for health. An economical source of active substances can be found by using logical methods for extracting fruit byproducts.

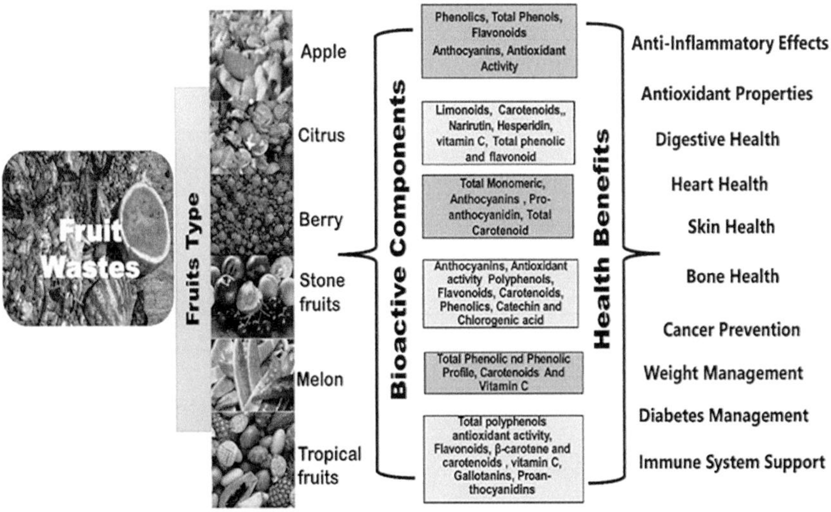

FIGURE 5.1 A systematic review of bioactive compounds of fruit wastes and their health benefits.

5.4 EXTRACTION TECHNIQUES FOR CONSTITUENTS OF COMMERCIAL AND NUTRACEUTICAL IMPORTANCE

Extraction techniques play a vital role in isolating constituents of commercial and nutraceutical importance from various natural sources, that is, plants, fruits, herbs, and marine organisms. These techniques help to obtain bioactive compounds, flavors, fragrances, essential oils, and other valuable components for use in industries like pharmaceuticals, food and beverages, cosmetics, and more. Extraction techniques form the cornerstone of modern industries that rely on obtaining valuable substances from various natural sources. These techniques encompass a diverse array of methods designed to isolate specific components of interest, ranging from bioactive compounds and essential oils to flavors, fragrances, and pharmaceutical agents. With applications spanning pharmaceuticals, nutraceuticals, food and beverages, cosmetics, and beyond, extraction techniques are pivotal in harnessing the potential of nature's bounty for commercial purposes.

By employing innovative methods that combine scientific knowledge with engineering precision, extraction processes continue to evolve, enabling industries to access the rich tapestry of chemicals and compounds that nature offers. This introduction delves into the world of extraction techniques, exploring their principles, methodologies, and the vast scope of their applications in the contemporary landscape. In this chapter, we will thoroughly discuss the most common techniques used in the extraction of the bio-active compound, as follows (Figure 5.2).

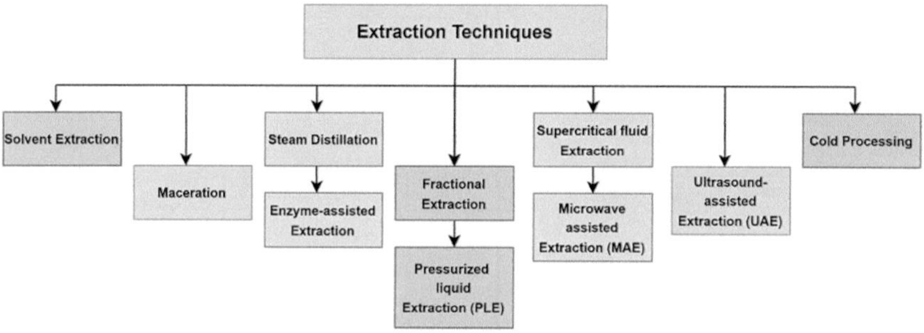

FIGURE 5.2 Different techniques for solvent extraction.

5.4.1 SOLVENT EXTRACTION

Solvent extraction, also known as liquid-liquid extraction or partitioning, hinges on the principle of selectively dissolving target compounds from a solid or liquid source material into a suitable solvent. The process capitalizes on differences in solubility, polarity, and other physicochemical properties to achieve separation. The solvent acts as a mediator, extracting the desired constituents while leaving behind impurities or unwanted compounds (Figure 5.3).

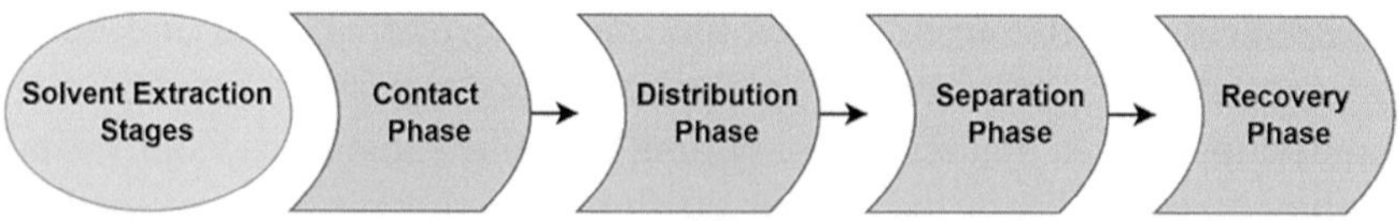

FIGURE 5.3 Stages of solvent extraction.

One of the defining characteristics of solvent extraction is the choice of solvent. The selection is influenced by the nature of the compounds being targeted and the desired end product. A diverse range of solvents, each with its unique solubility characteristics, are employed. For instance, polar solvents like water and ethanol are often used for extracting water-soluble compounds, while non-polar solvents like hexane are effective in extracting lipophilic substances such as essential oils. The stages of solvent extraction are as follows (Figure 5.2):

- **Contact Phase:** The source material is mixed or soaked with the solvent. This initiates the process of solubilization, where the solvent interacts with the target compounds.
- **Distribution Phase:** As the solvent and the source material interact, the target compounds move from the solid or liquid matrix into the solvent. This distribution is driven by factors such as concentration gradients and partition coefficients.
- **Separation Phase:** The solvent containing the extracted compounds, referred to as the extract or raffinate, is separated from the remaining source material.
- **Recovery Phase:** In some cases, further steps may be employed to recover the target compounds from the solvent, leading to the production of a concentrated extract.

Applications:

- **Natural Flavors and Fragrances:** Essential oils and natural flavors are obtained through solvent extraction from herbs, spices, and fruits. These extracts are used in food, beverages, and perfumes.
- **Food Additives:** Solvent extraction is used to isolate natural colors, antioxidants, and other food additives from plant sources.

5.4.2 STEAM DISTILLATION

The roots of steam distillation trace back to ancient civilizations such as Egypt, where it was used for perfumery, rituals, and medicinal practices. The technique evolved over time, with the Persian polymath Ibn Sina (Avicenna) refining the process during the Islamic Golden Age. Eventually, the knowledge spread to Europe during the Middle Ages, contributing to the development of modern distillation techniques. In the world of extraction techniques, where science meets artistry, steam distillation stands as a mesmerizing process that has been captivating minds and senses for centuries. As a method rooted in ancient traditions yet refined by modern science, steam distillation is the key to unlocking the ethereal essences and aromatic wonders hidden within plants and botanicals. Steam distillation is an age-old technique that harnesses the power of steam to extract essential oils and aromatic compounds from plant material. It is a method that marries the principles of heat, vaporization, and condensation to separate volatile components from the rest of the plant matrix. The result is a distilled liquid known as an essential oil, which holds the concentrated aroma and therapeutic properties of the source material (Figure 5.4).

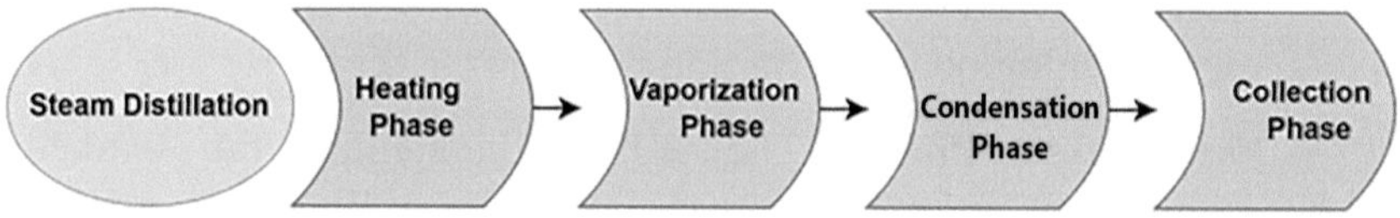

FIGURE 5.4 Stages of steam distillation.

- **Heating Phase**: Plant material, often flowers, leaves, or other aromatic parts, is placed in a distillation apparatus. Steam is then introduced, gently heating the plant material.
- **Vaporization Phase:** The heat causes the volatile compounds within the plant to vaporize. These compounds rise with the steam, carrying their precious aromas with them.
- **Condensation Phase:** The vaporized steam and essential oil mixture enters a cooling system, typically a condenser. Here, it undergoes a transformation from vapor to liquid as it cools down.
- **Collection Phase:** The condensed liquid, which now contains the essential oil and water, is collected. Since oil and water do not mix, they naturally separate, making it easy to extract the essential oil.

Application of Steam Distillation

- The extracted essential oils are revered for their therapeutic properties, contributing to holistic well-being and relaxation.
- In the culinary world, steam distillation is used to extract natural Flavors and fragrances from herbs, spices, and botanicals, enhancing the sensory experience of food and beverages.

- The medicinal properties of certain plants are harnessed through steam distillation, resulting in essential oils that find their way into natural remedies.
- From skincare products to perfumes, steam-distilled essential oils add a touch of nature's elegance to cosmetics and personal care items.

5.4.3 SUPERCRITICAL FLUID EXTRACTION (SFE)

Supercritical fluid extraction (SFE) stands as a remarkable method that has revolutionized the way we harness nature's treasures. With its unique ability to dissolve a wide range of compounds, SFE has become a cornerstone in industries ranging from pharmaceuticals to food and cosmetics. This chapter embarks on a journey through the fascinating world of SFE, exploring its principles, applications, and the transformative impact it has on our quest for valuable substances.

SFE lies in a state of matter that defies traditional classification—the supercritical state. A supercritical fluid exhibits properties that bridge the gap between a gas and a liquid, resulting from subjecting a substance to specific temperature and pressure conditions. In the case of SFE, carbon dioxide (CO_2) is often used as the supercritical solvent due to its relatively low critical temperature and pressure (Figure 5.5).

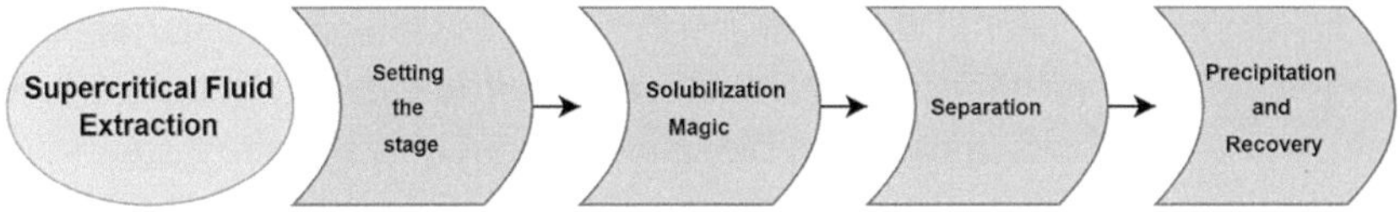

FIGURE 5.5 Steps of supercritical fluid extraction.

Stages of Extraction:

- **Setting the Stage:** The source material, which can range from botanicals to polymers, is placed in an extraction vessel. Supercritical CO_2 is then introduced and pressurized to its critical point, where it becomes a supercritical fluid.
- **Solubilization Magic:** In its supercritical state, CO_2 acts as a solvent with extraordinary solubility properties. It permeates the source material, selectively dissolving the target compounds.
- **Separation:** The supercritical fluid containing the dissolved compounds is separated from the residual material. By adjusting

the pressure and temperature, the solubilized compounds can be selectively collected.

- **Precipitation and Recovery:** When the pressure is reduced, the supercritical CO_2 returns to its gaseous state, leaving behind the extracted compounds. The result is a pure, solvent-free extract.

Applications:

- **Food and Beverages:** In decaffeination, flavor extraction, and lipid separation, SFE enhances the quality and purity of food products.

5.4.4 COLD PROCESSING

As a process that values quality over haste, cold processing extraction has gained prominence in industries seeking to retain the nutritional integrity, flavors, and aromas of botanicals and seeds. This chapter embarks on a journey through the world of cold processing extraction, exploring its principles, benefits, applications, and the art of capturing nature's essence while preserving its vitality.

Cold processing extraction is defined by its adherence to low temperatures throughout the extraction process. Unlike methods that involve heat and intense processing, cold extraction prioritizes maintaining the structural and chemical integrity of the source material. This technique is particularly well-suited for substances prone to degradation under high temperatures, such as essential oils and certain bioactive compounds (Figure 5.6).

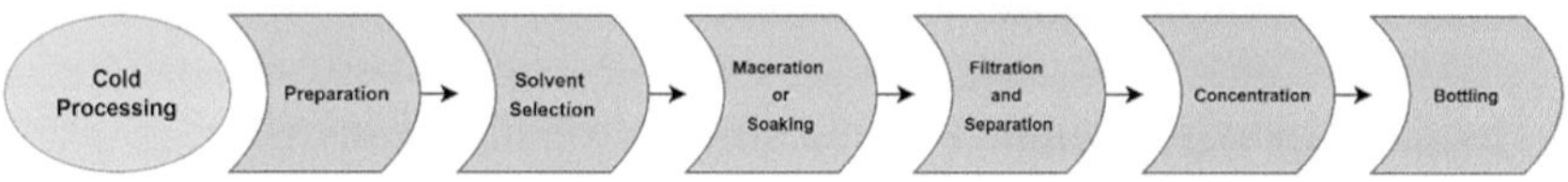

FIGURE 5.6 Steps of extraction.

- **Preparation:** The botanicals, seeds, or other materials to be extracted are carefully selected and prepared, ensuring optimal freshness and quality.

- **Solvent Selection:** Depending on the target compounds, an appropriate solvent is chosen. Common solvents include cold-pressed oils, glycerin, and vinegar.

- **Maceration or Soaking:** The source material is immersed in the chosen solvent and left to steep over a period of time. During this phase, the solvent gently draws out the desired compounds.
- **Filtration and Separation:** After the soaking period, the mixture is filtered to separate the extract from the solid plant material. This step ensures a pure, clean extract.
- **Concentration:** In some cases, the extract may undergo a concentration process to increase its potency. This step involves removing excess solvent.
- **Bottling:** The final extract, rich in the essence of the source material, is carefully bottled and labelled, ready for use.

Applications that Embrace Purity:

- **Essential Oils:** Cold extraction is ideal for obtaining essential oils from delicate flowers and herbs, ensuring that their aromatic compounds remain intact.
- **Nutritional Supplements:** Cold-processed extracts are used for the production of vitamins, minerals, and other nutritional supplements, maintaining the potency of active compounds.
- **Culinary Creations:** In gourmet cooking and fine dining, cold-pressed oils and extracts enhance dishes with unadulterated flavors and aromas.

15.4.5 MACERATION

Maceration is a technique that involves soaking plant material in a liquid—often alcohol, oil, or vinegar—to extract its essential components. This slow and patient process allows the liquid to absorb the flavors, colors, aromas, and medicinal properties of the botanicals, resulting in an enriched extract. Maceration traces its roots back to ancient civilizations where herbs, fruits, and flowers were steeped in liquids to create elixirs, tinctures, and remedies. The practice was an art passed down through generations, with different cultures and regions infusing their unique traditions into the process (Figure 5.7).

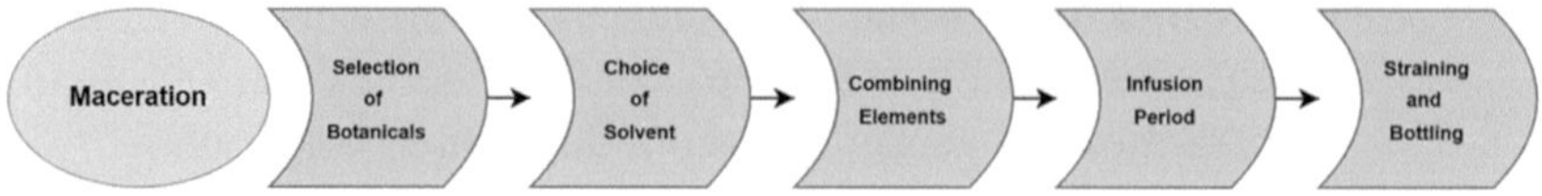

FIGURE 5.7 Stages of extraction.

Stages of Maceration:

- **Selection of Botanicals:** High-quality plant material is selected, cleaned, and sometimes chopped to increase surface area and facilitate extraction.
- **Choice of Solvent:** The chosen liquid solvent—alcohol, oil, or vinegar—acts as a medium to extract the compounds from the botanicals.
- **Combining Elements:** The plant material is placed into the chosen solvent, and the mixture is sealed in a container. The solvent's hydrophilic or lipophilic properties determine the components extracted.
- **Infusion Period:** The container is left undisturbed in a cool, dark place for a predetermined period, allowing the botanicals to slowly release their essence into the solvent.
- **Straining and Bottling:** After the infusion period, the mixture is strained to remove the solid botanicals, leaving behind the infused liquid—a potent extract capturing the botanical's characteristics.

Applications:

- **Culinary Arts:** Chefs and culinary enthusiasts employ maceration to create infused oils, vinegars, and spirits that enhance flavors in cooking and mixology.
- **Aromatherapy and Cosmetics:** Macerated oils serve as the base for many natural skincare products, perfumes, and aromatherapy blends.
- **Traditional Medicine:** Maceration has a place in traditional herbal medicine, as it's used to extract and preserve the medicinal properties of herbs for tinctures and herbal remedies.

- **Flavoring and Fragrance:** The technique contributes to the production of natural flavors, fragrances, and extracts used in the food, beverage, and cosmetic industries.

5.4.6 ENZYME-ASSISTED EXTRACTION (EAE)

Enzyme-assisted extraction (EAE) relies on a remarkable class of biological molecules known as enzymes. These proteins act as molecular architects, catalyzing specific biochemical reactions. In the context of extraction, enzymes are employed to break down the complex structures within plant cells, making it easier to extract valuable compounds. The process of EAE consists of the following sequence of steps (Figure 5.8).

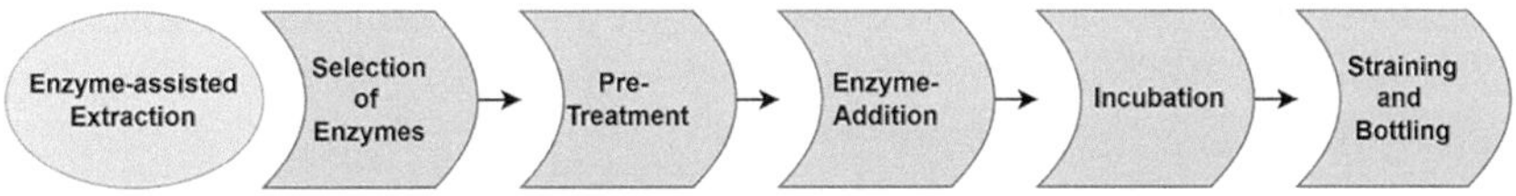

FIGURE 5.8 Sequence of extraction.

- **Selection of Enzymes:** Depending on the target compounds and source material, specific enzymes are chosen. Enzymes like cellulases, pectinases, and proteases are commonly used to degrade cell walls and facilitate extraction.
- **Pre-Treatment:** The source material is subjected to pre-treatment, which might involve crushing, cutting, or grinding to increase the surface area and facilitate enzyme penetration.
- **Enzyme Addition:** The selected enzymes are added to the pre-treated material. These enzymes work like microscopic demolition experts, breaking down the structural barriers that encase valuable compounds.
- **Incubation:** The enzyme-treated mixture is left to incubate under controlled conditions. During this time, the enzymes continue to work, breaking down cell walls and releasing target compounds.
- **Separation:** After the incubation period, the mixture is processed to separate the liquid extract from the solid residue, often using techniques like filtration or centrifugation.

- **Concentration and Purification:** The extracted liquid may undergo further processing to concentrate and purify the target compounds, resulting in a potent and valuable extract.

Applications:

- **Food and Beverages:** It is used to extract flavors, colors, and nutrients from fruits, vegetables, and grains for the production of beverages, sauces, and processed foods.

5.4.7 FRACTIONAL EXTRACTION

In the realm of extraction techniques, where precision meets complexity, fractional extraction emerges as a strategic method that allows us to delve deep into the intricate world of natural compounds. With its ability to selectively separate and isolate specific groups of compounds, fractional extraction has become a vital tool in industries ranging from pharmaceuticals to flavor and fragrance production.

Fractional extraction, often referred to as selective extraction, is an advanced technique that takes advantage of varying solubility characteristics of compounds within a source material. Unlike single-solvent extractions, fractional extraction involves multiple stages, each employing different solvents of varying polarities to target different groups of compounds. This intricate approach allows for the isolation of a broader spectrum of valuable substances (Figure 5.9).

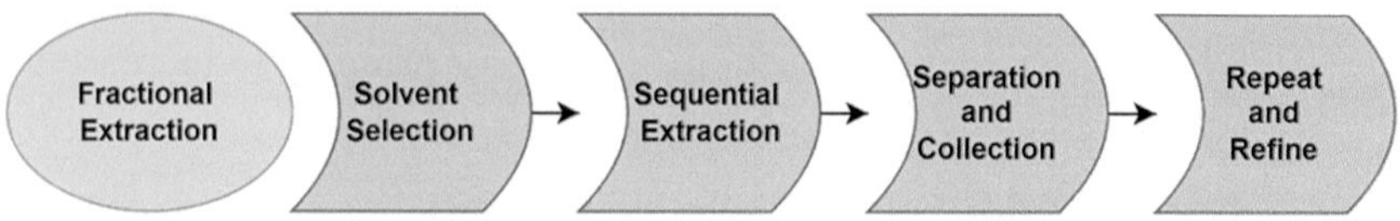

FIGURE 5.9 Sequence of extraction.

The process of fractional extraction is a meticulously orchestrated sequence of steps:

- **Solvent Selection:** Different solvents with varying polarities are chosen to target specific classes of compounds. Polar solvents are used to extract polar compounds, while non-polar solvents are used for non-polar compounds.

- **Sequential Extraction:** The source material undergoes multiple extraction stages. In each stage, a solvent is chosen to selectively dissolve a particular group of compounds.
- **Separation and Collection:** After each extraction stage, the mixture is separated, often through filtration or centrifugation, into the solvent-extract mixture. The solvent is then evaporated to obtain the extracted compounds.
- **Repeat and Refine:** Multiple extraction stages are performed, each with a different solvent, to target specific compounds of interest. The fractions obtained can be further refined or combined based on their intended applications.

Applications Across Industries:

- **Pharmaceuticals:** Fractional extraction is used to isolate specific classes of bioactive compounds from plant materials for drug development and research.
- **Flavor and Fragrance Production:** The technique aids in capturing the delicate flavors and aromas present in botanicals for use in perfumes, essential oils, and culinary extracts.
- **Nutraceuticals:** Fractional extraction helps obtain specific bioactive compounds with potential health benefits for use in dietary supplements.

5.4.8 MICROWAVE-ASSISTED EXTRACTION (MAE)

Microwave-assisted extraction (MAE) emerges as a powerful method that harnesses the energy of microwaves to unlock valuable compounds from various natural sources. MAE revolves around utilization of microwave energy, a type of electromagnetic radiation. This energy interacts with the source material and solvent, causing molecular vibrations, friction, and localized heating. This leads to accelerated extraction and enhanced solute-solvent interactions, resulting in more efficient compound extraction (Figure 5.10).

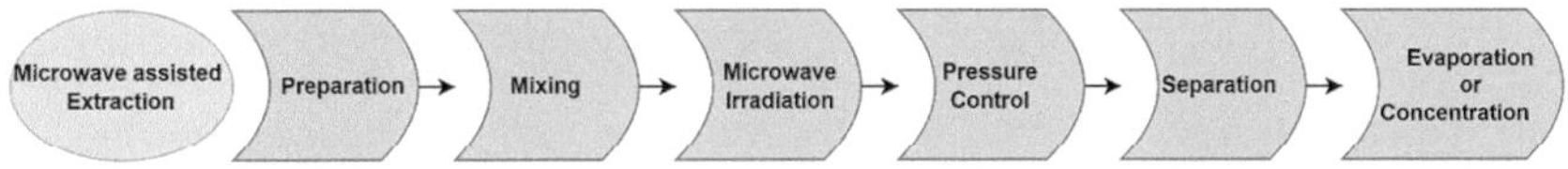

FIGURE 5.10 Steps of extraction.

The process of MAE unfolds in a series of carefully orchestrated steps:

- **Preparation:** The source material is selected, cleaned, and sometimes ground or chopped to optimize surface area and enhance extraction efficiency.
- **Solvent Selection:** A suitable solvent is chosen based on the target compounds' solubility properties.
- **Mixing:** The source material is mixed with the solvent in an extraction vessel.
- **Microwave Irradiation:** The mixture is exposed to microwave energy, which heats the solvent and creates localized conditions that facilitate the release of compounds from the source material.
- **Pressure Control:** In closed systems, pressure may build up due to vaporization. Pressure control mechanisms ensure safety and optimal extraction.
- **Separation:** After irradiation, the mixture is separated using traditional techniques such as filtration or centrifugation.
- **Evaporation or Concentration:** In some cases, the solvent is evaporated or concentrated to yield the desired compound-rich extract.

Advantages Over Traditional Methods:

- **Speed:** The rapid heating induced by microwaves significantly reduces extraction times compared to conventional methods.
- **Efficiency:** Microwaves promote efficient heat transfer, ensuring uniform heating and better solute-solvent interactions.
- **Preservation of Compounds:** The shorter extraction times help preserve thermally sensitive compounds that might degrade under prolonged heat exposure.
- **Reduced Solvent Usage:** The enhanced extraction efficiency often leads to reduced solvent usage, contributing to sustainability.

Applications:

- **Food and Beverages:** The technique is used to extract flavors, colors, and nutrients from food ingredients.

5.4.9 PRESSURIZED LIQUID EXTRACTION (PLE)

Pressurized liquid extraction (PLE) involves the application of pressure to enhance the extraction process. By increasing the pressure within a closed system, the solvating power of the solvent is improved, leading to higher solubility and increased interaction between the solvent and the target compounds. This results in more efficient extraction in a shorter period of time (Figure 5.11).

The process of PLE consists of the following steps:

- **Preparation:** The source material is selected, cleaned, and sometimes ground to optimize surface area and exposure to the solvent.
- **Sample Packing:** The prepared source material is placed into an extraction cell or cartridge.
- **Solvent Selection:** A suitable solvent is chosen based on the target compounds' solubility properties.

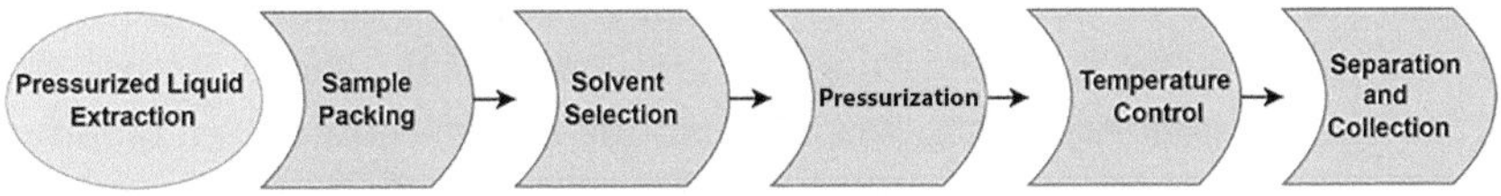

FIGURE 5.11 Steps of pressurized liquid extraction.

- **Pressurization:** The solvent is pumped into the extraction cell under controlled pressure. The increased pressure promotes enhanced solute-solvent interactions and solubility.
- **Temperature Control:** In some cases, temperature can be controlled to further enhance the extraction process by adjusting the equilibrium between the solid matrix and the solvent.
- **Separation and Collection:** After the pressurized extraction, the solvent is collected, often through a separate collection vessel, ready for analysis or further processing.

Advantages Over Traditional Methods:

- **Speed and Efficiency:** The increased pressure leads to higher solubility and faster extraction, resulting in shorter extraction times.
- **Reduced Solvent Usage:** The enhanced solubility means that less solvent is required, making the process more environmentally friendly.

- **Selective Extraction:** By controlling pressure and temperature, selective extraction of specific compounds can be achieved.
- **Consistency:** The controlled parameters in PLE lead to more consistent extraction results, reducing variability.

Applications:

- **Food and Beverages:** PLE enhances the extraction of flavors, antioxidants, and nutrients from food ingredients.
- **Natural Products:** Compounds for cosmetics, nutraceuticals, and herbal remedies are efficiently extracted using PLE.

5.4.10 ULTRASOUND-ASSISTED EXTRACTION (UAE)

Ultrasound-assisted extraction (UAE) revolves around the application of high-frequency sound waves to accelerate the extraction process. These sound waves generate microscopic bubbles through a phenomenon called cavitation. As the bubbles collapse, they create tiny shock waves that help disrupt cell structures, enhance solvent penetration, and facilitate the release of compounds from the source material. The process of UAE consists of the following step sequence (Figure 5.12):

- **Preparation:** The source material is selected, cleaned, and sometimes ground to increase surface area and facilitate the interaction between the solvent and compounds.
- **Solvent Selection:** A suitable solvent is chosen based on the target compounds' solubility properties.
- **Ultrasonic Bath:** The prepared source material is immersed in the chosen solvent within an ultrasonic bath, which generates sound waves.

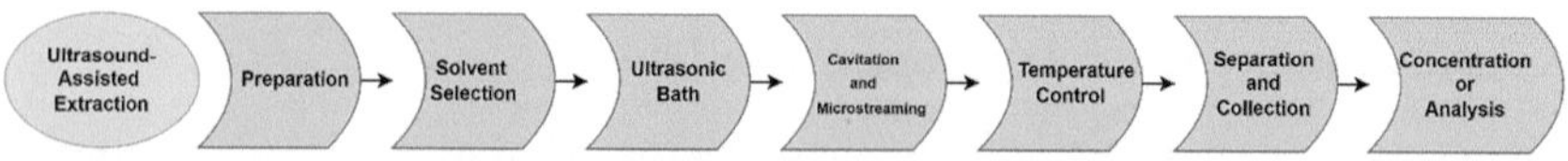

FIGURE 5.12 Sequence of ultrasound-assisted extraction.

- **Cavitation and Microstreaming:** The sound waves induce cavitation, creating tiny bubbles that expand and collapse. These implosions generate shock waves and microstreaming that promote cell disruption and compound release.

- **Temperature Control:** The process generates localized heating due to the cavitation process. In some cases, temperature control mechanisms ensure that delicate compounds are not degraded.
- **Separation and Collection:** After ultrasonic treatment, the solvent is separated from the source material, often through filtration or centrifugation.
- **Concentration or Analysis**: The collected solvent containing the extracted compounds can be concentrated or directly analyzed depending on the intended application.

Advantages Over Traditional Methods:

- **Speed and Efficiency:** The cavitation-induced disruptions accelerate the extraction process, resulting in reduced processing times.
- **Enhanced Mass Transfer:** The microstreaming created by cavitation enhances mass transfer, leading to improved solute-solvent interactions.
- **Preservation of Delicate Compounds:** The relatively lower temperatures and shorter extraction times help preserve thermally sensitive compounds.
- **Reduced Solvent Usage:** The enhanced efficiency often requires less solvent, contributing to sustainability.

Applications:

- **Food and Beverages:** The technique is used to extract flavors, antioxidants, and nutrients from food ingredients.

5.5 FOOD WASTE AND BIO ACTIVE COMPOUNDS

With the increased consciousness of sustainability and waste reduction, a significant focus has been placed on the management and utilization of food and agricultural byproducts. Among these, the waste generated from fruits and vegetables often remains an untapped resource with immense potential. Beyond their outward appearance, these scraps and peels harbor a treasure trove of bioactive components that can contribute to human health and well-being in unexpected ways (Figure 5.13).

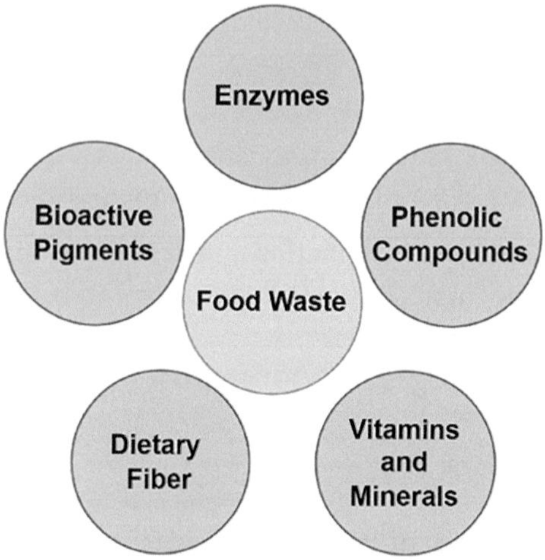

FIGURE 5.13 Food waste and bioactive components.

Fruits and vegetables are renowned for their nutritional richness, packed with vitamins, minerals, and dietary fiber. However, their value does not end with what is typically consumed. The outer layers, cores, and seeds that are often discarded contain a plethora of bioactive compounds, which are compounds that have the potential to positively influence human health beyond basic nutrition. These compounds have antioxidant, anti-inflammatory, and other health-promoting properties. In order to extract these bioactive components, various extraction techniques are followed by the food industries. These extracted bio-components are then used to develop nutraceuticals and dietary supplements, offering targeted health supplements (Table 5.3).

TABLE 5.3 Bioactive Compounds of Fruits and Vegetable Waste and Its Extraction Methods

Bioactive Compounds	Fruit and Vegetables Waste	Extraction Method
Phenolic Acid, Anthocyanins	Blackberry Pomace	Pressurized Liquid extraction (PLE) Super Critical Fluid Extraction
	Raspberry Pomace	Solid Phase Extraction
	Dragon Fruit peel	Methanol Extraction
Carotenoids, Phenolic acids, and Flavonoids	Tomato Skin	Supercritical CO_2 Extraction
	Pumpkin seeds, peel, fresh	Aqueous phase extraction
	Papaya Peel	Ultrasound-assisted extraction (UAE)
Phenolic acid, flavonoids, betaines	Red Beetroot peel extract	Maceration Technique
	Red beetroot juice	Ultrasound-assisted extraction (UAE)
	Pulp and peel of pricky pear	Single extraction

TABLE 5.3 *(Continued)*

Bioactive Compounds	Fruit and Vegetables Waste	Extraction Method
Phenolic acids, Anthocyanins, flavanol	Blueberry pomace extract	Ultrasound-assisted extraction (UAE)
Polyphenols, Phenolic and flavonoids	Broccoli and artichokes waste	Methanolic extraction
	Onion skin	Pressurized hot water extraction
	Potato waste	Maceration, hot water extraction
	Red pepper, carrots, and cucumber waste	Ultrasonic processor
Phenolic flavonoids, anthocyanins	Eggplant peel	Microwave-assisted extraction (MAE)

Source: Massimo et al. (2021).

In the pursuit of consumer-centric innovation, extraction techniques are at the forefront of creating products that resonate with health-conscious individuals. The extraction of bioactive compounds from plants, the concentration of vitamins and minerals, and the preservation of natural flavors enable the development of functional foods and nutraceuticals that offer not just sustenance, but genuine benefits for well-being.

Quality assurance is another cornerstone of the food industry, and extraction techniques play a crucial role in achieving consistency and safety. Through precise extraction methods, producers can control the concentration of active compounds, ensuring that products meet desired standards in flavor, color, aroma, and health attributes.

Extraction techniques illuminate the intricate dance between science and gastronomy within the food industry. They bridge the gap between raw ingredients and the culinary wonders that grace our tables. These techniques bring forth the hidden treasures of nature, empower sustainability efforts, facilitate innovation, and elevate the quality of products we consume. As the food industry continues to evolve, the art and science of extraction techniques will undoubtedly remain a cornerstone, enriching our lives one delectable creation at a time.

5.6 BIOACTIVE COMPOUNDS OF COMMERCIAL IMPORTANCE EXTRACTED FROM FRUIT WASTE

5.6.1 PECTIN

Pectin is a heteropolysaccharide that constitutes about 30% of cell walls, mainly the primary cell walls along with the middle lamella. Being considered as soluble fiber consisting of a long chain of indigestible sugar units, it has been utilized in diverse applications from pharmaceutical bases, cosmetics, nutritional, and functional food formulations. Pectin possesses an average molecular weight of 50 to 150 kDa and consists of galacturonic acid monomer units joined together by α-(1→4)-glycosidic linkages having acidic groups esterified with methoxy groups throughout the chain. It is mainly used in food applications as gelling agents, emulsifiers, thickeners, viscosities, and stabilizers (Min et al., 2010). It is used to increase the consistency of jams, jellies, and marmalades in food industries due to its ability in minimizing syneresis and binding water molecules. Pectin can be extracted from diverse sources of fruit wastes, but commercially apple pomace and citrus peels are mainly used for pectin production. Pectin has been found by various researchers to slow down the absorption of glucose in the human body due to its property of forming a mucilage layer that entraps the glucose molecules and hinders their digestion by minimizing contact with enzymes. It also has the property of binding with cholesterol and thus has been used to limit its absorption in the intestinal tract, and has thus been used in the treatment of patients suffering from cardiovascular diseases and obesity complications. The pectin from apple and citrus peels has been validated to enhance the viscosity of bulk and restrict the action of bile salts on cholesterol absorption (Sriamornsak and Pornsak, 2003). It is also used as medication for the treatment of constipation and diarrhea due to its ability to increase the bulk and viscosity of stools. Pectin has been found to possess prebiotic properties as it is degraded in the colon and large intestines of the human body, causing the formation of short-chain fatty acids and thus is used in the treatment of diarrheal diseases (Gomez et al., 2014). Pectin is being used for the encapsulation of medicines and antioxidant compounds for the controlled release of bioactive components (Liu et al., 2003). Pectin has the ability to form a gel in acidic conditions; it is due to this property of pectin that it is being used as a carrier for drugs to enhance contact time in gastric and ocular ailments (Sriamornsak et al., 2010). Pectin has the ability to heal wounds and ulcers due to its hydrophilicity, which maintains a moist environment with acidic conditions and prevents bacterial growth, along with its efficacy to bind drugs for healing wounds (Munarin et al., 2012). Pectin has been found to possess an anticarcinogenic effect by activating cell death pathways in cancerous cells. It also has the ability to bind and remove toxic heavy metals from the human body, especially lead and mercury. Pectin is used throughout the world in the formulation of low-calorie and sugar-free soft drinks due to its ability to increase viscosity and mouthfeel at low concentrations of sugars.

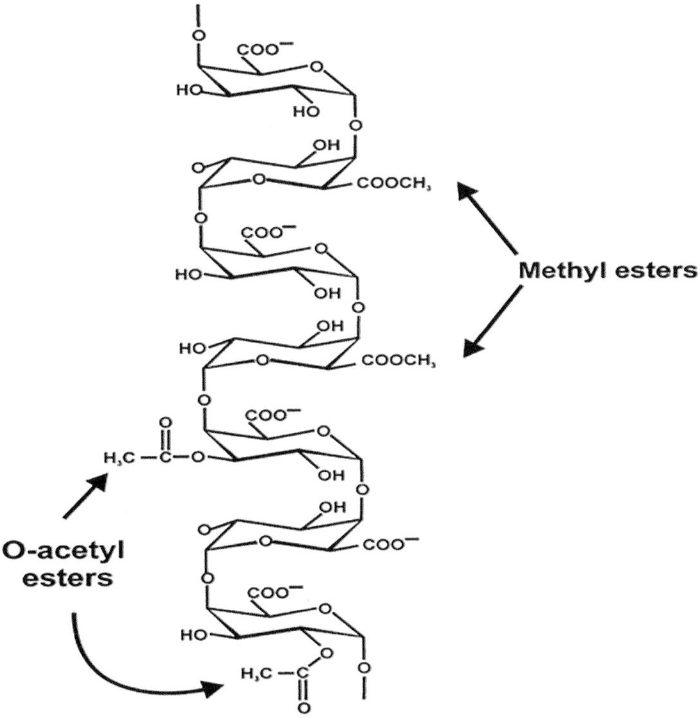

5.6.2 NARINGIN

Naringin ($C_{27}H_{32}O_{14}$) is a flavonoid glycoside occurring in the citrus fruit family, contributing bitterness, especially in the skin of grapefruit and orange. Naringin has the ability to act as an antioxidant and reduce fat deposition in adipose tissues. It has been discovered that naringin upregulates the mRNA levels of the enzymes involved in β-oxidation, reducing diet-induced obesity (Fukuchi et al., 2008). Naringin supplementation has been validated in lowering the elevation of plasma lipid concentrations, thus reducing plasma lipids and cholesterol (Pu et al., 2012). Naringenin has been found to possess a vasodilatation effect by activating large conductance of Ca^{2+}-activated K^{+} currents; thus, its supplementation can be used in improving hypertension in obese persons (Saponara et al., 2006).

Naringin supplementation has also been found to reduce the peroxidation of lipids, accelerate antioxidant enzymes, and reduce inflammatory cell and fibrosis (Rajadurai et al., 2006). Naringin has been proven to continue to have several pharmacological advantages, such as antioxidant, anti-aging, antihyperglycemic, antiosteoporosis, antitumor, antiapoptotic, anti-ulcer, antibacterial, and antiviral properties (Bharti et al., 2014). It has been discovered that taking naringin supplements can lower your risk of getting cancers like lung cancer (Le Marchand et al., 2000). The therapeutic effects of naringin have been found to be due to its ability to modulate enzymes and protein expressions, along with its effect on osteogenic differentiation and cell proliferation (Dai et al., 2009). Numerous signaling molecules implicated in the inflammatory response, such as tumor necrosis factor-α, interleukin-6, interleukin-8, and inducible nitric oxide synthase, have been shown to be downregulated by naringin (Habauzit et al., 2011). Naringin has been acknowledged to prevent cancer cell formation by interfering with their development and progression by modulating several unregulated signaling pathways manifested mainly with apoptosis, inflammation, angiogenesis, proliferation, autophagy, and metastasis (Zhang et al., 2018).

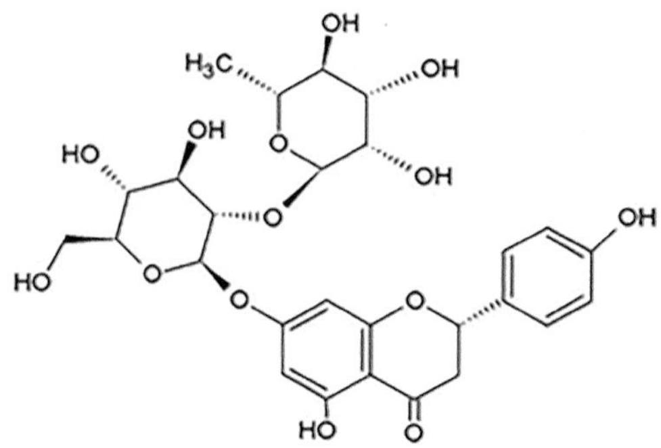

5.6.3 LIMONIN

Limonin ($C_{26}H_{30}O_8$), a natural hexacyclic triterpenoid bioactive compound, is a secondary metabolite possessing higher biological activities and is a main component of volatile oil in citrus fruits. Limonin, along with other limonoids, are the main contributors of bitterness in citrus fruits, with seeds retaining the maximum concentration of this component. Limonin has received an increasing interest by researchers due to its valuable pharmacological benefits like antioxidative, anticancer, anti-tumor, anti-parasitic, anti-bacterial, anti-virus, analgesic, anti-inflammatory, antihyperalgesic, and immunomodulatory activities. Besides, it has been discovered that limonin plays a significant impact in protecting nerve damage, hepatitis, liver protection, enteritis, hemorrhoids, and regulation of blood lipid profiling (You et al., 2007). There is an emerging trend in development of limonene-based phytomedicines in treatment of different kinds of ailments and diseases. These have been reported to inhibit the replication of HIV-1 and HTLV-1 in cells infected with such viruses. Limonin has been depicted to exhibit significant effect on cholesterol metabolism and results raising excellent HDL cholesterol while decreasing harmful LDL cholesterol and triglycerides. Limonin has been found to improve micro structural integrity in bones by increasing bone mass density and thus prevents the chances of bone loss and osteoporosis. Limonoids are having the ability to induce the activity of glutathione S-transferase, which is one among the major detoxifying enzyme system in human body and thus marked limonin as ant carcinogenic agent (Matheyambath and Paliyath, 2016).

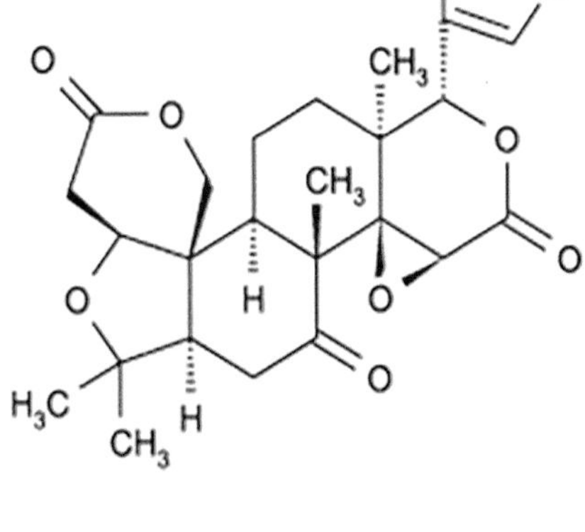

5.6.4 CATECHIN

A polyphenolic antioxidant found in a variety of fruits, mostly in their peels, that includes grapes, seeds, kiwis, blueberries, strawberries, raspberries, and cherries. The dietary catechins that are present in most food items include epicatechin, gallocatechol, and epigallocatechin. The antioxidant activity of catechin has been found to be effective in the treatment of some cancer types, neurological disorders, diabetes, inflammation, and cardiovascular diseases (Macakova et al., 2014). According to reports, epigallocatechin gallate inhibits the proliferative ability of carcinogenic cells by preventing the action of the enzyme telomerase. The reduction in coronary heart diseases by catechins has been attributed to their ability in minimizing cytotoxicity in fibroblast cells of the lungs. Catechins have the ability to reduce atherosclerosis, hypertension, cardiomyopathy, and cardiovascular dysfunction by reducing inflammation and myocardial cell infiltration, along with platelet aggregation (Khan and Mukhtar, 2018). The greater scavenging ability of catechins is thought to be the cause of their therapeutic effects as compared to ascorbic acid and tocopherols. It has the potential of reducing degradation of the extracellular matrix induced by pollution and exposure to ultraviolet rays (Shi et al., 2016). Catechins have the ability to activate the production of collagen in the skin, acting as an anti-aging agent (Arct et al., 2003).

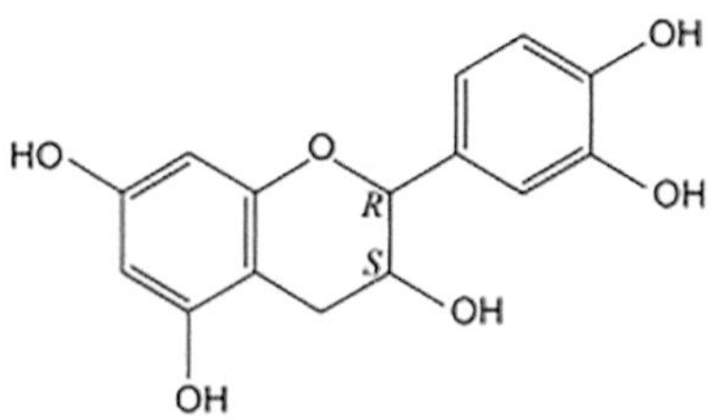
Catechin

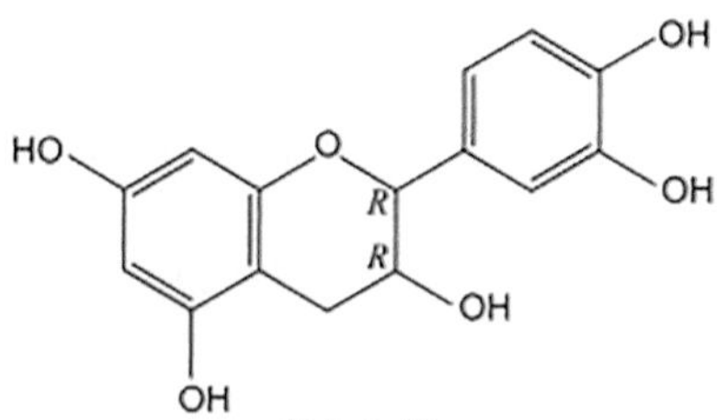
Epicatechin

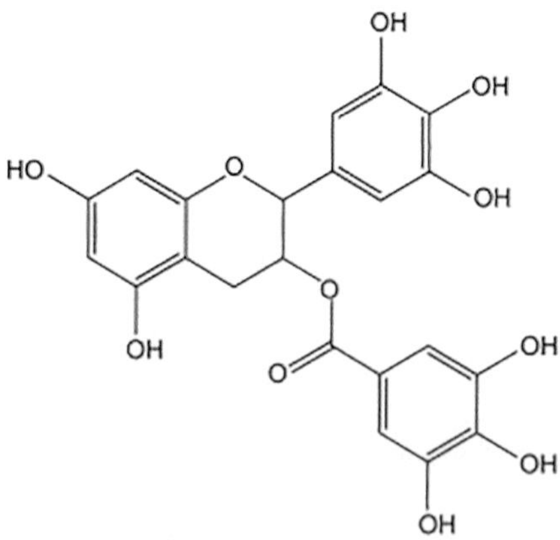
Epigallocatechin gallate

5.6.5 *CAROTENOIDS*

The carotenoids are the most important phytochemicals present in both edible and non-edible parts of fruits and vegetables that account for their red, yellow, and orange colors. Carotenoids are categorized into carotenes that consist of lutein, zeaxanthin, and β-cryptoxanthin, as well as the carotenoids β-carotene, α-carotene, and lycopene. The carotenoids are important precursors of vitamin A and have been found to protect the body from the development of degenerative diseases like macular degeneration, eye diseases, and cataracts (Granado et al., 1992). Carotenoids have the property of enhancing immunity and maintaining skin health by preventing the degradation of collagen, elastin, and fibrillin, thus improving skin healing and protecting from harmful UV radiations.

5.7 CONCLUSION

In conclusion, the exploration of bioactive compounds present in fruit wastes has revealed a treasure trove of potential health benefits that extend far beyond their initial role as discarded byproducts. The diverse array of compounds, including polyphenols, dietary fiber, vitamins, and minerals, offers a multifaceted approach to improving human health and well-being. The antioxidant properties of polyphenols found in fruit peels and seeds hold the promise of mitigating oxidative stress, a common denominator in the development of chronic diseases. As potent scavengers of free radicals, these compounds have the potential to reduce the risk of conditions such as cancer, cardiovascular diseases, and neurodegenerative disorders. The contribution of dietary fiber to digestive health, blood sugar regulation, weight management, and heart disease prevention underscores the importance of incorporating fruit wastes into dietary considerations. This natural fiber source serves not only as a remedy for gastrointestinal concerns but also as a proactive measure against metabolic disorders. Vitamins and minerals in fruit wastes play vital roles in maintaining immune function, supporting bone health, and sustaining overall physiological balance. Harnessing these nutrients from what was once considered waste material could lead to significant advancements in nutritional supplementation and preventive healthcare strategies. The anti-inflammatory, antimicrobial, and antiviral attributes of certain fruit waste bioactive compounds open doors to innovative solutions in both the pharmaceutical and food industries. As the demand for natural and effective treatments rises, these compounds could pave the way for novel therapies and preservatives. While promising, the field of fruit waste bioactive compounds also calls for continued research. Gaining a deeper understanding of their mechanisms of action, potential side effects, and optimal extraction methods is crucial for realizing their full therapeutic potential. In harnessing these compounds, transforming what was once discarded into valuable resources, we embark on a journey toward sustainable health solutions. Fruit wastes, once overlooked, emerge as reservoirs of bioactivity that can positively impact human health and the environment. By recognizing the potential within these discarded remnants, we are presented with an opportunity to enhance well-being, promote sustainability, and redefine the boundaries of resource utilization in the pursuit of a healthier future.

KEYWORDS

- **black gram byproducts**
- **gallic acid**
- **glucosidase inhibitor activity**
- **phenolic acid composition**
- **type 2 diabetes**
- **vanillic acid**

REFERENCES

Ajila, C. M., & Prasada Rao, U. J. S. (2013). Mango peel dietary fiber: Composition and associated bound phenolics. *Journal of Functional Foods, 5*(1), 444–450.

Arct, J., Bielenda, B., Oborska, A., & Pytkowska, K. (2003). The tea and its cosmetic application. *Journal of Applied Cosmetology, 21*(3), 117–127.

Arshad, Z. I. M., Amid, A., Yusof, F., Jaswir, I., Ahmad, K., & Loke, S. P. (2014). Bromelain: An overview of industrial application and purification strategies. *Applied Microbiology and Biotechnology, 98*(15), 7283.

Banerjee, J., Singh, R., Vijayaraghavan, R., MacFarlane, D., Patti, A. F., & Arora, A. (2017). Bioactives from fruit processing wastes: Green approaches to valuable chemicals. *Food Chemistry, 225*, 10–22.

Berardini, N., Carle, R., & Schieber, A. (2004). Characterization of gallotannins and benzophenone derivatives from mango (*Mangifera indica* L. cv. 'Tommy Atkins') peels, pulp and kernels by high performance liquid chromatography/electrospray ionization mass spectrometry. *Rapid Communications in Mass Spectrometry, 18*(19), 2208.

Bharti, S., Rani, N., Krishnamurthy, B., & Arya, D. (2014). Preclinical evidence for the pharmacological actions of naringin: A review. *Planta Medica, 80*(6), 437.

Białek, A., Stawarska, A., Bodecka, J., Białek, M., & Tokarz, A. (2017). Pomegranate seed oil influences the fatty acids profile and reduces the activity of desaturases in livers of Sprague–Dawley rats. *Prostaglandins and Other Lipid Mediators, 131*, 9–16. https://doi.org/10.1016/j.prostaglandins.2017.05.004 (accessed on 11th August 2025).

Bilawal, A., Ishfaq, M., Gantumur, M. A., Qayum, A., Shi, R., Fazilani, S. A., Anwar, A., Jiang, Z., & Hou, J. (2021). A review of the bioactive ingredients of berries and their applications in curing diseases. *Food Bioscience, 44*, 101407.

Boyer, J., & Liu, R. H. (2004). Apple phytochemicals and their health benefits. *Nutrition Journal, 3*(15). https://doi.org/10.1186/1475-2891-3-5 (accessed on 11th August 2025).

Çaliskan Eleren, S., Özış Altıncekiç, Ş., & Altıncekiç, E. (2018). Biofuel potential of fruit juice industry waste. *Journal of Hazardous, Toxic, and Radioactive Waste, 22*(4), 05018002.

Cano, A., Medina, A., & Bermejo, A. (2008). Bioactive compounds in different citrus varieties: Discrimination among cultivars. *Journal of Food Composition and Analysis, 21*(5), 377.

Castro Coelho, S., Nogueiro Estevinho, B., & Rocha, F. (2021). Encapsulation in food industry with emerging electrohydrodynamic techniques: Electrospinning and electro spraying–A review. *Food Chemistry, 339*, 127850.

Chen, X. M., Tait, A. M., & Kitts, D. D. (2017). Flavonoid composition of orange peel and its association with antioxidant and anti-inflammatory activities. *Food Chemistry, 218*, 15–21. https://doi.org/10.1016/j.foodchem.2016.09.016 (accessed on 11th August 2025).

Cho, H., Jung, H., Lee, H., Yi, H. C., Kwak, H. K., & Hwang, K. T. (2015). Chemopreventive activity of ellagitannins and their derivatives from black raspberry seeds on HT-29 colon cancer cells. *Food & Function, 6*(5), 1675–1683. https://doi.org/10.1039/c5fo00274e (accessed on 11th August 2025).

Dai, K., Yan, S., Yan, W., Chen, D., & Xu, Z. (2009). Effects of naringin on the proliferation and osteogenic differentiation of human bone mesenchymal stem cell. *European Journal of Pharmacology, 607*(1–3), 1–5.

Deolindo, C. T. P., Monteiro, P. I., Santos, J. S., Cruz, A. G., Cristina da Silva, M., & Granato, D. (2019). Phenolic-rich Petit Suisse cheese manufactured with organic Bordeaux grape juice, skin, and seed extract: Technological, sensory, and functional properties. *LWT, 115*, 108493. https://doi.org/10.1016/j.lwt.2019.108493 (accessed on 11th August 2025).

Doulabi, M., Golmakani, M. T., & Ansari, S. (2020). Evaluation and optimization of microwave-assisted extraction of bioactive compounds from eggplant peel byproduct. *Journal of Food Processing and Preservation, 44*(6), e14853.

Duman, E., & Özcan, M. M. (2011). Fatty acid and mineral contents of kiwifruit seed and pulps. *Acta Horticulture, 913*, 647.

Femenia, A., Rosselló, C., Mulet, A., & Cañellas, J. (1995). Chemical composition of bitter and sweet apricot kernels. *Journal of Agricultural and Food Chemistry, 43*(2), 356.

Food and Agriculture Organisation (FAO). (2021). FAO STAT: Statistical database. https://www.fao.org/faostat (accessed on 11th August 2025).

Foster, R., Williamson, C. S., & Lunn, J. (2009). Briefing paper: Culinary oils and their health effects. *Nutrition Bulletin, 34*(1), 4–47.

Fukuchi, Y., Hiramitsu, M., Okada, M., Hayashi, S., Nabeno, Y., Osawa, T., & Naito, M. (2008). Lemon polyphenols suppress diet-induced obesity by up-regulation of mRNA levels of the enzymes involved in beta-oxidation in mouse white adipose tissue. *Journal of Clinical Biochemistry and Nutrition, 43*(3), 201–209.

Gbenga-Fabusiwa, F. J., Jeff-Agboola, Y. A., Ololade, Z. S., Akinrinmade, R., & Agbaje, D. O. (2022). Waste-to-wealth: Nutritional potential of five selected fruit peels and their health benefits: A review. *African Journal of Food Science, 16*(7), 172–183.

Gomez, B., Gullon, B., Remoroza, C., Schols, H. A., Parajo, J. C., & Alonso, J. L. (2014). Purification, characterization, and prebiotic properties of pectic oligosaccharides from orange peel wastes. *Journal of Agricultural and Food Chemistry, 62*(40), 9769–9782.

Granado, F., Olmedilla, B., Blanco, I., & Rojas-Hidalgo, E. (1992). Carotenoid composition in raw and cooked Spanish vegetables. *Journal of Agricultural and Food Chemistry, 40*(11), 2135–2140.

Habauzit, V., Sacco, S. M., Gil-Izquierdo, A., Trzeciakiewicz, A., Morand, C., Barron, D., et al. (2011). Differential effects of two citrus flavanones on bone quality in senescent male rats in relation to their bioavailability and metabolism. *Bone, 49*(5), 1108–1116.

Harbeoui, H., Hichami, A., Wannes, W. A., Lemput, J., Tounsi, M. S., & Khan, N. A. (2019). Anti-inflammatory effect of grape (*Vitis vinifera* L.) seed extract through the down regulation of NF-κB and MAPK pathways in LPS-induced RAW264.7 macrophages. *South African Journal of Botany, 125*, 1–8. https://doi.org/10.1016/j.sajb.2019.06.026 (accessed on 11th August 2025).

Hussain, A., Kausar, T., Din, A., Murtaza, M. A., Jamil, M. A., Noreen, S., Rehman, H. U., Shabbir, H., & Ramzan, M. A. (2021). Determination of total phenolic, flavonoid, carotenoid, and mineral contents in peel, flesh, and seeds of pumpkin (*Cucurbita maxima*). *Journal of Food Processing and Preservation, 45*(7), e15542.

Jäger, S., Trojan, H., Kopp, T., Laszczyk, M. N., & Scheffler, A. (2009). Pentacyclic triterpene distribution in various plants: Rich sources for a new group of multi-potent plant extracts. *Molecules, 14*(6).

Jara-Palacios, M. J., Santisteban, A., Gordillo, B., Hernanz, D., Heredia, F. J., & Escudero-Gilete, M. L. (2019). Comparative study of red berry pomaces (blueberry, red raspberry, red currant and blackberry) as source of antioxidants and pigments. *European Food Research and Technology, 245*(1), 1.

Jiang, L. Y., He, S., Pan, Y. J., & Sun, C. R. (2010). Bioassay-guided isolation and EPR-assisted antioxidant evaluation of two valuable compounds from mango peels. *Food Chemistry, 119*(4), 1285.

Joshi, V. K. (2020). Fruit and vegetable processing waste management–An overview. *International Journal of Food and Fermentation Technology, 10*(2), 67–94.

Kandemir, K., Piskin, E., Xiao, J., Tomas, M., & Capanoglu, E. (2022). Fruit juice industry wastes as a source of bioactives. *Journal of Agricultural and Food Chemistry, 70*(23), 6805–6832.

Kheirkhah, H., Baroutian, S., & Quek, S. Y. (2019). Evaluation of bioactive compounds extracted from Hayward kiwifruit pomace by subcritical water extraction. *Food and Bioproducts Processing, 115*, 143.

Kim, J. H., & Kim, M. Y. (2016). The potential use of citrus juice waste as sources of natural phenolic antioxidants. *Journal of Applied Pharmaceutical Science, 6*(12), 202.

Lebaka, V. R., Wee, Y. J., Ye, W., & Korivi, M. (2021). Nutritional composition and bioactive compounds in three different parts of mango fruit. *International Journal of Environmental Research and Public Health, 18*, 741.

Li, T., Shen, P., Liu, W., Liu, C., Liang, R., Yan, N., & Chen, J. (2014). Major polyphenolics in pineapple peels and their antioxidant interactions. *International Journal of Food Properties, 17*(8), 1805.

Liu, L., Fishman, M. L., Kost, J., & Hicks, K. B. (2003). Pectin-based systems for colon-specific drug delivery via oral route. *Biomaterials, 24*(19), 3333–3343.

Liu, Q. B., Cheng, Z. Y., Yan, Z. Y., Wang, D., Bai, M., Huang, X. X., et al. (2018). *Prunus tomentosa* seed waste as a source of aromatic glycosides: Valuable phytochemicals with α-glucosidase inhibitory and hepatoprotective properties. *Industrial Crops and Products, 111*, 590–596.

Lucarini, M., Durazzo, A., et al. (2021). Fruit wastes as a valuable source of value-added compounds: A collaborative perspective. *Molecules, 26*(21), 6338.

Luo, F., Fu, Y., Xiang, Y., Yan, S., Hu, G., Huang, X., Huang, G., Sun, C., Li, X., & Chen, K. (2014). Identification and quantification of gallotannins in mango (*Mangifera indica* L.) kernel and peel and their antiproliferative activities. *Journal of Functional Foods, 8*(1), 282.

Macakova, K., Koleckar, V., Cahlíkova, L., Chlebek, J., Host'alkova, A., Kuca, K., Jun, D., & Opletal, L. (2014). Tannins and their influence on health. In *Recent Advances in Medicinal Chemistry* (Vol. 1, Chapter 6, pp. 159–208). Elsevier: Amsterdam, Netherlands.

Makori, S. I., Mu, T. H., & Sun, H. N. (2022). Profiling of polyphenols, flavonoids and anthocyanins in potato peel and flesh from four potato varieties. *Potato Research, 65*, 193–208.

Mallek-Ayadi, S., Bahloul, N., & Kechaou, N. (2018). Chemical composition and bioactive compounds of *Cucumis melo* L. seeds: Potential source for new trends of plant oils. *Process Safety and Environmental Protection, 113*, 68–77.

Manners, G. D. (2007). Citrus limonoids: Analysis, bioactivity, and biomedical prospects. *Journal of Agricultural and Food Chemistry, 55*(21), 8285–8294.

Marago, E., Iacopini, P., Camangi, F., Scattino, C., Ranieri, A., Stefani, A., & Sebastiani, L. (2015). Phenolic profile and antioxidant activity in apple juice and pomace: Effects of different storage conditions. *Fruits, 70*(4), 213.

Matheyambath, A. C., & Paliyath, G. (2016). In *Encyclopedia of Food and Health.*

Maurya, A. K., Pandey, R. K., Rai, D., Porwal, P., & Rai, D. C. (2015). Waste product of fruits and vegetables processing as a source of dietary fiber: A review. *Trends in Biosciences, 8*, 5129–5140.

Milala, J., Kosmala, M., Sojka, M., Kołodziejczyk, K., Zbrzezniak, M., & Markowski, J. (2013). Plum pomaces as a potential source of dietary fiber: Composition and antioxidant properties. *Journal of Food Science and Technology, 50*(5), 1012.

Min, B., Bae, I. Y., Lee, H. G., Yoo, S. H., & Lee, S. (2010). Utilization of pectin-enriched materials from apple pomace as a fat replacer in a model food system. *Bioresource Technology, 101*(14), 5414–5418.

Monagas, M., Gómez-Cordovés, C., Bartolomé, B., Laureano, O., & Ricardo da Silva, J. M. (2003). Monomeric, oligomeric, and polymeric flavan-3-ol composition of wines and grapes from *Vitis vinifera* L. cv. Graciano, Tempranillo, and Cabernet Sauvignon. *Journal of Agricultural and Food Chemistry, 51*(22), 6475.

Montoya, C., González, L., Pulido, S., Atehortua, L., & Robledo, S. M. (2019). Identification and quantification of limonoid aglycones content of citrus seeds. *Revista Brasileira de Farmacognosia, 29*(6), 710.

Muhammad Suffi, N. S., Mohamed, E., Camalxaman, S. N., Rambely, A. S., & Haron, N. (2021). The medicinal benefits, phytochemical constituents and antioxidant properties of banana blossom: A mini review. *Healthscope: The Official Research Book of Faculty of Health Sciences, Universiti Teknologi MARA, 4*, 113–118.

Munarin, F., Tanzi, M. C., & Petrini, P. (2012). Advances in biomedical applications of pectin gels. *International Journal of Biological Macromolecules, 51*, 681–689.

Nasir Md Nur 'Aqlah, Kabun, R., Wen, X. L. F., & Joseph, M. V. (2023). Review on the potential bioactive components in fruits and vegetable wastes as value-added products in the food industry. *Molecules, 28*(6), 2631.

Park, J. H., Lee, M., & Park, E. (2014). Antioxidant activity of orange flesh and peel extracted with various solvents. *Preventive Nutrition and Food Science, 19*, 291–298.

Prabowo, I., Utomo, E. P., Nurfaizy, A., Widodo, A., Widjajanto, E., & Rahadju, P. (2019). Characteristics and antioxidant activities of anthocyanin fraction in red dragon fruit peels (*Hylocereus polyrhizus*) extract. *Drug Invention Today, 12*(4), 670–678.

Pu, P., Gao, D. M., Mohamed, S., Chen, J., Zhang, J., Zhou, X. Y., Zhou, N. J., Xie, J., & Jiang, H. (2012). Naringin ameliorates metabolic syndrome by activating AMP-activated protein kinase in mice fed a high-fat diet. *Archives of Biochemistry and Biophysics, 518*, 61–70.

Rajadurai, M., & Stanely Mainzen Prince, P. (2006). Preventive effect of naringin on lipid peroxides and antioxidants in isoproterenol-induced cardiotoxicity in Wistar rats: Biochemical and histopathological evidences. *Toxicology, 228*, 259–268.

Ran, X. L., Zhang, M., Wang, Y., & Adhikari, B. (2019). Novel technologies applied for recovery and value addition of high value compounds from plant byproducts: A review. *Critical Reviews in Food Science and Nutrition, 59*, 450–461.

Reissner, A. M., Al-Hamimi, S., Quiles, A., Schmidt, C., Struck, S., Hernando, I., Turner, C., & Rohm, H. (2019). Composition and physicochemical properties of dried berry pomace. *Journal of the Science of Food and Agriculture, 99*, 1284–1293.

Saad, A. M., El-Saadony, M. T., El-Tahan, A. M., Sayed, S., Moustafa, M. A. M., Taha, A. E., Taha, T. F., & Ramadan, M. M. (2021). Polyphenolic extracts from pomegranate and watermelon wastes as substrate to fabricate sustainable silver nanoparticles with larvicidal effect against *Spodoptera littoralis*: Polyphenolic extracts from pomegranate and watermelon wastes. *Saudi Journal of Biological Sciences, 28*, 5674.

Santos, P. H., Ribeiro, D. H. B., Micke, G. A., Vitali, L., & Hense, H. (2019). Extraction of bioactive compounds from feijoa (*Acca sellowiana* (O. Berg) Burret) peel by low and high-pressure techniques. *Journal of Supercritical Fluids, 145*, 219–227. https://doi.org/10.1016/j.supflu.2018.12.016 (accessed on 11th August 2025).
Saponara, S., Testai, L., Iozzi, D., Martinotti, E., Martelli, A., Chericoni, S., Sgaragli, G., Fusi, F., & Calderone, V. (2006). (+/−)-Naringenin as large conductance Ca^{2+}-activated K^+ (BKCa) channel opener in vascular smooth muscle cells. *British Journal of Pharmacology, 149*, 1013–1021.
Savic, I., Nikolic, V., Savic-Gajic, I., Kundakovic, T., Stanojkovic, T., & Najman, S. (2016). Chemical composition and biological activity of the plum seed extract. *Advanced Technologies, 5*(2), 38.
Senica, M., Stampar, F., Veberic, R., & Mikulic-Petkovsek, M. (2017). Fruit seeds of the Rosaceae family: A waste, new life, or a danger to human health? *Journal of Agricultural and Food Chemistry, 65*(48), 10621.
Shi, M., Nie, Y., Zheng, X. Q., Lu, J. L., Liang, Y. R., & Ye, J. H. (2016). Ultraviolet B (UVB) photosensitivities of tea catechins and the relevant chemical conversions. *Molecules, 21*, 1345.
Singh, B., Singh, J. P., Kaur, A., & Singh, N. (2020). Phenolic composition, antioxidant potential and health benefits of citrus peel. *Food Research International, 132*, 109114.
Singla, M., Singh, A., & Sit, N. (2022). Effect of microwave and enzymatic pretreatment and type of solvent on kinetics of ultrasound-assisted extraction of bioactive compounds from ripe papaya peel. *Journal of Food Process Engineering, 42*, e14119.
Soncu, E. D., Kolsarici, N., Cicek, N., Ozturk, G. S., & Akoglu, I. T., & Arici, Y. K. (2015). The comparative effect of carrot and lemon fiber as a fat replacer on physico-chemical, textural, and organoleptic quality of low-fat beef hamburger. *Korean Journal for Food Science of Animal Resources, 35*, 370–381.
Song, J., Pan, T., Wu, J., & Ren, F. (2016). The improvement effect and mechanism of citrus fiber on the water-binding ability of low-fat frankfurters. *Journal of Food Science and Technology, 53*, 4197–4204.
Soong, Y. Y., & Barlow, P. J. (2004). Antioxidant activity and phenolic content of selected fruit seeds. *Food Chemistry, 88*(3), 411.
Spilmont, M., Léotoing, L., Davicco, M. J., Lebecque, P., Miot-Noirault, R., Pilet, P., et al. (2015). Pomegranate peel extract prevents bone loss in a preclinical model of osteoporosis and stimulates osteoblastic differentiation *in vitro*. *Nutrients, 7*, 9265–9284.
Sriamornsak, P. (2003). Chemistry of pectin and its pharmaceutical uses: A review. *Silpakorn University International Journal, 3*(1, 2), 206.
Sriamornsak, P., Wattanakorn, N., & Takeuchi, H. (2010). Study on the mucoadhesion mechanism of pectin by atomic force microscopy and mucin-particle method. *Carbohydrate Polymers, 79*, 54–59.
Sudha, M. L., Baskaran, V., & Leelavathi, K. (2007). Apple pomace as a source of dietary fiber and polyphenols and its effect on the rheological characteristics and cake making. *Food Chemistry, 104*, 686–692.
Sun-Waterhouse, D., Luberriaga, C., Jin, D., Wibisono, R., Wadhwa, S. S., & Waterhouse, G. I. N. (2013). Juices, fibres and skin waste extracts from white, pink or red-fleshed apple genotypes as potential food ingredients: A comparative study. *Food and Bioprocess Technology, 6*(2), 377.

Szymanowska, U., & Baraniak, B. (2019). Antioxidant and potentially anti-inflammatory activity of anthocyanin fractions from pomace obtained from enzymatically treated raspberries. *Antioxidants, 8*, 299. https://doi.org/10.3390/antiox8080299 (accessed on 11th August 2025).

Tedesco, D. E. A., Scarioni, S., Tava, A., Panseri, S., & Zuorro, A. (2021). Fruit and vegetable wholesale market waste: Safety and nutritional characterization for their potential re-use in livestock nutrition. *Sustainability, 13*, 9478.

Theagarajan, R., Malur Narayanaswamy, L., Dutta, S., Moses, J. A., & Chinnaswamy, A. (2019). Valorization of grape pomace (cv. Muscat) for development of functional cookies. *International Journal of Food Science and Technology, 54*(4), 1299.

Tremocoldi, M. A., Rosalen, P. L., Franchin, M., Massarioli, A. P., Denny, C., & Daiuto, É. R. (2018). Exploration of avocado byproducts as natural sources of bioactive compounds. *PLoS One, 13*, e0192577. https://doi.org/10.1371/journal.pone.0192577 (accessed on 11th August 2025).

Valiente, C., Arrigoni, E., Esteban, R., & Amado, R. (1995). Grape pomace as a potential food fiber. *Journal of Food Science, 60*, 818–820.

Verhaar, M. C., Stroes, E., & Rabelink, T. J. (2002). Folates and cardiovascular disease. *Arteriosclerosis, Thrombosis, and Vascular Biology, 22*, 6–13.

Vieira, F. G. K., Borges, G. D. S. C., Copetti, C., Di Pietro, P. F., Nunes, E. da C., & Fett, R. (2011). Phenolic compounds and antioxidant activity of the apple flesh and peel of eleven cultivars grown in Brazil. *Scientia Horticulture, 128*(3), 261.

Viladomiu, M., Hontecillas, R., Lu, P., & Bassaganya-Riera, J. (2013). Preventive and prophylactic mechanisms of action of pomegranate bioactive constituents. *Evidence-Based Complementary and Alternative Medicine, 2013*, 1.

Vorobyova, V., Skiba, M., Vasyliev, G., & Chygyrynets, O. (2021). Component composition and antioxidant activity of the blackcurrant (*Ribes nigrum* L.) and apricot pomace (*Prunus armeniaca* L.) extracts. *Journal of Chemical Technology and Metallurgy, 56*(4), 710.

Wang, Y., Li, L., Liu, H., Zhao, T., Meng, C., Liu, Z., & Liu, X. (2018). Bioactive compounds and *in vitro* antioxidant activities of peel, flesh and seed powder of kiwi fruit. *International Journal of Food Science and Technology, 53*(9), 2239.

Wani, A. A., Sogi, D. S., Singh, P., & Shivhare, U. S. (2011). Characterization and functional properties of watermelon (*Citrullus lanatus*) seed protein isolates and salt assisted protein concentrates. *Food Science and Biotechnology, 20*, 877–887.

William, P. T. (2005). *Water Treatment and Disposal*. John Wally.

Xu, Y., Fan, M., Ran, J., Zhang, T., Sun, H., Dong, M., Zhang, Z., & Zheng, H. (2016). Variation in phenolic compounds and antioxidant activity in apple seeds of seven cultivars. *Saudi Journal of Biological Sciences, 23*(3), 379.

Yılmaz, F. M., Görgüç, A., Karaaslan, M., Vardin, H., Ersuş Bilek, S., Uygun, Ö., & Bircan, C. (2019). Sour cherry byproducts: Compositions, functional properties and recovery potentials–A review. *Critical Reviews in Food Science and Nutrition, 59*, 3549.

Yilmaz, I. (2018). The biological and pharmacological importance of apricot. *SOJ Pharmacy & Pharmaceutical Sciences, 5*(1), 1.

You, W. T., Wang, S., He, L., Li, Y. Y., Yang, S. Y., & Pan, J. C. (2017). Advances in pharmacological activities of limonin monomers. *Chinese Medical Materials, 40*, 242–246.

Yu, X., Van De Voort, F. R., Li, Z., & Yue, T. (2007). Proximate composition of the apple seed and characterization of its oil. *International Journal of Food Engineering, 3*(5). https://doi.org/10.2202/1556-3758.1283 (accessed on 11th August 2025).

Zayed, A., Badawy, M. T., & Farag, M. A. (2021). Valorization and extraction optimization of citrus seeds for food and functional food applications. *Food Chemistry, 355*, 129529.

Zeb, A. (2020). Concept, mechanism, and applications of phenolic antioxidants in foods. *Journal of Food Biochemistry, 44*, e13394.

Zhang, H. W., Hu, J. J., Fu, R. Q., Liu, X., Zhang, Y. H., Li, J., Liu, L., Li, Y. N., Deng, Q., & Luo, Q. S. (2018). Flavonoids inhibit cell proliferation and induce apoptosis and autophagy through downregulation of PI3Kγ mediated PI3K/AKT/mTOR/p70S6K/ULK signaling pathway in human breast cancer cells. *Scientific Reports, 8*, 11255.

CHAPTER 6

Citrus Fruits: Wastes, Byproducts, and Utilization

RUBY PANDEY,[1] MOIRANGTHEM KALPANA DEVI,[2] VIVEK KUMAR,[3,*] and JITHENDER BHUKYA[1]

[1]*Department of Agricultural Engineering, School of Agricultural and Bio-Engineering, Centurion University of Technology and Management, Paralakhemundi, Odisha, India*

[2]*Department of Agricultural and Food Engineering, Indian Institute of Technology Kharagpur, Kharagpur, West Bengal, India*

[3]*Department of Food Processing Technology, Ghani Khan Choudhury Institute of Engineering and Technology, Malda, West Bengal, India*

*Corresponding author

ABSTRACT

This chapter focused on presenting different methods to utilize the citrus waste residues obtained after juice extraction. Citrus species contain various bioactive components, including essential oils, flavonoids, phenolic compounds, pectin, and Vitamin C, which offer extensive potential for application in both food and non-food industries. These waste residues can serve as sources of dietary fiber, flavor, gelling agents, and essential oils in the food, feed, and pharmaceutical sectors. Additionally, they can be utilized as sources of biofuel, biofertilizer, and bioactive ingredients in the cosmetics industry. While some efforts have been made to explore their usage in the mentioned areas, further studies on industrial-scale applications are necessary. Encouraging research to discover novel

Utilization of Food Waste and Byproducts. Mudasir Ahmad Malik and Kshirod Kumar Dash (Eds.)

DOI: 10.1201/9781779641151-6

applications of citrus waste in various food and non-food sectors would also be beneficial.

6.1 INTRODUCTION

The total citrus fruit production worldwide is estimated to be 143 million tons (FAO, 2020), out of which one-third is utilized for processing. In terms of consumption, citrus fruits surpass all other fruit types globally; there are reports stating it to be the most consumed fruit in the world due to the health benefits it confers. There are more than 145 botanical species in the *Citrus* genus, out of which the most cultivated are sweet orange (*Citrus sinensis*), mandarin (*Citrus reticulata*), lemon (*Citrus limon*), and grapefruit (*Citrus paradise*). Biochemically, it is rich in bioactive components/secondary metabolites which enhances its nutritional value (Lv et al., 2015) and, in turn, increases its demand among consumers. More than half of the total production of citrus fruits in the world is constituted by Brazil, India, China, Mexico, Spain, and USA (Paggiola et al., 2016), which produces around 50–60% organic waste (Satari and Karimi, 2018).

The citrus pomace that is obtained after juice extraction constitutes pulp, seed, and peels, which have the potential to be valorized as different food additives such as pectin, essential oil, carotenoids, antioxidants, dietary fiber, organic acids, vitamins, and bioactive ingredients in functional foods and nutraceuticals (Suri et al., 2022). The commercial citrus juice production gives rise to citrus waste called citrus pomace, mainly constituting pulp (10–30%) and peels (25–45%), as shown in Figure 6.1.

6.2 TYPES OF CITRUS FRUIT WASTE

Citrus fruit waste refers to the byproducts generated during the processing of citrus fruits, such as oranges, lemons, limes, grapefruits, and mandarins. These byproducts may include the following types of waste.

6.2.1 PEEL WASTE

The outer layer of citrus fruits, also known as the peel or rind, is often discarded during fruit processing. It is further divided into two parts

– flavedo (the outer part of the peel) and albedo (the inner part of the peel). Citrus peel waste contains a high amount of pectin and other bioactive compounds. It is widely used for the extraction of limonene (essential oil) and dried for medicinal purposes. It is also used as a flavoring ingredient in chocolates and bakery products. Marmalade, which contains at least 5% of fruit peel, finds application as citrus peel. In addition, attempts to utilize citrus peel for making plastic products have also been carried out by some researchers (Shan, 2016). Citrus peels are also an attractive source of natural pigments, carotenoids, which are found in two forms – water soluble and lipid soluble and have potential application in pharmaceuticals and as a natural coloring agent.

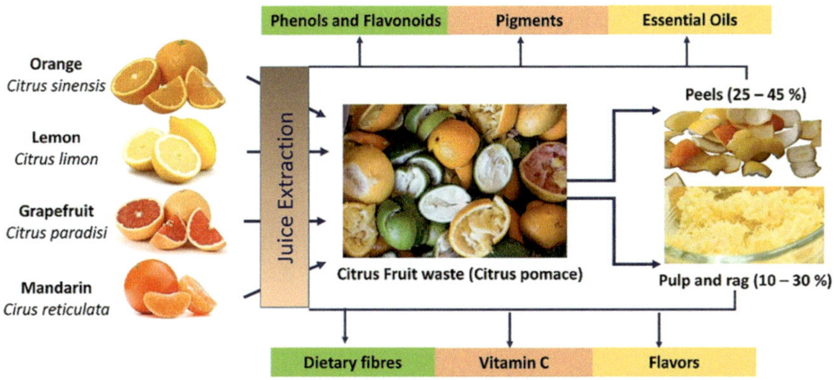

FIGURE 6.1 Citrus fruit processing waste and their utilization.

6.2.2 POMACE WASTE

The residue left after juice extraction is known as pomace. Citrus pomace is a very good source of pectin, polyphenols, Vitamin C, and cellulose (Espinosa-Pardo et al., 2017). The research findings have indicated up to 13% extraction of pectin from the lemon pomace (Azad et al., 2014) under optimum conditions. The findings of the study conducted by Fan et al. (2016) on orange peel reported that phenolic compounds yield from the citrus pomace exhibits good recovery with supercritical CO_2 extraction. The citrus pomace is also utilized for its non-food applications, mainly as a bio-fertilizer and bio adsorbent for improving soil fertility and industrial waste treatment, respectively (Panwar et al., 2021; Zema et al., 2018).

6.2.3 SEED WASTE

Citrus seeds are often removed during processing and are considered as waste product. However, they contain oils and other compounds that have potential uses in the food and cosmetic industries. Limonoids are the class of terpenoids that are present in citrus seeds and are known for having a positive effect in fighting different types of cancer (Quin et al., 2018). Citrus seeds are also rich in PUFA constituting around 34% linoleic acid and 8.7% linolenic acid that can be extracted by the method of solvent extraction (Yilmaz and Gunesar, 2017). As seeds are naturally rich in starch, citrus seeds could be potentially exploited for starch extraction and their use as a food additive (Zayed et al., 2021). The non-food applications of citrus seeds include preparation of medicinal and aromatic soaps and biodiesel production (Atolani et al., 2020; Ezekoye et al., 2019).

6.2.4 CLARIFIED JUICE WASTE

During the clarification of citrus juice either by enzyme or centrifugation or sedimentation, there is a certain amount of waste generated, such as pulp and other solids that are removed from the juice.

6.2.5 WHOLE FRUIT WASTE

Citrus fruits that do not meet quality standards for sale as fresh fruit could be used as a source of waste. These fruits are too small, too damaged, or too ripe for fresh consumption or juice extraction. They can be utilized for all kinds of applications discussed above.

These different types of citrus fruit waste have the potential for use in various applications, such as the extraction of pectin, production of animal feed, and the development of functional foods and nutraceuticals.

6.3 MAJOR BIOACTIVE COMPONENTS PRESENT IN CITRUS WASTE

Citrus fruits have an abundance of bioactive compounds, especially phenolics such as flavonoids, phenolic acids, and coumarins; terpenoids such as limonoids and carotenoids; and pectin (Cebadera-Miranda et al., 2019; Lado et al., 2018; Lu et al., 2021). In addition, they are also a rich source of ascorbic acid, tocopherols and tocotrienols, and minerals such as selenium, iron, zinc, copper, and manganese (Zou et al., 2016). The presence of these bioactive draws citrus waste to be utilized as an additive, encapsulant, flavor and fragrance agent, dietary fiber, and prebiotics. The non-food applications of citrus fruit include biofertilizer, biofuel, and so forth. Table 6.1 lists the review/research highlights of work on different bioactive compounds from citrus waste.

TABLE 6.1 Bioactive Components Review/Extraction from Citrus Species

Bioactive Compound	**Authors**	**Focus Points**
Vitamin C	Mena et al. (2009)	Stability of Vitamin C, Antioxidant activity, Factors responsible for Vitamin C degradation, Storability
	Mditshwa et al. (2017)	Post-harvest factors affecting Vitamin C content of citrus fruits, surface coating, thermal treatments, irradiation, and ethylene regreening lead to better vitamin C retention
	Elkhatim et al. (2018)	Vitamin C from citrus waste, Citrus peel as better source of Vitamin C, flavonoids, and phenolic compounds than other waste parts,
Phenolic compounds and Flavonoids	Hoshino et al. (2014)	Extraction of flavonoids from the Yuzu (*Citrus junos*) using supercritical CO_2 extraction
	Garcia-Castello et al. (2015)	Extraction of flavonoids from grapefruit using ultrasound-assisted extraction and conventional extraction
	Rafiq et al. (2018)	Citrus waste is a rich source of phenolic compounds and dietary fibers, inexpensive source of nutraceuticals
	Hwang et al. (2021)	Extraction of hesperidin and narirutin from mandarin peels using pulsed electric field aided subcritical water extraction method
	Li et al. (2022)	Extraction and characterization of total flavonoid content from grapefruit and lemon peel
Essential oils	Kamal et al. (2011)	Compared the chemical composition of essential oil from fresh, ambient- and oven-dried orange, grapefruit, and mandarin peels
	Ben Hsouna et al. (2017)	Evaluated the chemical composition, antioxidant activity, and antimicrobial activity of lemon essential oil against *Listeria monocytogenes* in minced beef meat
	Fakayode and Abobi (2018)	Optimized the extraction time and temperature for essential oil extraction from orange peels using response surface methodology. Reported oil yield of 3.38%
	González-Mas et al. (2019)	Comprehensive review on the essential oil composition of the rind part of the citrus species

TABLE 6.1 *(Continued)*

Bioactive Compound	Authors	Focus Points
	Hou et al. (2019)	Extraction of essential oils from mandarin peel using hydro distillation method and reported more than 3% oil yield
	Mahato et al. (2019)	Comprehensive review on citrus based essential oils, their chemical composition, and application in food preservation
	Thien Hien Tran et al. (2019)	Extraction of essential oil from lemon using microwave assisted hydro distillation method
Pectin	Kanmani et al. (2014)	Optimized the extraction time, temperature, and pH for highest pectin yield from orange, lemon, and sweet lime peels. Reported highest yield of 36.71%
	Ayora-Talavera et al. (2017)	Optimized extraction parameters for pectin extraction from lime residues post juice extraction and characterized the physicochemical properties of pectin along with antioxidant activity and polyphenol profile
	Panchami and Gunasekaran (2017)	Extracted pectin from different fruit waste and reported highest pectin yield from citrus peels
	Benassi et al. (2021)	Extracted and compared pectin yield from orange peels using three different methods Hot-water," Rapid Solid-Liquid Dynamic (RSLD) and microwave-assisted extraction

6.3.1 FLAVONOIDS

Flavonoids are secondary metabolites synthesized in plants and are responsible for a significant contribution to the total antioxidant activity. The flavonoids present in *Citrus* species are mainly classified into three types: flavanones, flavones, and flavanols (Addi et al., 2022). Citrus fruits contain a substantial amount of flavanone-7-O-glycosides, polyethoxylated flavones, flavanols (quercetin, rutin, and kaempferol), and anthocyanin (cyanidin and peonidin glucosides). In citrus fruits, the flavonoid concentration is found to be highest during the middle stages (60–80 days after pollination) of development, and there is a gradual decline during complete maturation. Different researchers have extracted phenolic compounds from various kinds of citrus waste and have reported good amounts of recovery (Cebadera-Miranda et al., 2019; Multari et al., 2020). Although flavonoids come under the category of non-nutritive compounds, they are linked to the treatment of some chronic illnesses such as cancer. Therefore, the focus point of researchers for long has been to extract different flavonoids such as hesperidin and naringin (belonging to a class of glycosides), and tangerine and nobiletin (polyethoxylated flavones) from citrus waste (Li et al., 2014).

The reported studies have concluded the presence of flavonoids in almost all parts of citrus fruits, but many studies have emphasized that peels and seeds are the main parts of citrus fruits from which phenolic compounds and flavonoids could be extracted in significant quantities. A detailed flavonoid composition of citrus fruits and the health benefits conferred by them, such as antimicrobial, antioxidant, anti-inflammatory, anti-aging, and cardioprotective properties, have been reviewed by Addi et al. (2021). The potential of citrus flavonoids from peels in the treatment of cancer has been recommended by researchers with supporting evidence (Koolaji et al., 2022). In addition, citrus peel extracts rich in flavonoids are good for gut health as well. They show positive effects on intestinal microecology by supporting the growth of *Bifidobacterium* species and short-chain fatty acids (Li et al., 2022). The therapeutic potential of citrus flavonoids against obesity and diabetes has also been advocated by researchers (Lu and Yip, 2023).

6.3.2 ESSENTIAL OILS

Citrus essential oils are commonly used in perfumes, cosmetics, and aromatherapy products. Research has shown that citrus essential oils exhibit antimicrobial, anti-inflammatory, and antioxidant properties. They are generally recognized as safe (GRAS). Almost all the parts of citrus plants, including leaves, flowers, fruits, and peels, are a rich source of essential oils, but extraction of essential oil from citrus waste leads to its valorization. The major bioactive components present in citrus essential oils are α-/β-pinene, *d*-limonene, linalool, β-myrcene, sabinene, α-humulene, and α-terpineol (Bora et al., 2020; Hou et al., 2019). The oil is intensively present in the oil glands of the citrus peel and constitutes around 1–3% of the total peel weight (Njoroge et al., 2005). It has shown its potential to be used as an effective natural antimicrobial against bacteria and yeasts (Dosoky and Setzer, 2018; Guo et al., 2018; Reyes-Jurado et al., 2019). A recent study published by Koochi et al. (2022) has reported high antioxidant activity of citrus peel essential oil, ranging from 375–643 mg Trolox equivalent/g. Findings of the experiments conducted by Fakayode and Abobi (2018) on orange peel have shown around 3.2% essential oil yield.

The extraction of citrus essential oil can be carried out by the method of hydro distillation (water or steam distillation), mechanical pressing, or solvent extraction (Kademi and Garba, 2017; Zeleke, 2022). The quality of oil obtained from various sources and methods is analyzed by equipment such as gas chromatography (GC) – flame ionization detector (FID). The applications of citrus essential oils are taking place in insect/pest control for stored food grains, antimicrobial packaging, nano-emulsified preservatives, flavoring agents in juice/concentrates, soft drinks, and so forth (Mahato et al., 2017).

6.3.3 PECTIN

Pectins are natural heteropolysaccharides that find wide application as gelling and thickening agents. They also fall under the category of soluble fiber and stabilizers in the food industry. They are mainly located in the albedo part of citrus peels. Citrus pomace and apple pomace are the most sought-after raw materials for commercial extraction, where around 20–30% more yield (dry basis) is obtained from the former as compared to the latter. The physical appearance of extracted pectin is that of a dry powder with a light cream/tan color (Kanmani et al., 2014). Pectin present in citrus waste also encourages its utilization in the preparation of edible/ biodegradable films. Attempts have been made for the production of biofilm from orange waste, utilizing the gelling ability of pectin and the strength of cellulose fibers (Jabbari et al., 2017).

Various ypes of extraction methods have been tried for the extraction of pectin from citrus waste, the major ones being conventional extraction, microwave-assisted extraction (MAE), ultrasound-assisted extraction (UAE), and enzyme-assisted extraction (EAE) (Singhal and Hulle, 2022). Hydro-cavitation assisted extraction of pectin is also one method, but it is relatively new (Meneguzzo et al., 2020; Nuzzo et al., 2021; Presentato et al., 2020). The general process of pectin extraction involves extraction in water/acidic medium followed by precipitation using alcohols. The major unit operations involved in pectin extraction and purification are washing, drying, grinding, extraction, centrifugation, filtration, and precipitation (Durán-Aranguren et al., 2021). The yield and rate of extraction of pectin depend upon several factors such as pH, temperature, extraction time, and type of solvent (Patil et al., 2022).

6.3.4 VITAMIN C

Vitamin C, also known as ascorbic acid, is a water-soluble vitamin and is found in abundance in citrus fruits. It is vital for the survival of humans and animals and shows preventive action against cancer, scurvy, oxidative stress, cold and flu, and boosts immunity as well. In citrus species, ascorbic acid is generally present in the range of 25–40 mg/100 mL juice (Choudhury et al., 2022), whereas citrus peels contain up to 100 mg ascorbic acid/100 g (dry weight basis) (El Khatim et al., 2018). Various research studies have claimed the successful extraction of vitamin C from citrus waste (Abdurrahman Isa et al., 2019; Papoutsis et al., 2016).

6.4 APPLICATIONS OF CITRUS WASTE IN FOOD INDUSTRY

The bioactive components extracted from citrus residues find their application in one or another kind of product in the food, animal feed, biofuel, flavors, and fragrances, pharmaceutical, and cosmetics industries. Concentrated citrus waste could be used for fermentation for the production of lactic acid, yeast, industrial alcohol/ethanol, and vinegar (Di Donna et al., 2020). One of the recent applications of citrus waste has been suggested for the development of nanoparticles with antimicrobial potential (Ahmed et al., 2018; Wei et al., 2016), nanocellulose (Marino et al., 2015), and production of pectinase enzyme.

Utilization of citrus residues in bakery products as a fiber and flavor-rich ingredient has been an area of research for over a decade. As per the context, studies have been conducted to incorporate dietary fiber from orange peels in muffins, resulting in a reduced glycemic index, and to increase the nutritional value of biscuits by incorporating mandarin peel powder (Ojha and Thapa, 2017; Romero-Lopez et al., 2011).

A recent study tried to use debittered fiber from sweet orange residues as a fat replacer in bakery products (Caggia et al., 2020).

The high potential of citrus residues to be used as a source of prebiotics has also been explored by evaluating the effect of citrus pectic oligosaccharides as a prebiotic agent (Foti et al., 2021). The presence of a good amount of pectin in citrus residues indicates its utilization as an encapsulating agent, which has also been supported by studies (Huang et al., 2019). Inclusion of citrus waste in animal feed leads to improvement in their overall health, indicated by their improved blood profile and weight gain (Ciftci et al., 2016). The potential of citrus waste in the production of biodegradable film is one of its most useful applications. Research focused on the production of edible/biodegradable film from citrus parts has shown positive findings in terms of tensile strength and antimicrobial activity of the packaging material (Jridi et al., 2020).

6.5 CONCLUSION

This chapter was focused on presenting the various ways in which we can utilize the citrus waste residues that are obtained after juice extraction. There are multiple bioactive components such as essential oils, flavonoids, phenolic compounds, pectin, and Vitamin C present in the citrus species, which opens a wide scope of its utilization in various food and non-food sectors. It can be used as a source of dietary fiber, flavor, gelling agent, and essential oils in the food, feed, and pharma sectors, and as a source of biofuel, biofertilizer, and bioactive ingredients in the cosmetics industry. Several attempts have already been made to evaluate its usage in all the above-mentioned areas; however, more studies related to industrial-scale applications are needed in the future. Also, finding the utilization of citrus waste in novel areas related to the food or non-food sector could be encouraged.

KEYWORDS

- **bioactive compounds**
- **biofuel**
- **citrus fruits**
- **gallic acid**
- **glucosidase inhibitor activity**
- **waste utilization**

REFERENCES

Abdurrahman Isa, A., Samsuri, S., & Aini Amran, N. (2019). Integration of maceration and freeze concentration for recovery of vitamin C from orange peel waste. *IOP Conference Series: Earth and Environmental Science, 268*(1), 12101. https://doi.org/10.1088/1755-1315/268/1/012101 (accessed on 11th August 2025).

Addi, M., Elbouzidi, A., Abid, M., Tungmunnithum, D., Elamrani, A., & Hano, C. (2022). An overview of bioactive flavonoids from citrus fruits. *Applied Sciences (Switzerland), 12*(29), 1–15. https://doi.org/10.3390/app12010029 (accessed on 11th August 2025).

Ahmed, S., Kaur, G., Sharma, P., Singh, S., & Ikram, S. (2018). Fruit waste (peel) as bio-reductant to synthesize silver nanoparticles with antimicrobial, antioxidant and cytotoxic activities. *Journal of Applied Biomedicine, 16*(3), 221–231. https://doi.org/10.1016/j.jab.2018.02.002 (accessed on 11th August 2025).

Atolani, O., Adamu, N., Oguntoye, O. S., Zubair, M. F., Fabiyi, O. A., Oyegoke, R. A., Adeyemi, O. S., Areh, E. T., Tarigha, D. E., Kambizi, L., & Olatunji, G. A. (2020). Helion and antimicrobial potentials of the *Citrus sinensis* seed oil for sustainable cosmeceutical production. *Helion, 6*, e03399. https://doi.org/10.1016/j.heliyon.2020.e03399 (accessed on 11th August 2025).

Ayora-Talavera, T. del R., Ramos-Chan, C. A., Covarrubias-Cárdenas, A. G., Sánchez-Contreras, A., García-Cruz, U., & Pacheco L., N. A. (2017). Evaluation of pectin extraction conditions and polyphenol profile from *Citrus* × *lantifolia* waste: Potential application as functional ingredients. *Agriculture (Switzerland), 7*(3). https://doi.org/10.3390/agriculture7030028 (accessed on 11th August 2025).

Azad, A. K. M., Ali, M. A., Akter, M. S., Rahman, J., & Ahmed, M. (2014). Isolation and characterization of pectin extracted from lemon pomace during ripening. *Journal of Food and Nutrition Sciences, 2*(2), 30–35. https://doi.org/10.11648/j.jfns.20140202.12 (accessed on 11th August 2025).

Bátori, V., Jabbari, M., Åkesson, D., Lennartsson, P. R., Taherzadeh, M. J., & Zamani, A. (2017). Production of pectin-cellulose biofilms: A new approach for citrus waste recycling. *International Journal of Polymer Science, 2017*, 1–9. https://doi.org/10.1155/2017/9732329 (accessed on 11th August 2025).

Ben Hsouna, A., Ben Halima, N., Smaoui, S., & Hamdi, N. (2017). *Citrus limon* essential oil: Chemical composition, antioxidant and antimicrobial activities with its preservative effect against *Listeria monocytogenes* inoculated in minced beef meat. *Lipids in Health and Disease, 16*(1), 1–11. https://doi.org/10.1186/s12944-017-0487-5 (accessed on 11th August 2025).

Benassi, L., Alessandri, I., & Vassalini, I. (2021). Assessing green methods for pectin extraction from waste orange peels. *Molecules, 26*(6). https://doi.org/10.3390/molecules26061766 (accessed on 11th August 2025).

Bora, H., Kamle, M., Mahato, D. K., Tiwari, P., & Kumar, P. (2020). Citrus essential oils (CEOs) and their applications in food: An overview. *Plants, 9*(3), 1–25. https://doi.org/10.3390/plants9030357 (accessed on 11th August 2025).

Caggia, C., Palmeri, R., Russo, N., & Timpone, R. (2020). Employ of citrus byproduct as fat replacer ingredient for bakery confectionery products. *Frontiers in Nutrition, 7*, 1–9. https://doi.org/10.3389/fnut.2020.00046 (accessed on 11th August 2025).

Cebadera-Miranda, L., Morales, P., & Cámara, M. (2020). Bioactive compounds in oranges from the Mediterranean climate area. In V. R. Preedy & R. R. Watson (Eds.), *The Mediterranean Diet* (2nd ed., pp. 293–309). Academic Press. https://doi.org/10.1016/B978-0-12-818649-7.00027-8 (accessed on 11th August 2025).

Choudhury, L., & Monika. (2022). Current applications of citrus and its underutilized parts: A review. *The Pharma Innovation Journal, 11*(7), 3489–3497. http://www.thepharmajournal.com (accessed on 11th August 2025).

Çiftçi, M., Şimşek, Ü. G., Dalkiliç, B., Azman, M. A., Yilmaz, Ö., Mutlu, S. I., Özçelik, M., Baykalir, Y., Tonbak, F., & Bahşi, M. (2016). Effect of dietary orange peel extract on physiological, biochemical, and metabolic responses of Japanese quail reared under low ambient temperature. *Turkish Journal of Veterinary & Animal Sciences, 40*(3), Article 5. https://doi.org/10.3906/vet-1504-11 (accessed on 11th August 2025).

Di Donna, L., Bartella, L., De Vero, L., Gullo, M., Giuffrè, A. M., Zappia, C., Capocasale, M., Poiana, M., D'Urso, S., & Caridi, A. (2020). Vinegar production from *Citrus bergamia* byproducts and preservation of bioactive compounds. *European Food Research and Technology, 246*(10), 1981–1990. https://doi.org/10.1007/s00217-020-03549-1 (accessed on 11th August 2025).

Dosoky, N. S., & Setzer, W. N. (2018). Biological activities and safety of *Citrus* spp. essential oils. *International Journal of Molecular Sciences, 19*(7), 1966. https://doi.org/10.3390/ijms19071966 (accessed on 11th August 2025).

Durán-Aranguren, D. D., Ramírez, C. J. A., Díaz, L. C. V., Valderrama, M. A., & Sierra, R. (2021). Production of pectin from citrus residues: Process alternatives and insights on its integration under the biorefinery concept. In M. A. Masuelli (Ed.), *Pectin: Chemical Properties, Uses and Health Benefits* (pp. 1–24). IntechOpen. https://doi.org/10.5772/intechopen.100153 (accessed on 11th August 2025).

Espinosa-Pardo, F. A., Nakajima, V. M., Macedo, G. A., Macedo, J. A., & Martínez, J. (2017). Extraction of phenolic compounds from dry and fermented orange pomace using supercritical CO_2 and cosolvents. *Food and Bioproducts Processing, 101*, 1–10. https://doi.org/10.1016/j.fbp.2016.10.002 (accessed on 11th August 2025).

Ethridge, E. C., & Kaukler, W. (2012). Finite element analysis of three methods for microwave heating of planetary surfaces. *50th AIAA Aerospace Sciences Meeting Including the New Horizons Forum and Aerospace Exposition*. https://doi.org/10.2514/6.2012-801 (accessed on 11th August 2025).

Ezekoye, V., Adinde, R., Ezekoye, D., & Ofomatah, A. (2019). Syntheses and characterization of biodiesel from *Citrus sinensis* seed oil. *Scientific African, 6*, e00217. https://doi.org/10.1016/j.sciaf.2019.e00217 (accessed on 11th August 2025).

Fakayode, O. A., & Abobi, K. E. (2018). Optimization of oil and pectin extraction from orange (*Citrus sinensis*) peels: A response surface approach. *Journal of Analytical Science and Technology, 9*(1), 1–10. https://doi.org/10.1186/s40543-018-0151-3 (accessed on 11th August 2025).

Fan, X., Gao, Y., He, W., Hu, H., Tian, M., Wang, K., & Pan, S. (2016). Production of nano bacterial cellulose from beverage industrial waste of citrus peel and pomace using *Komagataeibacter xylinus*. *Carbohydrate Polymers, 151*, 1068–1072. https://doi.org/10.1016/j.carbpol.2016.06.062 (accessed on 11th August 2025).

FAOSTAT. (2020). Food and agriculture data. *Food and Agriculture Organization*. Rome: FAO. https://www.fao.org/faostat/en/#data/QCL (accessed on 11th August 2025).

Foti, P., Ballistreri, G., Timpanaro, N., Rapisarda, P., & Romeo, F. V. (2022). Prebiotic effects of citrus pectic oligosaccharides. *Natural Product Research, 36*(12), 3173–3176. https://doi.org/10.1080/14786419.2021.1948845 (accessed on 11th August 2025).

Garcia-Castello, E. M., Rodriguez-Lopez, A. D., Mayor, L., Ballesteros, R., Conidi, C., & Cassano, A. (2015). Optimization of conventional and ultrasound-assisted extraction of flavonoids from grapefruit (*Citrus paradisi* L.) solid wastes. *LWT–Food Science and*

Technology, 64(2), 1114–1122. https://doi.org/10.1016/j.lwt.2015.07.024 (accessed on 11th August 2025).

González-Mas, M. C., Rambla, J. L., López-Gresa, M. P., Blázquez, M. A., & Granell, A. (2019). Volatile compounds in citrus essential oils: A comprehensive review. *Frontiers in Plant Science, 10*, 1–18. https://doi.org/10.3389/fpls.2019.00012 (accessed on 11th August 2025).

Hien Tran, T., Tien Nguyen, V., Phat Dao, T., Duc Lam, T., Quoc Toan, T., Duy Nguyen, T., Vo, D. V. N., Anh Vy, T., & Bui, L. M. (2019). New direction in research on extraction of *Citrus aurantifolia* (lemon fruit) essential oil grown in Mekong Delta–Vietnam via microwave-assisted hydro distillation. *IOP Conference Series: Materials Science and Engineering, 542*(1), 012038. https://doi.org/10.1088/1757-899X/542/1/012038 (accessed on 11th August 2025).

Hoshino, M., Suetsugu, T., Iwai, H., Takamizu, A., Tanaka, M., Quitain, A., Sasaki, M., & Goto, M. (2014). Extraction of citrus flavonoids from peel of *Citrus junos* using supercritical carbon dioxide with polar solvent. *The Materials Research Society of Japan, 39*. https://doi.org/10.12691/ces-1-4-7 (accessed on 11th August 2025).

Hou, H. S., Bonku, E. M., Zhai, R., Zeng, R., Hou, Y. L., Yang, Z. H., & Quan, C. (2019). Extraction of essential oil from *Citrus reticulata* Blanco peel and its antibacterial activity against *Cutibacterium acnes* (formerly *Propionibacterium acnes*). *Helion, 5*(12), e02947. https://doi.org/10.1016/j.heliyon.2019.e02947 (accessed on 11th August 2025).

Huang, X., Liu, Y., Zou, Y., Liang, X., Peng, Y., McClements, D. J., & Hu, K. (2019). Encapsulation of resveratrol in zein/pectin core-shell nanoparticles: Stability, bioaccessibility, and antioxidant capacity after simulated gastrointestinal digestion. *Food Hydrocolloids, 93*, 261–269. https://doi.org/10.1016/j.foodhyd.2019.02.039 (accessed on 11th August 2025).

Hwang, H. J., Kim, H. J., Ko, M. J., & Chung, M. S. (2021). Recovery of hesperidin and narirutin from waste *Citrus unshiu* peel using subcritical water extraction aided by pulsed electric field treatment. *Food Science and Biotechnology, 30*(2), 217–226. https://doi.org/10.1007/s10068-020-00862-z (accessed on 11th August 2025).

Jridi, M., Abdelhedi, O., Salem, A., Kechaou, H., Nasri, M., & Menchari, Y. (2020). Physicochemical, antioxidant and antibacterial properties of fish gelatin-based edible films enriched with orange peel pectin: Wrapping application. *Food Hydrocolloids, 103*, 105688. https://doi.org/10.1016/j.foodhyd.2020.105688 (accessed on 11th August 2025).

Kademi, H. I., & Garba, U. (2017). Citrus peel essential oils: A review on composition and antimicrobial activities. *International Journal of Food Safety, 9*(5), 38–44.

Kamal, G. M., Anwar, F., Hussain, A. I., Sarri, N., & Ashraf, M. Y. (2011). Yield and chemical composition of citrus essential oils as affected by drying pretreatment of peels. *International Food Research Journal, 18*(4), 1275–1282.

Koolaji, N., Shammugasamy, B., Schindeler, A., Dong, Q., Dehghani, F., & Valtchev, P. (2020). Citrus peel flavonoids as potential cancer prevention agents. *Current Developments in Nutrition, 4*(5), nzaa025. https://doi.org/10.1093/cdn/nzaa025 (accessed on 11th August 2025).

Lado, J., Gambetta, G., & Zacarías, L. (2018). Key determinants of citrus fruit quality: Metabolites and main changes during maturation. *Scientia Horticulture, 233*, 238–248. https://doi.org/10.1016/j.scienta.2018.01.055 (accessed on 11th August 2025).

Li, P., Yao, X., Zhou, Q., Meng, X., Zhou, T., & Gu, Q. (2022). Citrus peel flavonoid extracts: Health-beneficial bioactivities and regulation of intestinal microecology *in vitro*. *Frontiers in Nutrition, 9*, 888745. https://doi.org/10.3389/fnut.2022.888745 (accessed on 11th August 2025).

Lu, K., & Yip, Y. M. (2023). Therapeutic potential of bioactive flavonoids from citrus fruit peels toward obesity and diabetes mellitus. *Future Pharmacology, 3*(1), 14–37. https://doi.org/10.3390/futurepharmacol3010002 (accessed on 11th August 2025).

Lu, X., Zhao, C., Shi, H., Liao, Y., Xu, F., Du, H., Xiao, H., & Zheng, J. (2023). Nutrients and bioactives in citrus fruits: Different citrus varieties, fruit parts, and growth stages. *Critical Reviews in Food Science and Nutrition, 63*(14), 2018–2041. https://doi.org/10.1080/10408398.2021.1969891 (accessed on 11th August 2025).

Lv, X., Zhao, S., Ning, Z., Zeng, H., Shu, Y., Tao, O., Xiao, C., Lu, C., & Liu, Y. (2015). Citrus fruits as a treasure trove of active natural metabolites that potentially provide benefits for human health. *Chemistry Central Journal, 9*(1), 68. https://doi.org/10.1186/s13065-015-0145-9 (accessed on 11th August 2025).

Mahato, N., Sharma, K., Koteswararao, R., Sinha, M., Baral, E. R., & Cho, M. H. (2019). Citrus essential oils: Extraction, authentication and application in food preservation. *Critical Reviews in Food Science and Nutrition, 59*(4), 611–625. https://doi.org/10.1080/10408398.2017.1384716 (accessed on 11th August 2025).

Mahato, N., Sharma, K., Sinha, M., & Cho, M. H. (2018). Citrus waste derived nutra-/pharmaceuticals for health benefits: Current trends and future perspectives. *Journal of Functional Foods, 40*, 307–316. https://doi.org/10.1016/j.jff.2017.11.015 (accessed on 11th August 2025).

Mariño, M., Lopes, L., Durán, N., & Tasic, L. (2015). Enhanced materials from nature: Nanocellulose from citrus waste. *Molecules, 20*(4), 5908–5923. https://doi.org/10.3390/molecules20045908 (accessed on 11th August 2025).

Martí, N., Mena, P., Cánovas, J. A., Micol, V., & Saura, D. (2009). Vitamin C and the role of citrus juices as functional food. *Natural Product Communications, 4*(5), 1934578X0900400506. https://doi.org/10.1177/1934578X0900400506 (accessed on 11th August 2025).

Mditshwa, A., Magwaza, L. S., Tesfay, S. Z., & Mbili, N. (2017). Postharvest quality and composition of organically and conventionally produced fruits: A review. *Scientia Horticulture, 216*, 148–159. https://doi.org/10.1016/j.scienta.2016.12.033 (accessed on 11th August 2025).

Meneguzzo, F., Brunetti, C., Fidalgo, A., Ciriminna, R., Delisi, R., Albanese, L., Zabini, F., et al. (2019). Real-scale integral valorization of waste orange peel via hydrodynamic cavitation. *Processes, 7*(9), 581–604. https://doi.org/10.3390/pr7090581 (accessed on 11th August 2025).

Multari, S., Licciardello, C., Caruso, M., & Martens, S. (2020). Monitoring the changes in phenolic compounds and carotenoids occurring during fruit development in the tissues of four citrus fruits. *Food Research International, 134*, 109228. https://doi.org/10.1016/j.foodres.2020.109228 (accessed on 11th August 2025).

Njoroge, S. M., Koaze, H., Karanja, P. N., & Sawamura, M. (2005). Essential oil constituents of three varieties of Kenyan sweet oranges (*Citrus sinensis*). *Flavor and Fragrance Journal, 20*(1), 80–85. https://doi.org/10.1002/ffj.1377 (accessed on 11th August 2025).

Nuzzo, D., Picone, P., Giardina, C., Scordino, M., Mudò, G., Pagliaro, M., Scurria, A., et al. (2021). New neuroprotective effect of lemon IntegroPectin on neuronal cellular model. *Antioxidants, 10*(5), 669–686. https://doi.org/10.3390/antiox10050669 (accessed on 11th August 2025).

Ojha, P., & Thapa, S. (2017). Quality evaluation of biscuit incorporated with mandarin peel powder. *Scientific Study & Research. Chemistry & Chemical Engineering, Biotechnology, Food Industry, 18*(1), 19–30.

Paggiola, G., van Stempvoort, S., Bustamante, J., Barbero, J. M. V., Hunt, A. J., & Clark, J. H. (2016). Can bio-based chemicals meet demand? Global and regional case-study around citrus waste-derived limonene as a solvent for cleaning applications. *Biofuels, Bioproducts and Biorefining, 10*(6), 686–698. https://doi.org/10.1002/bbb.1677 (accessed on 11th August 2025).

Palanisamy, K., Dhivya, J., & Kuppamuthu, K. (2014). Extraction and analysis of pectin from citrus peels: Augmenting the yield from *Citrus limon* using statistical experimental design. *Iranica Journal of Energy & Environment, 5*, 303–312. https://doi.org/10.5829/idosi.ijee.2014.05.03.10 (accessed on 11th August 2025).

Panchami, P. S., & Gunasekaran, S. (2017). Extraction and characterization of pectin from fruit waste. *International Journal of Current Microbiology and Applied Sciences, 6*(8), 943–948. https://doi.org/10.20546/ijcmas.2017.608.116 (accessed on 11th August 2025).

Panwar, D., Saini, A., Panesar, P. S., & Chopra, H. K. (2021). Unravelling the scientific perspectives of citrus byproducts utilization: Progress towards circular economy. *Trends in Food Science & Technology, 111*, 549–562. https://doi.org/10.1016/j.tifs.2021.03.018 (accessed on 11th August 2025).

Papoutsis, K., Pristijono, P., Golding, J. B., Stathopoulos, C. E., Scarlett, C. J., Bowyer, M. C., & Vuong, Q. V. (2016). Impact of different solvents on the recovery of bioactive compounds and antioxidant properties from lemon (*Citrus limon* L.) pomace waste. *Food Science and Biotechnology, 25*(4), 971–977. https://doi.org/10.1007/s10068-016-0158-8 (accessed on 11th August 2025).

Patil, B., Patel, D., Patil, J., & Nibe, R. L. (2022). Extraction of pectin from citrus peels – A review. *International Journal of Advanced Research in Science, Communication and Technology, 2*(1), 631–637. https://doi.org/10.48175/ijarsct-2590 (accessed on 11th August 2025).

Presentato, A., Piacenza, E., Scurria, A., Albanese, L., Zabini, F., Meneguzzo, F., Nuzzo, D., et al. (2020). A new water-soluble bactericidal agent for the treatment of infections caused by Gram-positive and Gram-negative bacterial strains. *Antibiotics, 9*(9), 586–591. https://doi.org/10.3390/antibiotics9090586 (accessed on 11th August 2025).

Qin, S., Lv, C., Wang, Q., Zheng, Z., Sun, X., Tang, M., & Deng, F. (2018). Extraction, identification, and antioxidant property evaluation of limonin from pummelo seeds. *Animal Nutrition, 4*(3), 281–287. https://doi.org/10.1016/j.aninu.2018.05.005 (accessed on 11th August 2025).

Rafiq, S., Kaul, R., Sofi, S. A., Bashir, N., Nazir, F., & Ahmad Nayik, G. (2018). Citrus peel as a source of functional ingredient: A review. *Journal of the Saudi Society of Agricultural Sciences, 17*(4), 351–358. https://doi.org/10.1016/j.jssas.2016.07.006 (accessed on 11th August 2025).

Reyes-Jurado, F., Navarro-Cruz, A. R., Ochoa-Velasco, C. E., Palou, E., López-Malo, A., & Ávila-Sosa, R. (2020). Essential oils in vapor phase as alternative antimicrobials: A review. *Critical Reviews in Food Science and Nutrition, 60*(10), 1641–1650. https://doi.org/10.1080/10408398.2019.1586641 (accessed on 11th August 2025).

Romero-Lopez, M. R., Osorio-Diaz, P., Bello-Perez, L. A., Tovar, J., & Bernardino-Nicanor, A. (2011). Fiber concentrate from orange (*Citrus sinensis* L.) bagasse: Characterization and application as bakery product ingredient. *International Journal of Molecular Sciences, 12*(4), 2174–2186. https://doi.org/10.3390/ijms12042174 (accessed on 11th August 2025).

Satari, B., & Karimi, K. (2018). Citrus processing wastes: Environmental impacts, recent advances, and future perspectives in total valorization. *Resources, Conservation and Recycling, 129*, 153–167. https://doi.org/10.1016/j.resconrec.2017.10.032 (accessed on 11th August 2025).

Shan, Y. (2016). Chapter 7: Biotransformation of citrus peel (Y. Shan, Ed.; pp. 85–90). Academic Press. https://doi.org/10.1016/B978-0-12-809785-4.00007-1 (accessed on 11th August 2025).

Singhal, S., & Swami Hulle, N. R. (2022). Citrus pectins: Structural properties, extraction methods, modifications and applications in food systems–A review. *Applied Food Research, 2*(2), 100215. https://doi.org/10.1016/j.afres.2022.100215 (accessed on 11th August 2025).

Sir Elkhatim, K. A., Elagib, R. A. A., & Hassan, A. B. (2018). Content of phenolic compounds and vitamin C and antioxidant activity in wasted parts of Sudanese citrus fruits. *Food Science and Nutrition, 6*(5), 1214–1219. https://doi.org/10.1002/fsn3.660 (accessed on 11th August 2025).

Suri, S., Singh, A., & Nema, P. K. (2022). Current applications of citrus fruit processing waste: A scientific outlook. *Applied Food Research, 2*(1), 100050. https://doi.org/10.1016/j.afres.2022.100050 (accessed on 11th August 2025).

Wei, Y., Fang, Z., Zheng, L., Tan, L., & Tsang, E. P. (2016). Green synthesis of Fe nanoparticles using *Citrus maxima* peels aqueous extracts. *Materials Letters, 185*, 384–386. https://doi.org/10.1016/j.matlet.2016.09.029 (accessed on 11th August 2025).

Yilmaz, E., & Güneşer, B. A. (2017). Cold pressed versus solvent extracted lemon (*Citrus limon* L.) seed oils: Yield and properties. *Journal of Food Science and Technology, 54*(7), 1891–1900. https://doi.org/10.1007/s13197-017-2622-8 (accessed on 11th August 2025).

Zayed, A., Badawy, M. T., & Farag, M. A. (2021). Valorization and extraction optimization of citrus seeds for food and functional food applications. *Food Chemistry, 355*, 129609. https://doi.org/10.1016/j.foodchem.2021.129609 (accessed on 11th August 2025).

Zeleke, Z. Z. (2022). Extraction of essential oil from lemon and orange peel by Clevenger apparatus: Comparative GCMS analysis of chemical composition, from Debre Berehan Market town Amahara Region Ethiopia. *Annals of Biotechnology, 5*(1), 1–4. https://doi.org/10.33582/annbiotechnol.2022.1026 (accessed on 11th August 2025).

Zema, D. A., Calabrò, P. S., Folino, A., Tamburino, V., Zappia, G., & Zimbone, S. M. (2018). Valorization of citrus processing waste: A review. *Waste Management, 80*, 252–273. https://doi.org/10.1016/j.wasman.2018.09.024 (accessed on 11th August 2025).

Zou, Z., Xi, W., Hu, Y., Nie, C., & Zhou, Z. (2016). Antioxidant activity of citrus fruits. *Food Chemistry, 196*, 885–896. https://doi.org/10.1016/j.foodchem.2015.09.072 (accessed on 11th August 2025).

CHAPTER 7

Sugarcane Waste, Byproducts, and Utilization

VIVEK KUMAR,[1] ANWESA SARKAR,[1] RUBY PANDEY,[2] and MOIRANGTHEM KALPANA DEVI[3,*]

[1]*Department of Food Processing Technology, Ghani Khan Choudhury Institute of Engineering and Technology, Malda, West Bengal, India*

[2]*Department of Agricultural Engineering, School of Agricultural and Bio-Engineering, Centurion University of Technology and Management, Paralakhemundi, Odisha, India*

[3]*Department of Agricultural and Food Engineering, Indian Institute of Technology, Kharagpur, West Bengal, India*

*Corresponding author

ABSTRACT

The sugarcane industry plays a significant role in global sugar production and generates considerable amounts of byproducts and waste materials. It is crucial to comprehend these byproducts and waste to explore sustainable opportunities and make efficient use of them. Each byproduct, such as bagasse, molasses, filter cake, press mud, and vinasse, has the potential for value-added applications beyond their traditional uses. By adopting innovative approaches, the sugarcane industry can move towards a more sustainable and resource-efficient future, promoting environmental stewardship and economic growth. Sustainable utilization is of paramount importance in the sugarcane industry. Embracing sustainable practices for handling byproducts and waste can minimize environmental impacts,

Utilization of Food Waste and Byproducts. Mudasir Ahmad Malik and Kshirod Kumar Dash (Eds.)

DOI: 10.1201/9781779641151-7

conserve resources, and contribute to a more circular economy. Additionally, sustainable utilization offers economic benefits, including revenue diversification, cost savings, and market advantages. By recognizing the value of byproducts and waste materials and transforming them into valuable resources, the sugarcane industry can strike a balance between environmental responsibility and economic prosperity.

7.1 INTRODUCTION

Sugarcane plays a vital role in India's agriculture, and Uttar Pradesh contributes to over 70% of its production. Despite sugarcane often being perceived as an unsuitable crop due to its reputation as unhealthy food, this perception is not universally true. In various other countries, sugarcane has been successfully used as an alternative fuel source, and its medicinal benefits are widely recognized worldwide (Quereshi et al., 2020). Additionally, it finds applications in industrial and commercial sectors. On the contrary, sugarcane is a flexible and rapidly growing plant that requires minimal maintenance and can flourish even in poor soil conditions (Raza et al., 2021).

7.2 SUGARCANE BYPRODUCTS AND THEIR APPLICATIONS

Sugarcane offers a variety of products, including food (such as sucrose, jaggery, and syrups), fiber (cellulose), fodder (green tops, bagasse, and molasses), fuel, and chemicals (such as bagasse, molasses, and alcohol). The primary byproducts of the cane sugar industry are bagasse, molasses, and press mud (Solomon, 2011). Additionally, the sugar industry and distillery generate low-value co-products and byproducts, such as green leaves, green tips, rubbish, boiler ash, and effluents. By diversifying and utilizing these co-products and byproducts, various other sectors can be established, reducing dependence solely on sugar production (Sarker et al., 2017). Sugarcane undergoes different processing methods based on its intended use. It can be processed into raw or white sugar crystals, providing sweetness to food and beverages. Moreover, various forms of molasses are produced for use in products like rum, beer, and animal feed. Additionally, sugarcane is dried and utilized as animal fodder. These processing methods involve multiple steps due to the numerous applications of sugarcane products.

Originally native to India and Asia, sugarcane thrives in warm, humid tropical environments. It is a high-yielding crop, producing an average of 165 pounds of refined sugar per acre (Sahu, 2018).

In tropical countries, a significant portion of commercial cane is utilized for ethanol fuel production or as a direct fuel source. However, some farmers have begun growing cane varieties that yield bagasse, the fibrous material left after juice extraction from cane stalks, which can serve as fuel or cattle fodder. According to Coca-Cola research, about 90% of sugarcane is processed into sucrose, utilized in various products like plastics, car tires, soft drinks, and pharmaceuticals. Sweeteners, accounting for approximately 6% of processed cane, are used in items such as jams, jellies, dairy products, soft drinks, and preserves. Among these sweeteners, around 70% is refined beet or cane sugar, while the remaining 30% is converted into fructose-rich HFCS. As for the remaining 25%, bagasse is a fibrous residue typically burned for energy at refineries, but it can also be processed to produce refined sugar. Desugarized molasses from this process can be used as animal feed (So et al., 2021).

To maximize the value of the sugarcane harvest, efforts should be made to integrate the use of sugarcane, its co-products, and byproducts to create a variety of value-added products. Currently, sugarcane serves multiple purposes, including food (sugar), fiber (cellulose), fuel (bagasse), and fodder (green tops, garbage, molasses, and other byproducts). Here is an overview of the economic significance and consumption of various sugarcane byproducts in India.

7.2.1 BAGASSE

Bagasse, a cellulose-rich waste product, holds significant importance in both agriculture and industry. It serves as a valuable raw material for producing paper, cardboard, and various other products. Comprising around 60% combustible material (cellulose), bagasse can be utilized for power generation, fertilizer manufacturing, and even for producing biogas at home. Additionally, bagasse proves beneficial as a biofertilizer and a biopesticide, demonstrating effectiveness against soil-borne diseases and pests. The remarkable aspect of bagasse lies in its eco-friendly nature, unlike synthetic fertilizers, as it leaves no harmful impact on the soil. On the contrary, bagasse enhances soil fertility by improving water retention capacity due to its rich nutritional content. This property aids in boosting crop productivity and quality while posing no harm to the environment or human health. Furthermore, bagasse tableware is gaining popularity as an ecological substitute for plastic tableware. Also known as sugarcane fiber dinnerware, it is manufactured using sugarcane leftovers and provides an environmentally friendly alternative to polystyrene and Styrofoam tableware (Deshwal et al., 2021).

7.2.1.1 USES OF SUGARCANE BAGASSE AND ITS BASED PRODUCTS

Once sugarcane stalks are crushed to extract juice, the leftover dry fibrous residue is known as bagasse. The amount of bagasse produced per 100 tons of crushed cane ranges from 25 to 32 tons, with an average of approximately 27–28%. Fresh mill bagasse contains about 48–50% water, 2.5–3.5% dissolved solids (Brix), and 46–48% fiber (Quereshi et al., 2020). The fiber in sugarcane comprises insoluble cellulosic materials. Bagasse finds various applications, and some of its uses include.

7.2.1.2 FUEL

Bagasse serves as a predominant fuel source in sugar mills, providing ample thermal energy to fulfill all the requirements of a typical sugar mill when utilized in significant quantities.

7.2.1.3 BIO-ELECTRICITY FROM BAGASSE COGENERATION

By utilizing high-pressure boilers and a specialized steam turbine, bagasse can be effectively converted into energy. Approximately 450 kWh of electricity can be generated from one metric tons of bagasse. Presently, India has 147 sugar mills with co-generation installations, collectively producing 3,067 MW of power. These mills not only meet their own captive power needs but also contribute up to 1,900 MW of surplus electric power to the National Grid. In the future, it is expected that sugar mills in India will have the capacity to generate up to 7,000 MW of electric power, and they are projected to provide around 3,000 MW of surplus electricity to the National Grid (Quereshi et al., 2020).

7.2.1.4 PAPER FROM BAGASSE

Bagasse serves as a widely used alternative to wood in tropical and subtropical countries like India, China, Colombia, Iran, Thailand, and Argentina for the production of pulp, paper, and board. Numerous bagasse-based facilities are dedicated to creating various paper products, including kraft paper, white writing and printing materials, newsprint, and more. Bagasse pulp finds application in the manufacturing of bags, corrugated wrapping and writing paper, as well as toilet paper. Tamil Nadu Newsprints Limited (TNPL) in India is known for producing high-quality newsprint from bagasse. Sugarcane proves to be a highly efficient source of pulp and paper fiber, with one hectare yielding about five tons per year, twice as much as one hectare of wood in the same conditions. Additionally, sugarcane has a significantly shorter rejuvenation period, making it a more rapid and sustainable resource (Chandel et al., 2017).

Utilizing bagasse for paper production offers several advantages over traditional tree pulp papers, with the most significant being the avoidance of tree-cutting. This approach is not only more cost-effective but also renewable. Sugar waste paper is particularly favored for napkins, tissue paper, disposable food containers, plates, bowls, trays, and other paper products.

7.2.1.5 BOARDS OR AGGLOMERATED -PRODUCTS

The board production industry often relies on forest woods, but the practice of cutting trees leads to reduced forest cover and climate change challenges. As a result, board manufacturing businesses are seeking alternatives to forest wood, and bagasse proves to be the most suitable and extensively usable raw material in this context (De Paoli et al., 2011).

Bagasse-based composites hold the potential to replace high-density and costly wood fiberboard as the core material for laminated floors. In India, considerable advancements have been made in recent years concerning the utilization of bagasse for creating agglomerated products, as highlighted in sub section.

7.2.1.6 PANEL OR INSULATING BOARD

Bagasse undergoes a treatment process in a rotary digester using hot water or steam under pressure to produce a panel or insulating board. Following this, the pulp is rinsed and fed into board machines, forming a continuous wet mat. The mat is then rolled to remove excess water before being cut into sheets and dried. The end product is non-bendable and does not readily accept paint. However, it serves as an efficient heat insulator and can be treated with chemicals to enhance bug and fire resistance. Bagasse finds application as a material for ceilings and wall partitions (Veana et al., 2014).

7.2.1.7 PARTICLE BOARD

Produced through a dry process that involves binding bagasse fiber with resin, this material possesses higher density and hardness compared to insulating boards. It can be made water-resistant and is suitable for crafting various types of furniture, including cabinets, cupboards, racks, almirahs, partition boards, and table tops. India houses 12 particle board plants that effectively utilize surplus bagasse from nearby sugar mills as the fibrous raw material for this purpose (de Sande et al., 2021).

7.2.1.8 BAGASSE-CEMENT BOARD

These types of boards are gaining popularity as construction materials due to their exceptional physical characteristics, including resistance to fire, fungi, insects, and extreme weather conditions. They are manufactured by blending ligno-cellulosic fibers with substances like Portland cement, calcinated magnesite, or gypsum (de Sande et al., 2021).

7.2.1.9 BIOGAS PRODUCTION FROM BAGASSE

Bagasse has the potential to undergo fermentation to yield sludge or biogas, a mixture of carbon dioxide and methane, achieved through inoculation or the addition of agricultural manure. The gas produced consists of approximately 60–65% methane, 30–35% carbon dioxide, and trace amounts of hydrogen, with a calorific value of 5500 kcal/m^3. Additionally, bagasse can serve as an energy source in gas-producing plants, where the calorific value of the gas produced from bagasse is 1200 kcal/kg, consisting of 60% nitrogen, 17% carbon monoxide, and 6% hydrogen (de Sande et al., 2021).

7.2.1.10 PRODUCTION OF FURFURAL

Furfural is a clear or yellow oily liquid with an almond-like scent in its pure form, but when exposed to air, it gradually changes color from yellow to brown and eventually black, earning the name furfuraldehyde. This significant organic compound is obtained from agro-industrial wastes and residues that contain pentosans, a type of carbohydrate. The production of furfural relies entirely on pentosan-containing residues collected during the processing of various agricultural (such as bagasse) and forest products, as there are no commercial synthetic pathways. In highly industrialized countries, furfural has a limited market, primarily used in the manufacturing of nylon and butadiene. The most important commercial intermediate derived from furfural is furfuryl alcohol. In Cuba, new medications derived from cane lignin have been developed, while India has the potential to explore opportunities in this industry (Jamir et al., 2021).

7.2.1.11 USE AS ANIMAL FEED

Raw bagasse is not highly digestible due to its low nitrogen content and high lignin content (around 20%), which limits its suitability as animal feed. However, there are methods to enhance its palatability and digestibility. By supplementing it with molasses and enzymes like bromelain, its energy, protein, and mineral value can be increased, making it a more cost-effective alternative feed option for small farmers and landless laborers (Aguilar, 2019).

Through the fermentation process involving molasses and enzymes, bagasse can be transformed into calf feed. Several countries, including Thailand, Japan, Malaysia, Korea, Taiwan, the Middle East, and Australia, offer this product for sale.

7.2.1.12 THANOL PRODUCTION

The cellulose-based residue, bagasse, from the sugarcane crop holds potential for ethanol production using a simultaneous saccharification-cum-fermentation (SSF) process, utilizing either enzymatic or acid hydrolysis technology. Researchers are currently working on developing a microbial consortia to optimize the process, aiming to produce approximately 200 liters of ethanol per tons of processed bagasse (Kumar et al., 2016).

7.2.2 SUGARCANE MOLASSES

Molasses is a thick syrup used for sweetening, obtained from crushed sugar cane or sugar beets during the sugar-making process. Despite limited research on its health effects, molasses offers a broad range of uses and applications that are often overlooked due to its appearance as a simple brown syrup. The process of manufacturing molasses involves two phases from sugar cane: first, sugar is extracted from the cane juice after boiling it down, and then further evaporation results in the thick, black molasses. Apart from molasses, there are several other components utilized for various purposes in sugar cane production.

The residue left after extracting juice from cane stalks, known as bagasse, serves as fuel in the plant. Additionally, beet pulp is used in the preparation of pet foods. From the dry residue, cane wax is extracted and utilized in the production of cosmetics, polish, and paper coatings (Shweta et al., 2010). The various application of sugar cane molasses are as follows.

7.2.2.1 MOLASSES AS A FERTILIZER

Due to its common status as a waste product, molasses finds primary usage as an organic fertilizer. Despite this, there is general consensus regarding the beneficial effects of molasses as a soil fertilizer. According to claims, for every 48 metric tons of molasses applied to the field, the land receives approximately 51.3 kg of K_2, 5.2 kg of N, and 2.5 kg of P_2O_5 as fertilizers (Dotaniya et al., 2016).

7.2.2.2 MOLASSES AS ANIMAL FEED

The importance of molasses as a livestock feed has been acknowledged since the discovery of sugar. The key feature of cane molasses as a feed is its abundant carbohydrate content, mainly in the form of sugar molasses, which also contains small amounts of protein and very low levels of calcium and phosphorus, but a high proportion of potassium.

In addition, molasses contains traces of B-complex vitamins and minerals such as cobalt, boron, iodine, copper, manganese, and zinc. It can also serve as a feed for pigs and poultry. The National Sugar Institute in Kanpur, India, produced the first molasses-based animal feed (UP).

7.2.2.3 MOLASSES AS A DISTILLING RAW MATERIAL

Molasses serves as the main raw material utilized in the distilling industry, where two primary processes are involved. The initial step is fermentation, followed by distillation as the second step.

7.2.2.4 FERMENTATION

Fermentation is a biochemical reaction that takes place in an organic substance due to the activity of microorganisms such as yeasts, molds, and bacteria acting as biochemical catalysts or enzymes. Traditionally, fermentation was understood as the process of converting sugar to ethyl alcohol and carbon dioxide.

Molasses finds diverse applications in various fermentation processes, particularly in producing vinegar and acetic acid, which are important components. Vinegar, made through the fermentation of sugar or starchy materials with alcohol and acetic acid, primarily contains acetic acid. Acetic acid has various derivatives, including cellulose acetate for clothing and home furnishings, vinyl acetate anhydride, acetamide, and others. Another significant fermentation product is butanol acetone, for which molasses is a valuable source. It is used in explosives manufacturing. Acetone, a highly flammable substance, can also be derived from the fermentation of cane molasses and is utilized in the electronics sector as a plasticizer, dyeing agent, detergent, and cleaning agent (Rodrigues and Hu, 2017).

Additionally, molasses fermentation produces lactic acid, a colorless, odorless, syrupy liquid used in various industries like food, pharmaceuticals, and plastics. Citric acid is another essential outcome of molasses fermentation. It finds widespread applications in the food and beverage industry (60%), pharmaceutical (16%), chemical (15%), cosmetics and toiletries (3%), and other industries (6%).

7.2.2.5 DISTILLATION INDUSTRIES

Distillation is a process that separates liquids based on their boiling points or boiling ranges through evaporation and re-condensation. This method yields one or more types of alcohol as the primary outcome.

- **Rum**: Rum, an alcoholic distillate, is produced by fermenting sugarcane juice or sugarcane molasses, with an alcohol content ranging from 43.15 to 52.50% by volume. The mashing process involves diluting fresh molasses and adding ammonium sulfate, sulfuric acid, and, occasionally, phosphorus. The mash substance undergoes basic distillation and condensation. Rum is typically aged to enhance its flavor, color, and aroma, and further treated with herbs, sugar, and oils (Rodrigues and Hu, 2017).

- **Ethyl Alcohol**: Ethyl alcohol is produced similarly to rum. Molasses is first diluted with water, ammonium sulfate, phosphoric acid, sulfuric acid, and sodium chloride, with a sugar concentration of 14 to 18%. Yeast culture is added to the mash at a rate of 5 to 8% by volume, and fermentation takes 30 to 72 hours, depending on other conditions. The fermented mash or beer, along with the resulting alcohol, undergoes further refinement in a rectifying column, resulting in rectified spirit. Industrial alcohol serves primarily as fuel, antiseptic and cleaning agent, dye, medicine, soap solvent, and is utilized in various chemical processes (Rodrigues and Hu, 2017).

Ordinary Denatured Spirit (ODS) is commonly used in hospitals, households, and for polishing purposes. From sugarcane, 72–75 liters of ethanol can be produced per tons, and a tons of molasses yields approximately 220–250 liters of ethanol.

7.2.2.6 MOLASSES BYPRODUCTS

Dextran, a byproduct of molasses, serves as a stabilizer in ice cream, sugar syrup, and other confectionery products. Additionally, it finds application in oil well drilling. Aconitic acid, derived from molasses, is mainly used as a plasticizer, wetting agent, and flavoring ingredient within the chemical industry. Itaconic acid, produced through molasses fermentation, is employed for manufacturing thermoplastics and transparent materials. Monosodium Glutamate (MSG) is another molasses byproduct utilized as a flavor enhancer. During fermentation of molasses by *Saccharomyces cerevisiae*, carbon dioxide and alcohol are generated. CO_2 is produced at a rate of 16% by weight of molasses, with approximately 70 to 75% recoverable. The produced carbon dioxide is utilized as a cooling agent and in the production of carbonated beverages (Devi et al., 2020).

7.2.2.7 BIOETHANOL PRODUCTION FROM SUGARCANE MOLASSES

One approach to reduce molasses waste is to utilize it as biomass for bioethanol production. Due to its high sugar content, molasses can be effectively used to produce bioethanol, which holds significant economic value.

The molasses-based bioethanol production technique called Simultaneous Saccharification and Fermentation (SSF) employs 10% (100 g/L molasses) and 20% (200 g/L molasses) substrates. To simplify the process and avoid bacterial contamination, commercial instant dry yeast was selected for this fermentation as it can be directly used as a starter. The sugar industry in West Java, Indonesia, is one of the sugarcane processing industries that produces molasses, which is then converted into bioethanol.

7.2.3 PRESSMUD OR FILTER MUD: USES AND PRODUCTS

During the sugar manufacturing process, sugarcane juice undergoes a purification procedure known as clarification, which can be achieved through carbonation or sulfuration methods. This process generates a bulky precipitate. The clarified juice is then filtered using vacuum filters or the filling process, resulting in waste material known as pressmud, which contains some sugar. On average, around 7 to 8% of the crushed cane remains as pressmud.

Pressmud is rich in fertilizers such as calcium sulfite and calcium phosphate, and it also contains small amounts of nitrogen, phosphate, and potassium. It consists of approximately 20 to 25% organic matter and about 45% lime on a dry basis. As a result, it serves as an excellent fertilizer for acidic soils.

Dried pressmud cakes can be used as animal feed when combined with cane molasses and toppings. Additionally, pressmud has other applications, including its use in the production of construction lime, although this may pose waste disposal challenges. It can also be utilized in the creation of metal polishing powder, board chalk, and tooth powder.

During sugarcane crop harvesting, the tops, leaves, and trash are removed, and the stems are cleaned of their leaves. On average, cane tops and trash account for 25 to 35% of the cane weight on the field. Cattle value cane tops as a source of feed (James, 2020).

7.2.4 VINASSE

Vinasse is a liquid waste that is generated as a byproduct of ethanol production. It is a residue left after the fermentation and distillation processes involved in the production of ethanol from various feedstocks, such as sugarcane, corn, or other crops rich in sugar or starch. Vinasse is a dark-colored liquid with a high organic content, and its composition may vary depending on the raw materials and production processes used.

During ethanol production, the fermentation process converts the sugars present in the feedstock into alcohol. After fermentation, the liquid mixture undergoes distillation to separate the ethanol from other components. The remaining liquid, known as vinasse, contains water, organic compounds, minerals, residual sugars, and other substances that were not converted into ethanol (Reis et al., 2020).

7.2.4.1 TRANSFORMATION OF VINASSE FROM WASTE TO VALUABLE RESOURCE

The transformation of vinasse from a waste product to a valuable resource represents a significant shift towards sustainability and resource efficiency. Through innovative approaches and advancements in technology, vinasse is being transformed into a valuable resource with economic, environmental, and agricultural benefits. This section explores the transformation of vinasse from waste to a valuable resource and highlights its potential applications.

7.2.4.2 FERTILIZER AND SOIL AMENDMENT

Vinasse is rich in organic matter and nutrients, making it an excellent fertilizer and soil amendment. Its application to agricultural fields improves soil fertility, enhances nutrient availability, and promotes the growth and productivity of crops.

7.2.4.3 BIOGAS AND BIOENERGY PRODUCTION

One of the significant transformations of vinasse is its utilization as a feedstock for biogas production. Through anaerobic digestion (AD), vinasse undergoes microbial fermentation, resulting in the production of biogas. Biogas, primarily composed of methane and carbon dioxide, serves as a renewable energy source for heat and electricity generation. This transformation not only helps in waste management but also reduces reliance on fossil fuels, contributing to greenhouse gas mitigation (Torgbo et al., 2021).

7.2.4.4 ANIMAL FEED SUPPLEMENT

Vinasse can be used as a feed supplement for livestock and poultry. It contains residual sugars, organic matter, and essential nutrients that serve as an energy source and contribute to animal nutrition. When properly balanced with other feed ingredients, vinasse can be incorporated into animal feed formulations, promoting efficient resource utilization and reducing feed costs.

7.2.4.5 INDUSTRIAL APPLICATIONS

Vinasse has potential applications in various industrial sectors. It can be utilized as a raw material in the production of bio-based chemicals, enzymes, and microbial products. Vinasse contains organic compounds that can serve as substrates for fermentation processes, leading to the production of value-added products. This transformation contributes to the development of a bio-based economy and reduces the dependence on fossil fuel-derived chemicals (Mohamed et al., 2021).

7.2.4.6 WATER MANAGEMENT AND ENVIRONMENTAL BENEFITS

The transformation of vinasse into a valuable resource also addresses water management and environmental concerns. Proper utilization of vinasse reduces the volume of waste requiring disposal, minimizing the environmental impact of ethanol production. Furthermore, incorporating vinasse into agricultural practices enhances water retention in soils, reducing water consumption and improving water use efficiency.

7.2.4.7 UTILIZATION OF VINASSE AS A BIOFERTILIZER IN AGRICULTURE

The utilization of vinasse as a biofertilizer in agriculture has gained significant attention due to its nutrient-rich composition and potential benefits for soil fertility and crop productivity. Vinasse, a byproduct of ethanol production, is a liquid residue containing organic matter, residual sugars, and essential nutrients. When properly managed and applied, vinasse serves as an effective biofertilizer, providing nutrients to plants and improving soil health (Reis et al., 2020). This section explores the utilization of vinasse as a biofertilizer in agriculture and highlights its advantages in sustainable farming practices (Waraczewshi et al., 2022).

7.2.4.8 ENVIRONMENTAL SUSTAINABILITY

Utilizing vinasse as a biofertilizer aligns with sustainable agricultural practices and environmental stewardship. By reducing reliance on synthetic fertilizers, vinasse application minimizes the potential for nutrient runoff into water bodies, mitigating water pollution and eutrophication. It also contributes to the reduction of greenhouse gas emissions associated with the production and use of chemical fertilizers, supporting climate change mitigation efforts (Reis et al., 2020).

7.3 INTEGRATED APPROACHES

Integrated approaches for sugarcane byproduct utilization involve the strategic combination of multiple utilization methods to maximize the value and benefits derived from different byproducts. These approaches aim to optimize resource utilization, minimize waste generation, and create a circular economy within the sugarcane industry. This section presents examples of integrated approaches for sugarcane byproduct utilization, highlighting their potential and advantages.

7.3.1 BIOENERGY PRODUCTION AND FERTILIZER PRODUCTION

One integrated approach involves the utilization of sugarcane byproducts for both bioenergy production and fertilizer production. Bagasse, the fibrous residue left after sugarcane juice extraction, can be used as a fuel for bioenergy generation, producing heat and electricity. The combustion of bagasse in boilers or the conversion of bagasse into biogas through AD provides renewable energy. The residual ash from the combustion process can then be processed to produce nutrient-rich organic fertilizers. This approach maximizes the energy potential of bagasse while also creating value-added fertilizer products (Hernández-Pérez et al., 2016).

7.3.2 BIOGAS PRODUCTION AND ORGANIC FARMING

Another integrated approach involves the utilization of sugarcane byproducts for biogas production and organic farming. Vinasse, a liquid residue from ethanol production, can be utilized as a substrate for biogas production through AD. The biogas generated can be used for electricity and heat generation. Additionally, the digestate, the residual material from the AD process, can be used as an organic fertilizer for sugarcane cultivation or other organic farming practices. This integrated approach combines renewable energy production with organic farming, promoting sustainable agriculture and resource efficiency (Khatri and Pandit, 2022).

7.3.3 BIO-BASED MATERIALS AND WASTE MANAGEMENT

An integrated approach for sugarcane byproduct utilization involves the production of bio-based materials and effective waste management. Bagasse, for example, can be utilized not only for energy production but also for the production of bio-based materials such as pulp and paper, bio-composites, or bio-based chemicals. The utilization of bagasse for these applications adds value to the byproduct while reducing waste and reliance on non-renewable resources. Additionally, the residual waste streams generated during the production processes can be further treated and utilized for biogas production or composting, closing the loop in waste management (Khatri and Pandit, 2022).

7.3.4 ANIMAL FEED AND BIOENERGY GENERATION

An integrated approach involves the utilization of sugarcane byproducts for both animal feed production and bioenergy generation. Molasses, a byproduct of sugar production, can be used as a feed supplement for livestock and poultry due to its nutrient content. The excess molasses or low-quality molasses can be utilized for bioenergy production through AD or fermentation processes. This approach maximizes the utilization of molasses, ensuring economic value through animal feed production while also harnessing its energy potential.

7.3.5 COMBINED HEAT AND POWER (CHP) SYSTEMS

A comprehensive integrated approach involves the implementation of combined heat and power (CHP) systems within the sugarcane industry. CHP systems utilize multiple byproducts, such as bagasse, filter cake, and press mud, for simultaneous heat and electricity generation. Bagasse can be used as a primary fuel source, while filter cake and press mud can serve as supplementary fuel sources in boilers or biogas production systems. The generated heat can be utilized for various industrial processes, and the electricity produced can be used on-site or fed into the grid, reducing the reliance on fossil fuels and maximizing the utilization of byproducts (Khatri and Pandit, 2022).

7.4 FUTURE PERSPECTIVES AND SUSTAINABILITY

7.4.1 POTENTIAL ADVANCEMENTS IN SUGARCANE BYPRODUCT UTILIZATION

Advancements in sugarcane byproduct utilization have the potential to revolutionize the sugar industry, making it more sustainable, resource-efficient, and economically viable. These advancements leverage innovative technologies, process optimization, and new product development to extract maximum value from sugarcane byproducts. This section explores some potential advancements in sugarcane byproduct utilization and their implications (Kumar and Kumar, 2018).

7.4.2 ADVANCED BIOGAS PRODUCTION

Advancements in AD technologies hold great potential for maximizing biogas production from sugarcane byproducts. Improved digester designs, optimization of process parameters, and the use of pretreatment techniques can enhance biogas yields and quality. Additionally, the integration of biogas upgrading technologies can enable the production of biomethane, a higher-quality gas suitable for injection into the natural gas grid or for use as a transportation fuel (Boyetey et al., 2023).

7.4.3 HIGH-VALUE BIOCHEMICALS

Advancements in biochemistry and biotechnology offer opportunities for the production of high-value biochemicals from sugarcane byproducts

(Khatri and Pandit, 2022). Biorefinery concepts can be applied to extract and convert specific compounds, such as sugars, lignocellulose, or organic acids, into valuable products. For example, bio-based chemicals, bioplastics, and biopolymers can be synthesized from the sugars present in molasses or other sugarcane byproducts, contributing to the development of a bio-based economy.

7.4.4 CIRCULAR ECONOMY AND WASTE VALORIZATION

Advancements in waste valorization technologies can enhance the circular economy aspects of sugarcane byproduct utilization. Innovative processes, such as hydrothermal treatment, pyrolysis, or gasification, can convert lignocellulosic residues like bagasse into biochar, bio-oils, or syngas. These products can then be used as inputs for energy generation, biofuel production, or as precursors for the synthesis of chemicals and materials, creating a closed-loop system for waste utilization.

7.4.5 NUTRIENT RECOVERY AND SUSTAINABLE FERTILIZERS

Advancements in nutrient recovery technologies can improve the sustainability of sugarcane byproduct utilization. Techniques such as struvite precipitation, ion exchange, or membrane filtration can extract valuable nutrients, such as nitrogen, phosphorus, and potassium, from vinasse or other liquid effluents (Khatri and Pandit, 2022). The recovered nutrients can be transformed into slow-release fertilizers or used as nutrient supplements, reducing the reliance on synthetic fertilizers and minimizing nutrient runoff.

7.4.6 VALUE-ADDED ANIMAL FEED

Advancements in processing and formulation can enhance the quality and value of sugarcane byproducts used in animal feed. Advanced techniques for drying, pelleting, and extrusion can improve the nutritional composition and palatability of byproducts like press mud or bagasse, making them more suitable for livestock and poultry diets (Khatri and Pandit, 2022). This can result in better animal performance, reduced feed costs, and increased profitability for the agricultural sector.

7.4.7 INTEGRATED BIOREFINERIES

Advancements in integrated biorefinery concepts can transform the sugar industry by diversifying its product portfolio and revenue streams. Integrated biorefineries aim to extract value from multiple feedstocks and byproducts, optimizing their utilization through a combination of biochemical, thermochemical, and separation processes. This approach enables the production of a wide range of products, including biofuels, bio-based chemicals, biopolymers, and bio-based materials, fostering a more sustainable and economically viable industry.

7.5 ACTION FOR FURTHER RESEARCH AND DEVELOPMENT IN THE FIELD

The field of sugarcane waste utilization presents a wealth of opportunities for research and development, which can drive innovation, improve existing practices, and unlock the full potential of sugarcane byproducts. This call to action highlights the importance of further research and development in this field and encourages stakeholders to invest in exploring new avenues and addressing key challenges.

7.5.1 ENHANCING CONVERSION TECHNOLOGIES

There is a need to focus on developing and improving conversion technologies for sugarcane byproduct utilization. Advancements in AD, pyrolysis, fermentation, and other conversion processes can optimize resource recovery, increase energy efficiency, and produce higher-value products. Research should explore novel approaches, such as the integration of multiple technologies, process intensification, and the use of catalysts or enzymes to enhance conversion efficiencies.

7.5.2 VALORIZATION OF LIGNOCELLULOSIC RESIDUES

Lignocellulosic residues, such as bagasse and press mud, offer significant potential for value creation. Further research is needed to develop cost-effective and scalable processes for converting these residues into bio-based materials, biofuels, or specialty chemicals. Exploring innovative pretreatment techniques, enzymatic hydrolysis, and fermentation strategies can improve lignocellulosic biomass conversion and maximize its utilization potential.

7.5.3 SUSTAINABLE NUTRIENT MANAGEMENT

Research efforts should focus on developing sustainable nutrient management strategies for vinasse and other liquid byproducts. This includes exploring efficient methods for nutrient recovery, optimizing application rates, and timing, and assessing the environmental impact of byproduct utilization in agricultural systems. The development of precision nutrient management technologies and decision support tools can enable farmers to effectively utilize these byproducts as fertilizers while minimizing nutrient loss and environmental risks.

7.5.4 VALORIZATION OF SECONDARY BYPRODUCTS

Secondary byproducts, such as filter cake and molasses, offer untapped potential for value-added applications. Research should explore innovative uses for these byproducts, such as the production of bio-based chemicals, specialty products, or food ingredients. Understanding their chemical composition, functionality, and exploring new processing methods can unlock new markets and revenue streams (Reis et al., 2020).

7.6 CONCLUSION

The sugarcane industry is not only a major global player in sugar production but also generates significant quantities of byproducts and waste materials. Understanding these byproducts and waste is essential for exploring sustainable opportunities and utilizing them effectively. From bagasse and molasses to filter cake, press mud, and vinasse, each byproduct holds potential for value-added applications beyond their traditional uses. By embracing innovative approaches, the sugarcane industry can move towards a more sustainable and resource-efficient future, contributing to environmental stewardship and economic growth. The importance of sustainable utilization in the sugarcane industry cannot be overstated. By embracing sustainable practices for utilizing byproducts and waste, the industry can minimize environmental impacts, conserve resources, and contribute to a more circular economy. Simultaneously, sustainable utilization offers economic benefits such as revenue diversification, cost savings, and market advantages. By recognizing the value of byproducts and waste materials and transforming them into valuable resources, the sugarcane industry can achieve a balance between environmental stewardship and economic prosperity.

KEYWORDS

- **bagasse**
- **molasses**
- **pressmud**
- **sustainability**
- **valorization**
- **vinasse**

REFERENCE

Aguilar-Rivera, N. (2019). A framework for the analysis of socioeconomic and geographic sugarcane agroindustry sustainability. *Socio-Economic Planning Sciences, 66*, 149–160.

Boyetey, M. J. B., Torgbo, S., Sukyai, P., Watthanasakphuban, N., & Kamonsutthipaijit, N. (2023). Filter cake-derived calcium carbonate polymorphs from sugar refinery for hydroxyapatite production as a sustainable material for biomedical application. *Ceramics International, 49*(14), 23417–23425.

Chandel, A. K., & Silveira, M. H. L. (2017). *Advances in Sugarcane Biorefinery: Technologies, Commercialization, Policy Issues and Paradigm Shift for Bioethanol and byproducts*. Elsevier.

De Paoli, F., Bauer, A., Leonhartsberger, C., Amon, B., & Amon, T. (2011). Utilization of byproducts from ethanol production as substrate for biogas production. *Bioresource Technology, 102*(11), 6621–6624.

de Sande, V. T., Sadique, M., Pineda, P., Bras, A., Atherton, W., & Riley, M. (2021). Potential use of sugar cane bagasse ash as sand replacement for durable concrete. *Journal of Building Engineering, 39*, 102277.

Deshwal, G. K., Alam, T., Panjagari, N. R., & Bhardwaj, A. (2021). Utilization of cereal crop residues, cereal milling, sugarcane and dairy processing byproducts for sustainable packaging solutions. *Journal of Polymers and the Environment, 29*, 2046–2061.

Devi, G. K., Vignesh, K., & Chozhavendhan, S. (2020). Effective utilization of sugarcane trash for energy production. In *Refining Biomass Residues for Sustainable Energy and Bioproducts* (pp. 259–273). Academic Press.

Dotaniya, M. L., Datta, S. C., Biswas, D. R., Dotaniya, C. K., Meena, B. L., Rajendiran, S., & Lata, M. (2016). Use of sugarcane industrial byproducts for improving sugarcane productivity and soil health. *International Journal of Recycling of Organic Waste in Agriculture, 5*, 185–194.

Hernández-Pérez, A. F., Costa, I. A. L., Silva, D. D. V., Dussán, K. J., Villela, T. R., Canettieri, E. V., & Felipe, M. G. A. (2016). Biochemical conversion of sugarcane straw hemicellulosic hydrolyzate supplemented with co-substrates for xylitol production. *Bioresource Technology, 200*, 1085–1088.

James, J. (2020). Sugarcane press mud modification of expansive soil stabilized at optimum lime content: Strength, mineralogy and microstructural investigation. *Journal of Rock Mechanics and Geotechnical Engineering, 12*(2), 395–402.

Jamir, L., Kumar, V., Kaur, J., Kumar, S., & Singh, H. (2021). Composition, valorization and therapeutical potential of molasses: A critical review. *Environmental Technology Reviews, 10*(1), 131–142.

Khatri, P., & Pandit, A. B. (2022). Systematic review of life cycle assessments applied to sugarcane bagasse utilization alternatives. *Biomass and Bioenergy, 158*, 106365.

Kumar, K. S., Praveen, U. M., Prathyusha, A., Akhila, V., & Sasidhar, P. (2016). A comprehensive study on partial replacement of cement with sugarcane bagasse ash, rice husk ash & stone dust. *International Journal of Civil Engineering and Technology, 7*(3), 163–172.

Kumar, R., & Kumar, M. (2018). Upgradation of jaggery production and preservation technologies. *Renewable and Sustainable Energy Reviews, 96*, 167–180.

Mohamed, R. M., Ali, M. R., Smuda, S. S., & Abedelmaksoud, T. G. (2021). Utilization of sugarcane bagasse aqueous extract as a natural preservative to extend the shelf life of refrigerated fresh meat. *Brazilian Journal of Food Technology, 24*, e2020167.

Quereshi, S., Naiya, T. K., Mandal, A., & Dutta, S. (2020). Residual sugarcane bagasse conversion in India: Current status, technologies, and policies. *Biomass Conversion and Biorefinery*, 1–23.

Raza, Q. U. A., Bashir, M. A., Rehim, A., Sial, M. U., Ali Raza, H. M., Atif, H. M., & Geng, Y. (2021). Sugarcane industrial byproducts as challenges to environmental safety and their remedies: A review. *Water, 13*(24), 3495.

Reis, C. E. R., Valle, G. F., Bento, H. B., Carvalho, A. K., Alves, T. M., & de Castro, H. F. (2020). Sugarcane byproducts within the biodiesel production chain: Vinasse and molasses as feedstock for oleaginous fungi and conversion to ethyl esters. *Fuel, 277*, 118064.

Rodrigues Reis, C. E., & Hu, B. (2017). Vinasse from sugarcane ethanol production: Better treatment or better utilization? *Frontiers in Energy Research, 5*, 7.

Sahu, O. (2018). Assessment of sugarcane industry: Suitability for production, consumption, and utilization. *Annals of Agrarian Science, 16*(4), 389–395.

Sarker, T. C., Azam, S. M. G. G., & Bonanomi, G. (2017). Recent advances in sugarcane industry solid byproducts valorization. *Waste and Biomass Valorization, 8*, 241–266.

Shweta, Kumar, R., Singh, B. L., & Deepshikha, V. (2010). Integrating microbial composting and vermicomposting for effective utilization of byproducts of sugar cane–processing industries. *Bioremediation Journal, 14*(3), 158–167.

So, S., Wanapat, M., & Cherdthong, A. (2021). Effect of sugarcane bagasse as industrial by-products treated with *Lactobacillus casei* TH14, cellulase and molasses on feed utilization, ruminal ecology and milk production of mid-lactating Holstein Friesian cows. *Journal of the Science of Food and Agriculture, 101*(11), 4481–4489.

Solomon, S. (2011). Sugarcane byproducts-based industries in India. *Sugar Tech, 13*, 408–416.

Torgbo, S., Quan, V. M., & Sukyai, P. (2021). Cellulosic value-added products from sugarcane bagasse. *Cellulose, 28*(9), 5219–5240.

Veana, F., Martínez-Hernández, J. L., Aguilar, C. N., Rodríguez-Herrera, R., & Michelena, G. (2014). Utilization of molasses and sugar cane bagasse for production of fungal invertase in solid state fermentation using *Aspergillus niger* GH1. *Brazilian Journal of Microbiology, 45*, 373–377.

Waraczewski, R., Muszyński, S., & Sołowiej, B. G. (2022). An analysis of the plant- and animal-based hydrocolloids as byproducts of the food industry. *Molecules, 27*(24), 8686.

CHAPTER 8

Brewer's Spent Grain: Wastes, Byproducts, and Utilization

NEHA SHARMA, MANMEET KAUR,* and MOHINI PRABHA SINGH

Department of Microbiology, Punjab Agricultural University, Ludhiana, Punjab, India

*Corresponding author

ABSTRACT

The most typical byproduct in beer production is called Brewer's Spent Grain (BSG) and accounts for about 85% of all byproducts. Most of this substance consists of the barley grain husks, which remain as a solid after wort production. BSG is commonly used as an inexpensive source of animal feed as it is a good source of nutrients such as protein, fiber, minerals, and vitamins. On the other hand, the developments of the circular economy and technological advances in the processing of BSG have recently stimulated interest in alternative uses of the substance. In addition, due to its low cost, year-round availability, and valuable chemical composition, BSG is used as a raw material in a wide range of industrial applications. Investigating alternative uses of BSG is relevant, not only from the perspective of the brewer who can benefit from valorizing this byproduct, but also from an environmental point of view, as the recycling and reuse of industrial waste and byproducts have become increasingly important. This chapter demonstrates the attempts that have been made to utilize the readily available byproduct of brewing.

Utilization of Food Waste and Byproducts. Mudasir Ahmad Malik and Kshirod Kumar Dash (Eds.)

DOI: 10.1201/9781779641151-8

8.1 INTRODUCTION

Brewer's waste grain (BSG) is the most common brewery byproduct, accounting for 85% of all byproducts produced (Aliyu and Bala, 2011). BSG is a solid, insoluble residue from after-wort production (Tang et al., 2009). According to Ainsworth (2008), 3.4 million tons of BSG are produced by brewing in the European Union each year. Although BSG is readily available, it receives little attention from industry and is typically only used in animal feed (Xiros et al., 2008). It is undervalued due to its complex composition and high moisture content, which makes storage and transportation difficult. Another factor contributing to its frequent classification as a waste product is the potential for rapid microbial degradation. Due to its high-water content of around 70%, BSG has a short shelf life of 7–10 days in warm climates. Because of its readily available and affordable biomass, BSG continues to be an important resource for industrial growth (Robertson et al., 2010). BSG is a lignocellulosic compound with 70% fiber and 20% protein.

8.2 GENERATION OF BREWER'S SPENT GRAIN (BSG)

Malted barley is the main ingredient in the production of beer. The three main components of the barley grain are the germ (embryo), the endosperm (aleurone and starchy endosperm), and the grain husks. The grain envelopes (from the inside out) can be further classified into the pericarp layers, the husk, and the seed coat (Mussatto et al., 2006). The fermentable (maltose and maltotriose) and non-fermentable (dextrins) carbohydrates, soluble proteins, polypeptides, and amino acids all arise during the enzymatic breakdown of the starchy endosperm of malted barley during brewing. Wort is the resulting liquid that yeast ferments to make beer. The insoluble grain components, which mainly consist of grain hulls, form the BSG. The lauter tun used in traditional BSG brewing forms the bed through which the mash is filtered. When malt is first ground, it is important to ensure that the grain hulls are intact to ensure adequate filtering (Santos et al., 2003).

8.3 BREWER SPENT GRAIN COMPOSITION

The layers of the pericarp, husk, and seed coat of the original barley grain form BSG. The amount of starchy endosperm and the walls of empty aleurone cells that remain after mashing can vary depending on how effectively the process has been carried out. Depending on the brewing method used, the beer may contain low levels of starch and some residual hops from the mashing process (Mussatto et al., 2006). BSG is a diverse substance, especially considering the differences between breweries. This is caused by a number of factors such as the type of grain used, the time of harvest, the type of hops used, the malting and mashing routine, and the use of additives during the brewing process. BSG is essentially a lignocellulosic substance, and its main components are fiber (cellulose and hemicellulose), protein, and lignin (Figure 8.1) (Mussatto and Teixeira, 2010). Protein can make up 30% of BSG's dry weight, with fiber making up about 50% of that. Due to its high protein and fiber content, BSG is an interesting raw material for culinary and non-food applications. Hemicellulose, which accounts for up to 40% of dry weight and consists largely of arabinoxylan (AXE), is the main component of the BSG. According to Mandalari et al. (2005) it is assumed that hydrogen bonds bind AX to the cellulose fibers. The arabinose residue can then be esterified with ferulic acid in place of the β-(1, 4)-linked xylose residues that form the backbone of AX. AX chains can be crosslinked via diferulic acid bridges (Mendis et al., 2014).

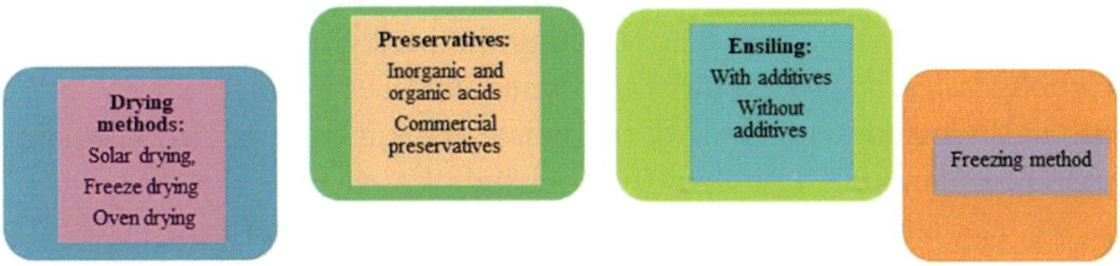

FIGURE 8.1 Schematic representation of preservation techniques for BSG.

Cellulose is another important polysaccharide in spent grain. The most abundant monosaccharides in brewer spent grain are glucose, xylose, and arabinose, but traces of rhamnose and galactose have also been found (Mandalari et al., 2005). Lignin, which makes up around 10–28% of the total dry weight of spent grain, is another important component. Lignin, a complex structured polyphenolic macromolecule, maintains the rigidity and structural integrity of plant cell walls. The complex lignin consists of three monomers, namely coniferyl alcohol, sinapyl alcohol, and coumaroyl alcohol, which are linked into a branched network structure by radical-induced condensation processes (Niemi et al., 2013). Although it varies significantly, the protein content of BSG is typically around 20% of dry weight. The most common are glutelin's, hordeins, albumins, and globulins.

Essential amino acids make up about 30% of total protein composition, with lysine (14.3%) being the most abundant. This is important as lysine is often absent from grain products (Blandino et al., 2003). The minerals most commonly found in BSG are Ca, P, Si, and Mg (Meneses et al., 2013).

8.4 CHARACTERISTIC'S OF BREWER'S SPENT GRAIN (BSG)

Brewer's Spent Grain (BSG) consists primarily of the pericarp, seed coat, and layers of skin that originally covered the grain for protection. Most of the BSG is therefore found in the walls of the skin, pericarp, and seed coat, which are high in cellulose, non-cellulose polysaccharides, and lignin, and may also contain some protein and fat. In addition to silica, most of the polyphenolic compounds contained in barley grain are also present in significant amounts in the peel (Macleod, 1979). The type of barley used and the quality and type of chemicals added during the brewing process all have an impact on the chemical composition of BSG. In general, however, BSG is considered to be a lignocellulosic substance rich in protein and dietary fiber, accounting for around 20% and 70% of its content, respectively. The elemental composition of the brewery's waste grains is shown in Table 8.1. Mashing the barley eliminates most of its carbohydrates, concentrating the proteins and fiber in the remaining grain. BSG also includes amino acids, vitamins, and minerals. Mineral elements present in amounts less than 0.5% include Ca, Co, Cu, Fe, Mg, Mn, P, K, Se, Na, and S. Also, the vitamins included are pantothenic acid (8.5), biotin (0.1), choline (1800), folic acid (0.2), niacin (44), riboflavin (1.5), thiamine (0.7), and pyridoxine (0.7). Amino acids that bind to proteins at lower levels include tyrosine, proline, threonine, arginine, and lysine. Tryptophan, methionine, phenylalanine, cystine, isoleucine, and histidine are also present (Mussatto et al., 2006).

TABLE 8.1 Chemical Composition of Dry Brewer's Spent Grain (BSG)

Proteins	Lipids	Fibers	Carbohydrates	Lignin	Ashes	Lyzine	References
14.2–31%	3–13%	59.1–74.1%	–	–	–	–	Martins et al. (2017)
18–35.4%	–	–	–	–	–	14.30%	Nazzaro et al. (2018)
24.7%	–	–	–	–	4.18%	–	Meneses et al. (2013)
15–26%	4–10%	70%	–	–	–	–	Nigam et al. (2017)
31%	9%	–	–	16%	4%	–	El-Shafey et al. (2004)
22.44%	5.3%	–	46.52%	19.57%	3.54%	–	Qin et al. (2018)

8.5 PRESERVATION TECHNIQUES FOR BARLEY SPENT GRAIN

Due to the high moisture content of BSG, as shown in Figure 8.1, several strategies have been proposed to extend its shelf life. Spent grain is difficult to transport and cannot be stored for a long time. To address these issues, BSG is neither an economically viable nor a commercially successful product. Various preservation techniques have been used, as explained below in sub section.

8.5.1 DRYING

Drying is the most common preservation method for BSG. Many breweries have BSG processing plants that use a two-stage drying process, where the water content is first reduced by pressing (by about 60%) and then dried to ensure moisture content (by about 10%) (Lynch et al., 2016). Factory drying is the most effective method of preserving BSG. The preservation of BSG through oven drying, freeze drying, and solar drying reduces product volume while lowering transport and storage costs without changing the composition of the product (Faccenda et al., 2021). The freezing technique should not be used to preserve BSG as it can damage the arabinose sugar. In order to avoid adverse effects, BSG must be dried in an oven at temperatures of 60°C. There is a possibility that during oven drying, the grain temperature will approach that of the dryer, and the dried grain will burn. After 180 days of dry storage and 16 hours of sun exposure, the nutritional and microbiological quality of BSG was unchanged (Faccenda et al., 2021).

Using superheated steam as a drying alternative has several advantages, including higher drying efficiency, reduced environmental impact, and improved recovery of priceless organic components. Autoclaving is the effective long-term storage of BSG (Lynch et al., 2016). Due to its chemical composition, BSG is very susceptible to microbial attack and chemical deterioration. Rapid colonization of microorganisms and loss of nutrients occurred for BSG at a storage temperature of 20°C; cooling produced a comparable response at a lower level. The chemical composition of BSG is not affected by freezing, although autoclaving BSG did cause polysaccharides and phenolic chemicals to become soluble. The temperature profile recorded during autoclaving and the reduction of any remaining starch influence changes. Superheated steam offers several advantages over other drying methods, including higher recovery of valuable organics, lower environmental impact, and higher drying efficiency. BSG can be stored for a very long time by autoclaving (Lynch et al., 2016). Due to its chemical composition, BSG is extremely susceptible to microbial attack and chemical degradation. For BSG, a storage temperature of 20°C led to rapid microbial colonization and nutrient loss; cooling had a comparable effect, but at a lower level. Although autoclaving BSG solubilized polysaccharides and phenolic compounds, freezing BSG did not affect its chemical composition. Changes are influenced by the temperature profile of the autoclave and the disappearance of any residual starch. Stabilization of discarded grains in breweries for storage and the processes used can greatly change grain properties (Aulds and Aldron, 2010).

8.5.2 ENSILING

One of the best and cheapest alternatives for preservation is silage. This approach is difficult due to the low dry matter concentration (between 12 and 25%) and the low sugar content, which can quickly ferment into lactic acid in the silage. When ensiling, the material must be compacted and sealed as quickly as possible to minimize losses through aerobic degradation by unfavorable bacteria and molds. In addition, early lactic acid production is required to lower pH and stop clostridial development. If the feed is not stored anaerobically, it decomposes quickly and encourages the growth of mold and mycotoxins. They observed a significant increase in the *lactobacilli* population during the first two days after ensiling the wet brewery spent grains, followed by a decrease in the remaining 16 days of the experiment, suggesting that bacterial inoculants could increase the availability of BSG. By lowering pH, increasing early lactate concentrations, and decreasing initial acetate and butyrate concentrations, the inclusion of lactic acid bacteria supported fermentation (Terefe, 2022). Silage made from grain residues from breweries contains more NH_3-N, more lactic acid bacteria, and a lower pH. Brewery residue silage has more NH_3-N, lower pH, and more lactic acid bacteria (Dai et al., 2022). It guarantees pH reduction, lactic acid fermentation under anaerobic conditions, and preservation of silage quality (Souza et al., 2012).

8.5.3 ORGANIC AND INORGANIC ACIDS

Both organic and inorganic acids are among the best additives for BSG silage. BSG preservatives require acids such as lactic, acetic, formic, benzoic, hydrochloric, and sulfuric acids. A small amount (0.5%) of sulfuric acid can be added to ensiled BSG to increase its nutritional value (Kazemi et al., 2014). Formic acid as an ensiling additive reduces the amount of organic acid and prevents fermentation, since many types of bacteria, such as lactic acid bacteria, have an antibacterial effect on formic acid (Coskuntuna et al., 2010).

BSG silages produce a small amount of propionic acid, but no isobutyric acid, butyric acid, or isovaleric acid is detectable in any sample. Without additives, BSG silage is poorly maintained and has high concentrations of acetic, butter, and nitrogenous ammonia. It is important to keep BSG preserved with a higher formic acid-propionic acid combination concentration (0.75%) to avoid microbial contamination and nutrient degradation. Weak acid preservatives excel at preventing rather than destroying microbial development, resulting in significantly longer lag periods. Low pH values, where the concentration of undissociated acid in solutions is higher, make preservatives more effective. Weak acids arrest growth by rapidly diffusing undissociated molecules across the plasma membrane. When these molecules break down inside the cells, free protons are produced, which acidify the cytoplasm and stop growth (Terefe, 2022).

8.5.4 COMMERCIAL PRESERVATIVES AND OTHERS

Storage of BSG requires commercial preservatives and other chemicals to prevent degradation. However, ESR needs to be salt treated to promote protein degradation, enhance *in situ* dry matter degradation, and limit microbial development (fungus, yeast, and mold counts) (Terefe et al., 2022). The dry, pourable Silo-King GPX Preservative also contains fermentation extracts of *A. oryzae*, *B. subtilis*, and *T. longibrachiatum*. In addition, it contains the lactic acid-producing microorganisms *Lactobacillus plantarum*, *Pediococcus pentosaceus*, and *Enterococcus faecium*. This product also contains the antifungal ingredients such as sodium benzoate, propionic acid, potassium sorbate, benzoic acid, acetic acid, and sorbic acid, and antioxidant preservative such as butylated hydroxytoluene. Monosodium phosphate serves as a nutrient and acidifier, unlike sodium chloroaluminate, which absorbs moisture. When sodium format is added, the pH is lowest and the levels of acetic acid, butyric acid, ammoniacal nitrogen, and lactic acid are highest. After fermentation, sodium format is added to the BSG to increase the amount of dry matter, water-soluble carbohydrates, and neutral detergent fibers. To improve the silage quality of BSG for short-term storage, both calcium propionate and sodium format are used as supplements (Lv et al., 2020). Commercial carbohydrase (Econase) and protease (Alcalase) enzymes are required to convert the protein and carbohydrate components of BSG into more useful products (Terefe, 2022).

8.6 MICROORGANISM GROWTH AND FERMENTATION OF BSG

The ability of BSG to rapidly degrade after removal from the lauter tun due to microbial activity if not handled and stored properly is evidence that it is a good substrate for microbial development. The fact that it contains proteins and carbohydrates, in addition to a high water content (70–80%), leads to such rapid degradation (Mussatto, 2014). BSG has received particular attention as a medium for the growth of fungi for the production of lignocellulosic enzymes. Many filamentous fungi, including those from the phylum Ascomycota (including species of *Fusarium*, *Aspergillus*, *Neurospora*, and *Trichoderma*), grow naturally in the environment and act as organic matter decomposers.

Bacteria such as *Lactobacillus*, *Streptomyces*, *Bifidobacterium*, and *Bacillus* species were also cultivated on BSG for the production of enzymes. Examples include using *A. fumigatus*, *S. malaysiensis*, and *F. oxysporum* to produce cellulases, *A. fumigatus* and *Bacillus* to produce amylase, *Humicola grisea* and *P. janczewskii* to produce hemicelluloses (Lynch et al., 2016). The majority of research has shown that no additional food sources are required for effective enzyme production. However, under certain circumstances, the addition of amino acids, vitamins, and inorganic substances increased enzyme productivity. BSG has also been used by LAB as a medium for lactic acid production and represents a cheaper alternative to the expensive raw materials (glucose, sucrose, or starch) used in the biotechnological production of lactic acid. The feasibility of BSG fermentation to lactic acid using *L. rhamnosus* was investigated. A unique method of steam explosion followed by an enzymatic treatment was used to generate supplemented (Tween 80) BSG hydrolysate which produced 19 g/L milk acid from a total sugar concentration of 40 g/L (Shindo and Tachibana, 2004).

8.7 EXTRACTION OF HIGHLY VALUE-ADDED COMPOUNDS FROM BSG

Due to the numerous compounds present in brewery waste grain, it is extracted to produce compounds with desired qualities. BSG is extracted using a variety of techniques including alkaline hydrolysis (Mussatto et al., 2007), enzymatic hydrolysis (Mussatto et al., 2008), microwave-assisted extraction (MAE) (Coelho et al., 2014), solvent-assisted extraction (Meneses et al., 2013), supercritical carbon dioxide extraction (Spinelli et al., 2016), and ultrasonically assisted extraction (Reis et al., 2015). The following products are available through extraction.

8.7.1 ARABINOXYLANS, POLYPHENOL, ANTIOXIDANTS AND GLUCOSE

Pentoses like xylose and arabinose are converted into a polysaccharide called arabinoxylan (Lukasiak et al., 2013). Along with cellulose, lignin, and other hemicelluloses, it is a form of hemicellulose found in BSG. It is able to bind to polyphenols such as ferulic acid and p-coumaric acid. Ultrasonic aid, microwave aid, or a mixture of HCl and ethanol can be used to extract arabinoxylans (Reis et al., 2015). Studies have shown that supercritical extraction of CO_2 with 60% ethanol at 35 MPa, 40°C, and an extraction time of 240 minutes can successfully produce phenolic or flavonoid fractions (Kitryte et al., 2015). Strong antioxidant properties are found in the extracted substance. Phenolic fractions can also be obtained by solvent extraction (acetone and water). Ferulic and p-coumaric acid can be recovered well by solvent extraction with acetone: water solution and alkaline BSG extraction (Guido et al., 2017). Microwave and ultrasonic extraction did not increase the yield of polyphenols compared to the traditional maceration method (Birsan et al., 2019). The research by investigated the use of *B. subtilis* WX-17 to improve the nutritional value of BSG in solid-state fermentation. Total antioxidants and total unsaturated fatty acids (UFAs) increased 1.7– and 5.8-fold, respectively (Tan et al., 2019).

8.7.2 PROTEINS

BSG has a high protein concentration (about 20% in dry matter), which makes it a promising potential source of vegetable protein for the food sector. The selectivity of the extraction process is crucial for protein extraction. The proteins from BSG had 21.4% extraction yield and 60.2% purity after alkaline treatment. When dilute acid and an alkaline pretreatment were combined, a very high degree of extraction (95%) was achieved (Du et al., 2020). A high selectivity was achieved with reduced horizontal extraction (65%) through a hydrothermal pretreatment that required a significantly lower temperature and required no use of chemicals. Carboxylate salt, urea, and DES have been used to successfully extract protein from brewer's spent grain up to 80% (Wahlstrom et al., 2017). Although this method has the disadvantage that residual DES is present in the protein product, this method could be attractive for human food production if a urea substitute can be found. Another interesting strategy involves the enzymatic destruction of spent grain proteins using ultrasound. Yu et al. (2020) reported the efficiency of protein separation is increased (61.6 to 69.8%), the enzymatic reaction time is shortened (by 56%), and the cost of using enzymes can be reduced by 73%.

8.8 UTILIZATION OF BREWER'S SPENT GRAIN

Since BSG is the main byproduct in brewing, little thought has been given to its marketability, and its disposal often leads to environmental problems. Nevertheless, due to its chemical composition, it can be used as a raw material. These are some potential uses for this agricultural byproduct.

8.8.1 BREWER'S SPENT GRAIN IN FOOD

Due to the comparatively high fiber and protein content, the potential use of BSG in the manufacture of various food products for inclusion in the human diet has been investigated. Mussatto (2014) investigated that brewer's spent grain was mainly used to make baked goods such as cookies, bread, cakes, muffins, waffles, tortillas, pancakes, snacks, brownies, and donuts. Table 8.2 provides some non-bakery applications and examples of studies on the use of BSG in these commodities. After removal from the brewery, BSG is often dried and crushed to make it more palatable for use in food preparations. Particle size is reduced, improving customer acceptance and ease of analysis (Mussatto, 2014). Extrusion cooking has also been used to introduce BSG into various food products (Ainsworth et al., 2007). Although it is normal for extruded snack foods to rise, it has been shown that increasing BSG levels can minimize expansion during extrusion cooking. A drop in starch concentration could be related, as starch is primarily what creates the dough in the extruder barrel (Lynch et al., 2016). It is generally accepted that the presence of fibers often makes products tougher, darker, and denser with less bulk. BSG protein hydrolysates can also be of importance when considering their use in food, particularly in view of their technological-functional properties, where solubility is a key component. Protein hydrolysis alters amino acid chains, charge, molecular weight, hydrophobic groups, and amino acid side chains, thus affecting sensory properties, emulsification, solubility, and viscosity (Celus et al., 2007).

TABLE 8.2 Application of BSG Component in Food Products

Sl. No.	Food Product	Proportion BSG Incorporated (%w/w)	Processing Form of Incorporated BSG	Findings	References
1.	Fruit beverages	2.5, 10 (% v/v)	Hydroxycinnamic acid	Addition of BSG phenolic extracts to juices did not significantly enhance their total phenolic content.	Celus et al. (2007)
2.	Dough (wheat)	15,25,35	Flour	Increased water absorption and decreased viscosity	Szwajgier et al. (2010)
3.	Cookies	10–40	Oven dried at various temperatures and fractionated into specific flour components	Cookies prepared with BSG had high protein, fiber, and lysine content	Ozvural et al. (2009)

TABLE 8.2 *(Continued)*

Sl. No.	Food Product	Proportion BSG Incorporated (%w/w)	Processing Form of Incorporated BSG	Findings	References
4.	Frankfurters	1,3,5	Milled and sieved to various fractions	Water holding capacity and dietary fibers increased whereas texture reduced	Ainsworth et al. (2007)
5.	Bread (wheat)	15	Flour fermented with sourdough starter and xylanase	Xylanase reduced firmness, antioxidant activity increased and fermentation reduced the phytic content	Ktenioudaki et al. (2015)
6.	Baked snacks	10,15,25	Flour	BSG increased the fiber and the protein content	Ktenioudaki et al. (2013)
7.	Breadsticks	15,25,35	Flour	Sensory scores increased as the BSG content decreased	Ktenioudaki et al. (2015)
8.	Extruded snack	10,20,25,30	Extruded	Increased fiber and protein content, Phytic acid increased	Stojceska et al. (2008)
9.	Ready to eat snacks	10,20,30	Flour	Increased protein and fiber content	Stojceska et al. (2009)
10.	Ready to eat extruded snacks	10	Dried and milled	Increased total dietary fiber	Steinmacher et al. (2012)

8.8.2 BSG AS A SOIL AMENDMENT AND FERTILIZER

Brewery residues are a great source of nutrients for crops, as they contain important components such as potassium and phosphorus (Dessalew et al., 2017). For soil insects such as *G. mellonella* larvae, BSG application has been shown to be both a fertilizer and a pesticide (Qiu et al., 2019). In addition to adding nutrients, BSG to the soil can help increase the amount of organic matter present. This makes it a particular possibility as a raw material for the production of biochar with the intention of using it in soil. Improved soil quality, more efficient fertilizer use, and improved water holding capacity are benefits of applying biochar in soil (Nair et al., 2017). In addition, it promotes a change in the microbiome associated with the roots, thus expanding the diversity of the microbial community in the soil. In addition, it allows the soil to function well as a carbon sink. Reportedly, adding water char derived from spent brewer's yeast to the soil contributed to better aggregation. The use of BSG in hop cultivation significantly improved the growth of the plant's root system (Amorielloa et al., 2020).

8.8.3 BSG AS ADSORBENT

BSG has been developed for the removal of volatile chemicals from exhaust gases and for its use as an adsorbent for heavy metals from aqueous solutions. Similar to charcoal, pyrolyzed BSG can absorb volatile chemical substances from coconut shells. With an adsorption capacity of 30.5 mg of orange acid 7 per gram of BSG at 30°C, BSG has also been used successfully as an adsorbent for orange acid dye 7 in wastewater. BSG is a raw material from which activated carbon can be made, which is used to purify gasoline and water. It is produced by treating BSG with an alkaline solution to extract the lignin, precipitating it with sulfuric acid, and then impregnating or activating the extracted lignin with 3 g/g phosphoric acid at 60°C. It can adsorb metal ions and phenolic chemicals in liquid media (Mussatto, 2014).

8.8.4 BSG AS A SOURCE OF BIOACTIVE PEPTIDES, PROTEINS AND PROTEIN HYDROLYSATES

Technical-functional additives with emulsification and gasification capabilities, nutrients for fortified foods and dietary supplements, and functional effects on the human body are just a few of the uses for plant-based proteins (Pojic et al., 2018). BSG proteins are employed in the production of complex proteins, fermented beverages, cookies, antioxidant and antimicrobial packaging materials, edible films, complex proteins, and edible proteins. Lynch et al. stated that proteins and protein hydrolysates produced from BSG have immuno-modulatory capabilities that could be useful in the treatment of inflammatory, hypertensive, and metabolic illnesses (Lynch et al., 2016). The two most common proteins in BSG are hordein, which accounts for 43% of the total protein composition, and glutelin, which accounts for 21.5%. Amino acids are essential to human health, and BSG is a good source of these nutrients. The essential amino acids of the BSG are histidine, methionine, lysine, phenylalanine, and tryptophan, while the non-essential amino acids are alanine, serine, proline, and glycine (Ikram et al., 2017), while less common in the BSG are leucine, glutamine, vanillin, and other amino acids identical to those in sprouted barley (Wen et al., 2019). Inactive amino acids found in the basic structure of proteins are called bioactive peptides. These inactive amino acids can later be converted to active forms by fermentation, food processing with enzymes (Toldra et al., 2018). The BIOPEP-UWM Bioactive Peptides Database lists ACE inhibitors altering blood pressure control, antioxidant action, antibacterial activity, opioid activity, antithrombotic, and immunomodulating activity as some of the biological roles of bioactive peptides. Bioactive peptides are added to functional foods to benefit from the amino acids they contain (Minkiewicz et al., 2019).

8.8.5 BSG AS A SOURCE OF PHENOLIC COMPOUNDS

The components of phenolic compounds in plants include tannins, flavonoids, lignans, phenolic acids, stilbene, and phenolic alcohols. Hydroxycinnamic acids (HCA), which are widespread in BSG and accumulate in cell walls, are the most important source of antioxidants in cereals, both in free and bound form (Guido et al., 2017). Ikram et al. (2017) considered phenolic chemicals to be natural antioxidants associated with a number of chronic diseases such as cancer, neurological diabetes, and cardiovascular disease. In the human body, ferulic acid acts similarly to vitamin C in skin defense and regeneration (Ravindran et al., 2016).

8.8.6 BSG AS A SOURCE OF FOOD PACKAGING

The development of biological and biodegradable materials is influenced by the ability of BSG proteins to bind to polypeptide chains. Plasticizers such as sorbitol, polyethylene glycol, and glycerin are needed to add flexibility and handling to protein-based films as they make the film less fragile and give it some plastic properties. Film formation is influenced by protein-protein interactions, which are stronger at pH values near the isoelectric point. Cellulose is used for food packaging because of its fine network, biodegradability, and strong water resistance. It can be used as a thickener, stabilizer, or as a source of fiber in functional meals and beverages. In addition, nanocomposite materials can be made from cellulose (Mussatto et al., 2016). Organic polyurethane, which has emerged as the most promising filler, can be made from natural cellulosic fibers such as BSG (Formela et al., 2017).

8.9 SUSTAINABLE MATERIALS

The next interesting application for brewery grain waste is in building materials (Olacia et al., 2020). As a filling and reinforcement, grain from the brewery seems to be a good choice. In addition, this technique reduces the cost of bio composites. Another way to recycle brewery biowaste is to modify building materials with beer waste. Ferraz et al. (2013) studied ceramic bricks made from powder mixes enhanced by dry BSG additions to the brick raw material at 5, 10, and 15%. The incorporation of 5% BSG by mass seems to be the best compromise between high mechanical flexural strength and low thermal conductivity. With a strength of 15 MPa, which is 12% less than the strength of the unmodified brick, the product could withstand mechanical bending. However, the thermal conductivity of the improved brick fell by 28% and was now 0.46 W/mK. Ferraz et al. (2019) used hot presses to produce shells from starch and BSG. The product was compared to typical expanded polystyrene trays. The results showed that all samples produced with BSG values of 40 to 80% had higher flexural strengths than their non-biodegradable equivalents.

8.10 BIOTECHNOLOGICAL PROCESSES

8.10.1 SUBSTRATE FOR CULTIVATION OF MICROORGANISMS

The number of enzymes (alpha-amylases, cellulases, and hemicelluloses) produced by growing fungi is increased by supplementing BSG with amino acids, vitamins, and inorganic substances (Lynch et al., 2016). Microorganisms that can thrive in BSG medium include *Pleurotus*, *Lentinus*, *Agrocybe*, *Trichoderma*, and *Streptomyces* (Mussatto, 2009). BSG is particularly well suited for the isolation, maintenance, and production of both established and new biologically active chemical screening strains (Guido et al., 2017).

8.10.2 UTILIZATION OF SUBSTRATES IN FERMENTATION PROCESSES

8.10.2.1 PREBIOTICS

Xylan makes up most of the hemicellulose found in plant cell walls. It has been shown that xylo-oligosaccharides (XOS), which are largely formed by the hydrolysis of Xylan, have prebiotic properties. Incorporation of XOS into foods results in acceptable organoleptic properties and shows heat and pH stability. It has beneficial health effects, particularly in the treatment of microencapsulation, neurotoxicity, intestinal inflammation, and constipation. Most of the lignocellulosic biomass consists of cellulose, hemicellulose, and lignin. The crystalline substance cellulose is made up of many layers of polysaccharides and a chain of glucose molecules connected by hydrogen bonds. The integrity and strength of the cell walls are supported by the phenolic components of lignin (p-coumaryl alcohol, coniferyl alcohol, and sinapyl alcohol). BSG contains 16.5% hemicellulose, which consists mainly of Xylan (10.3%), arabinan (5.1%), and a very small amount of acetyl groups (1.1%) (Amorim et al., 2019).

8.10.2.2 ETHANOL PRODUCTION

The production of ethanol, xylitol, and other products can be done using lignocellulose hydrolysates as fermentation medium. However, in order to maximize the efficiency of the process, a detox must first be performed. Hemicelluloses break down to form xylose, arabinose, glucose, acetic acid, and galactose, all of which can be used by microorganisms to produce ethanol. Glucose is formed when cellulose is broken down. The two methods used are simultaneous saccharification and fermentation (SSF) and separate hydrolytic fermentation (SHF) (Ravindran et al., 2016). The process to produce combustible ethanol involves five steps: pretreatment of BSG with acidic heat solutions, hydrolysis of BSG with enzymes to produce simple sugars from starch and cellulose, fermentation of sugars to produce ethanol (using microorganisms such as *Z. mobilis*, *S. cerevisiae*, *E. coli*, *B. subtilis*, or *P. pastoris*) and distillation of the resulting ethanol (Mussatto, 2014).

8.10.2.3 LIGNOCELLULOSIC YEAST CARRIER

A high-density population of yeasts is used for the continuous fermentation of beer, often resulting from the attachment of biofilms to solid surfaces. Solid substrates exhibit a number of important properties, including improved adhesion, physical durability, high availability, and low production costs (Moreira et al., 2013). The major obstacle to the industrial adoption of immobilized cell continuous fermentation systems is the high cost of immobilization support (Dragone et al., 2008). Because lignocellulosic yeast (LCYC) is a byproduct of the beer industry and can be easily incorporated into the continuous fermentation of beer, BSG is an excellent source of LCYC.

8.10.2.4 XYLITOL PRODUCTION

Xylitol has a lower caloric value than sucrose; therefore, insulin-deficient patients can also consume it as a powdered sweetener. The downside is the increased cost of sucrose, which can be eliminated by an inexpensive manufacturing process (Da Silva et al., 2012). Xylitol can be produced by microbial fermentation, and the BSG is cheaper than other lignocellulosic materials (Guido et al., 2017). The BSG substrate must first be fermented with *Candida guilliermondii* to produce xylitol. CO_2 is created during the fermentation process and is removed from the bioreactor. The broth is then evaporated at a temperature of 40°C, and the insolubility of xylitol is increased to 5°C by ethanol crystallization. Because it is lower in calories than sucrose, xylitol can be used as a powdered sweetener by people with insulin resistance. The downside is the higher price of sucrose, which can be reduced by using a cheap manufacturing process (Da Silva et al., 2012). The BSG is less expensive than other lignocellulosic materials and can produce xylitol through microbial fermentation (Guido et al., 2017). To produce xylitol, the BSG substrate must first be fermented with *C. guilliermondii*. During the fermentation process, CO_2 is produced and expelled from the bioreactor. Then the broth is evaporated at a temperature of 40°C, and the crystallization of ethanol increases the insolubility of the xylitol to a temperature of 5°C. The efficiency of centrifugation in removing the crystallized xylitol from molasses is 87–92% (Mussatto et al., 2013). Diabetics can consume xylitol because of its low calorie content.

In addition, it helps with osteoporosis, otitis media, and lung infections. Xylitol has significant economic value due to its anticariogenic properties, which stop the development of harmful acids that attack tooth enamel and promote tooth remineralization (Guido et al., 2017).

8.11 CONCLUSION

The use of agro-industrial byproducts is becoming increasingly popular from an economic and ecological point of view. Brewers spent grains are homogeneous and abundant in biowaste. They are commonly made by large breweries but are expensive to ship due to their high moisture content. The best way to take advantage of the above byproduct is to process it close to the brewery, as it saves on transportation costs. Despite all the possible uses listed, the use of this technology is still limited. The development of new methods of utilizing this agro-industrial byproduct is of paramount importance due to the large-scale production of spent grain all year round. More attention needs to be paid to finding financially viable alternative drying methods.

KEYWORDS

- **adsorption and wastewater treatment**
- **bioactive compounds**
- **biotechnological processes**
- **brewers' spent grain**
- **soil amendment and fertilizer**
- **sustainable materials**

REFERENCES

Ainsworth, P., İbanoğlu, S., Plunkett, A., İbanoğlu, E., & Stojceska, V. (2007). Effect of brewers' spent grain addition and screw speed on the selected physical and nutritional properties of an extruded snack. *Journal of Food Engineering, 81*, 702–709.

Aliyu, S., & Bala, M. (2011). Brewer's spent grain: A review of its potentials and applications. *African Journal of Biotechnology, 10*, 324–331.

Amoriello, T., Fiorentino, S., Vecchiarelli, V., & Pagano, M. (2020). Evaluation of spent grain biochar impact on hop (*Humulus lupulus* L.) growth by multivariate image analysis. *Applied Sciences, 10*, 533.

Amorim, C., Silvério, S. C., Silva, S. P., Coelho, E., Coimbra, M. A., Prather, K. L. J., & Rodrigues, L. R. (2018). Single-step production of arabino-xylooligosaccharides by recombinant *Bacillus subtilis* 3610 cultivated in brewers' spent grain. *Carbohydrate Polymers, 199*, 546–554.

Aulds, C. R. B. F., & Aldron, K. E. W. W. (2010). Effect of storage conditions on the microbial ecology and biochemical stability of cell wall components in brewers' spent grain. *Journal of Agricultural and Food Chemistry*. https://doi.org/10.1021/jf1001099

Birsan, R. I., Wilde, P., Waldron, K. W., & Rai, D. K. (2019). Recovery of polyphenols from brewer's spent grains. *Antioxidants, 8*, 380.

Blandino, A., Al-Aseeri, M., Pandiella, S., Cantero, D., & Webb, C. (2003). Cereal-based fermented foods and beverages. *Food Research International, 36*, 527–543.

Celus, I., Brijs, K., & Delcour, J. A. (2007). Enzymatic hydrolysis of brewers' spent grain proteins and techno-functional properties of the resulting hydrolysates. *Journal of Agricultural and Food Chemistry, 55*, 8703–8710.

Coelho, E., Rocha, M. A. M., Saraiva, J. A., & Coimbra, M. A. (2014). Microwave superheated water and dilute alkali extraction of brewers' spent grain arabinoxylans and arabinoxylo-oligosaccharides. *Carbohydrate Polymers, 99*, 415–422.

Coskuntuna, L., Koç, F., Özduven, M. L., & Coskuntuna, A. (2010). Effects of organic acid on silage fermentation and aerobic stability of wet brewer's grain at different temperatures. *Bulgarian Journal of Agricultural Science, 16*, 651–658.

Da Silva, S. S., & Chandel, A. K. (2012). *D-Xylitol: Fermentative Production, Application and Commercialization* (Vol. 1, p. 345). Springer.

Dai, T., Wang, J., Dong, D., Yin, X., Zong, C., & Jia, Y. (2022). Effects of brewers' spent grains on fermentation quality, chemical composition and *in vitro* digestibility of mixed silage prepared with corn stalk, dried apple pomace and sweet potato vine. *Italian Journal of Animal Science, 21*, 198–207.

de Souza, L. C., Zambom, M. A., & dos Pozza, M. S. (2012). Development of microorganisms during storage of wet brewery waste under aerobic and anaerobic conditions. *Revista Brasileira de Zootecnia, 41*, 188–193.

Dessalew, G., Beyene, A., Nebiyu, A., & Ruelle, M. L. (2017). Use of industrial diatomite wastes from beer production to improve soil fertility and cereal yields. *Journal of Cleaner Production, 157*, 22–29.

Dragone, G., Mussatto, S. I., Almeida, E., & Silva, J. B. (2008). Influence of temperature on continuous high gravity brewing with yeasts immobilized on spent grains. *European Food Research and Technology, 228*, 257–264.

Du, L., Arauzo, P. J., Meza Zavala, M. F., Cao, Z., Olszewski, M. P., & Kruse, A. (2020). Towards the properties of different biomass-derived proteins via various extraction methods. *Molecules, 25*, 488.

El-Shafey, E. I., Gameiro, M. L. F., Correia, P. F. M., & De Carvalho, J. M. R. (2004). Dewatering of brewer's spent grain using a membrane filter press: A pilot plant study. *Separation Science and Technology, 39*, 3237–3261.

Faccenda, A., Zambom, M. A., Avila, A. S., Castagnara, D. D., Dri, R., Fischer, M. L., Tinini, R. C. R., Dessbesell, J. G., Almeida, A. R. E., & Almeida, K. V. (2021). Influence of the storage period on the nutritional and microbiological value of sun-dried brewer's grains. *Revista Colombiana de Ciencias Pecuarias, 34*, 4.

Ferraz, E., Coroado, J., Gamelas, J., Silva, J., Rocha, F., & Velosa, A. (2013). Spent brewery grains for improvement of thermal insulation of ceramic bricks. *Journal of Materials in Civil Engineering, 25*, 1638–1646.

Ferreira, A. M., Martins, J., Carvalho, L. H., & Magalhães, F. D. (2019). Biosourced disposable trays made of brewer's spent grain and potato starch. *Polymers, 11*, 923.

Formela, K., Hejna, A., Zedler, Ł., Przybysz, M., Ryl, J., Saeb, M. R., & Piszczyk, L. (2017). Structural, thermal and physico-mechanical properties of polyurethane/brewers' spent grain composite foams modified with ground tire rubber. *Industrial Crops and Products, 108*, 844–852.

Guido, L. F., & Moreira, M. M. (2017). Techniques for extraction of brewer's spent grain polyphenols: A review. *Food and Bioprocess Technology, 10*, 1192–1209.

Ikram, S., Huang, L. Y., Zhang, H., Wang, J., & Yin, M. (2017). Composition and nutrient value proposition of brewers' spent grain. *Journal of Food Science, 82*, 2232–2242.

Kazemi, M., Tahmasbi, A. M., Valizadeh, R., Naserian, A. A., Haghayegh, G. H., & Jami, Y. E. (2014). Studies on the effects of different chemical additives on the nutritive value of ensiled barley distillers' grain (BDG) using *in vitro* techniques. *Direct Research Journal of Agriculture and Food Science, 2*, 19–24.

Kitrytė, V., Saduikis, A., & Venskutonis, P. R. (2015). Assessment of antioxidant capacity of brewer's spent grain and its supercritical carbon dioxide extract as sources of valuable dietary ingredients. *Journal of Food Engineering, 167*, 18–24.

Ktenioudaki, A., Alvarez-Jubete, L., Smyth, T. J., Kilcawley, K., Rai, D. K., & Gallagher, E. (2015). Application of bioprocessing techniques (sourdough fermentation and technological aids) for brewer's spent grain breads. *Food Research International, 73*, 107–116.

Ktenioudaki, A., O'shea, N., & Gallagher, E. (2013). Rheological properties of wheat dough supplemented with functional byproducts of food processing: Brewer's spent grain and apple pomace. *Journal of Food Engineering, 116*, 362–368.

Łukasiak, J., Olsen, K., Georgiou, C. A., & Georgakopoulos, D. G. (2013). Cereal arabinoxylans: Advances in structure and physicochemical properties. *European Food Research and Technology, 237*, 33–48.

Lv, J., Fang, X., Feng, G., Zhang, G., Zhao, C., Zhang, Y., & Li, Y. (2020). Effects of sodium formate and calcium propionate additives on the fermentation quality and microbial community of wet brewers' grains after short-term storage. *Molecules, 10*, 1–14.

Lynch, K. M., Steffen, E. J., & Arendt, E. K. (2016). Brewers' spent grain: A review with an emphasis on food and health. *Journal of the Institute of Brewing, 122*, 553–568.

MacLeod, A. M., & Pollock, J. R. A. (1979). The physiology of malting. In *Brewing Science* (Vol. 1, pp. 145–232). Academic Press.

Mandalari, G., Faulds, C. B., Sancho, A. I., Saija, A., Bisignano, G., Lo Curto, R., & Waldron, K. W. (2005). Fractionation and characterization of arabinoxylans from brewers' spent grain and wheat bran. *Journal of Cereal Science, 42*, 205–212.

Martinez, E. A., Villarreal, M. L. M., Almeida e Silva, J. B., Solenzal, A. I. N., Canilha, L., & Mussatto, S. I. (2002). Uso de diferentes materias primas para la producción biotecnológica de xilitol. *Ciencia y Tecnología Alimentaria, 3*, 295–301.

Martins, Z. E., Pinho, O., & Ferreira, I. M. P. L. V. O. (2017). Food industry byproducts used as functional ingredients of bakery products. *Trends in Food Science & Technology, 67*, 106–128.

McCarthy, A. L., O'Callaghan, Y. C., Piggott, C. O., FitzGerald, R. J., & O'Brien, N. M. (2013). Brewers' spent grain: Bioactivity of phenolic component, its role in animal nutrition and potential for incorporation in functional foods: A review. *Proceedings of the Nutrition Society, 72*, 117–125.

Mendis, M., & Simsek, S. (2014). Arabinoxylans and human health. *Food Hydrocolloids, 42*, 239–243.

Meneses, N. G., Martins, S., Teixeira, J. A., & Mussatto, S. I. (2013). Influence of extraction solvents on the recovery of antioxidant phenolic compounds from brewer's spent grains. *Separation and Purification Technology, 108*, 152–158.

Minkiewicz, P., Iwaniak, A., & Darewicz, M. (2019). BIOPEP-UWM database of bioactive peptides: Current opportunities. *International Journal of Molecular Sciences, 20*, 5978.

Moreira, M. M., Morais, S., Carvalho, D. O., Barros, A. A., Delerue-Matos, C., & Guido, L. F. (2013). Brewer's spent grain from different types of malt: Evaluation of the antioxidant activity and identification of the major phenolic compounds. *Food Research International, 54*, 382–388.

Mussatto, S. I. (2009). Biotechnological potential of brewing industry byproducts. In S. N. Nigam & A. Pandey (Eds.), *Biotechnology for Agro-Industrial Residues Utilization: Utilization of Agro-Residues* (pp. 313–326). Springer: Berlin/Heidelberg, Germany.

Mussatto, S. I. (2014). Brewer's spent grain: A valuable feedstock for industrial applications. *Journal of the Science of Food and Agriculture, 94*, 1264–1275.

Mussatto, S. I., Dragone, G., & Roberto, I. C. (2005). Kinetic behavior of *Candida guilliermondii* yeast during xylitol production from brewer's spent grain hemicellulosic hydrolysate. *Biotechnology Progress, 21*, 1352–1356.

Mussatto, S. I., Dragone, G., & Roberto, I. C. (2006). Brewers' spent grain: Generation, characteristics and potential applications. *Journal of Cereal Science, 43*, 1–14.

Mussatto, S. I., Dragone, G., & Roberto, I. C. (2007). Ferulic and p-coumaric acids extraction by alkaline hydrolysis of brewer's spent grain. *Industrial Crops and Products, 25*, 231–237.

Mussatto, S. I., Fernandes, M., Milagres, A. M. F., & Roberto, I. C. (2008). Effect of hemicellulose and lignin on enzymatic hydrolysis of cellulose from brewer's spent grain. *Enzyme and Microbial Technology, 43*, 124–129.

Mussatto, S. I., Moncada, J., Roberto, I. C., & Cardona, C. A. (2013). Techno-economic analysis for brewer's spent grains uses on a biorefinery concept: The Brazilian case. *Bioresource Technology, 148*, 302–310.

Nair, V. D., Nair, P. K. R., Dari, B., Freitas, A. M., Chatterjee, N., & Pinheiro, F. M. (2017). Biochar in the agro ecosystem climate-change-sustainability nexus. *Frontiers in Plant Science, 8*, 1051.

Nazzaro, F., Fratianni, F., Ombra, M. N., D'Acierno, A., & Coppola, R. (2018). Recovery of biomolecules of high benefit from food waste. *Current Opinion in Food Science, 22*, 43–54.

Niemi, P., Aura, A. M., Maukonen, J., Smeds, A. I., Mattila, I., Niemelä, K., Tamminen, T., Faulds, C. B., Buchert, J., & Poutanen, K. (2013). Interactions of a lignin-rich fraction from brewer's spent grain with gut microbiota *in vitro*. *Journal of Agricultural and Food Chemistry, 61*, 6754–6762.

Nigam, P. S. (2017). An overview: Recycling of solid barley waste generated as a byproduct in distillery and brewery. *Waste Management, 62*, 255–261.

Olacia, E., Pisello, A. L., Chiodo, V., Maisano, S., Frazzica, A., & Cabeza, L. F. (2020). Sustainable adobe bricks with seagrass fibers: Mechanical and thermal properties characterization. *Construction and Building Materials, 239*, 117669.

Ozvural, E. B., Vural, H., Gokbulut, İ., & Ozboy-Ozbaş, O. (2009). Utilization of brewer's spent grain in the production of frankfurters. *International Journal of Food Science & Technology, 44*, 1093–1099.

Pojic, M., Misan, A., & Tiwari, B. (2018). Eco-innovative technologies for extraction of proteins for human consumption from renewable protein sources of plant origin. *Trends in Food Science & Technology, 75*, 93–104.

Qin, F., & Mussatto, S. I. (2017, October 17–19). Protein and amino acids recovery from brewer's spent grains by different pretreatment technologies. In *Proceedings of the Brazilian Bioenergy Science and Technology Conference 2017: Designing a Sustainable Bioeconomy* (p. 193). Campos do Jordão, Brazil.

Qin, F., Johansen, A. Z., & Mussatto, S. I. (2018). Evaluation of different pretreatment strategies for protein extraction from brewer's spent grains. *Industrial Crops and Products, 125*, 443–453.

Qiu, L., Li, J. J., Li, Z., & Wang, J. J. (2019). Production and characterization of biocontrol fertilizer from brewer's spent grain via solid-state fermentation. *Scientific Reports, 9*, 480.

Ravindran, R., & Jaiswal, A. K. (2016). Exploitation of food industry waste for high-value products. *Trends in Biotechnology, 34*, 58–69.

Reis, S. F., Coelho, E., Coimbra, M. A., & Abu-Ghannam, N. (2015). Improved efficiency of brewer's spent grain arabinoxylans by ultrasound-assisted extraction. *Ultrasonics Sonochemistry, 24*, 155–164.

Robertson, J. A. I., Anson, K. J. A., Treimo, J., Faulds, C. B., Brocklehurst, T. F., Eijsink, V. G. H., & Waldron, K. W. (2010). Profiling brewers' spent grain for composition and microbial ecology at the site of production. *LWT–Food Science and Technology, 43*, 890–896.

Santos, M., Jimenez, J. J., Bartolome, B., Gomez-Cordoves, C., & del Nozal, M. J. (2003). Variability of brewer's spent grain within a brewery. *Food Chemistry, 80*, 17–21.

Shindo, S., & Tachibana, T. (2004). Production of L-lactic acid from spent grain, a byproduct of beer production. *Journal of the Institute of Brewing, 110*, 347–351.

Spinelli, S., Conte, A., Lecce, L., Padalino, L., & Del Nobile, M. A. (2016). Supercritical carbon dioxide extraction of brewer's spent grain. *The Journal of Supercritical Fluids, 107*, 69–74.

Steinmacher, N. C., Honna, F. A., Gasparetto, A. V., Anibal, D., & Grossmann, M. V. (2012). Bioconversion of brewer's spent grains by reactive extrusion and their application in bread-making. *LWT–Food Science and Technology, 46*, 542–547.

Stojceska, V., & Ainsworth, P. (2008). The effect of different enzymes on the quality of high-fiber enriched brewer's spent grain breads. *Food Chemistry, 110*, 865–872.

Stojceska, V., Ainsworth, P., Plunkett, A., & İbanoglu, S. (2008). The recycling of brewer's processing byproduct into ready-to-eat snacks using extrusion technology. *Journal of Cereal Science, 47*, 469–479.

Stojceska, V., Ainsworth, P., Plunkett, A., & İbanoglu, S. (2009). The effect of extrusion cooking using different water feed rates on the quality of ready-to-eat snacks made from food byproducts. *Food Chemistry, 114*, 226–232.

Szwajgier, D., Wasko, A., Targoński, Z., Niedzwiadek, M., & Bancarzewska, M. (2010). The use of a novel ferulic acid esterase from *Lactobacillus acidophilus* K1 for the release of phenolic acids from brewer's spent grain. *Journal of the Institute of Brewing, 116*, 293–303.

Tan, Y. X., Mok, W. K., Lee, J., Kim, J., & Chen, W. N. (2019). Solid state fermentation of brewers' spent grains for improved nutritional profile using *Bacillus subtilis* WX-17. *Fermentation, 5*, 52.

Tang, D., Yin, G., He, Y., Hu, S., Li, B., Li, L., Liang, H., & Borthakur, D. (2009). Recovery of protein from brewer's spent grain by ultrafiltration. *Biochemical Engineering Journal, 48*, 1–5.

Terefe. (2022). Preservation techniques and their effect on nutritional values and microbial population of brewer's spent grain: A review. *CABI Agriculture and Bioscience, 3*, 51.

Toldrá, F., Reig, M., Aristoy, M. C., & Mora, L. (2018). Generation of bioactive peptides during food processing. *Food Chemistry, 267*, 395–404.

Wahlstrom, R., Rommi, K., Willberg-Keyriläinen, P., Ercili-Cura, D., Holopainen-Mantila, U., Hiltunen, J., Makinen, O., Nygren, H., Mikkelson, A., & Kuutti, L. (2017). High yield protein extraction from brewer's spent grain with novel carboxylate salt–urea aqueous deep eutectic solvents. *ChemSelect, 2*, 9355–9363.

Wen, C., & Zhang, J., & Duan, Y. (2019). A mini-review on brewer's spent grain protein: Isolation, physicochemical properties, application of protein, and functional properties of hydrolysates. *Journal of Food Science, 84*, 3330–3340.

Xiros, C., Topakas, E., Katapodis, P., & Christakopoulos, P. (2008). Hydrolysis and fermentation of brewer's spent grain by *Neurospora crassa. Bioresource Technology, 99*, 5427–5435.

Yu, D., Sun, Y., Wang, W., O'Keefe, S. F., Neilson, A. P., Feng, H., Wang, Z., & Huang, H. (2020). Recovery of protein hydrolysates from brewer's spent grain using enzyme and ultrasonication. *International Journal of Food Science & Technology, 55*, 357–368.

CHAPTER 9

Meat: Wastes, Byproducts, and Utilization

PALAK MAHAJAN[1,*] and MUDASIR AHMAD MALIK[2]

[1]*Department of Biotechnology Engineering and Food Technology, University Institute of Engineering, Chandigarh University, Mohali, Punjab, India*

[2]*Department of Food Engineering and Technology, Ghani Khan Choudhury Institute of Engineering and Technology, Malda, West Bengal, India*

*Corresponding author

ABSTRACT

The emergence of waste from an industry negatively impacts the economy and environmental pollution of the nation. These impacts can be converted into the nation's strength by converting these wastes into useful byproducts. However, non- or under-utilization of the byproducts results in a loss of potential revenue as well as an increase in the cost of disposal. In many circumstances, wastes from the production of meat have the potential to be recycled into new raw materials or transformed into useful goods with a higher value. The majority of meat industry waste is generated during slaughtering. Depending on the kind of animals slaughtered, different waste materials are produced by the meat industries. Due to the abundance of organic compounds in the waste products of the meat processing sector, their disposal is rather complex. Meat industry wastes such as skin, blood, bones, meat trimmings, hooves, feet, and viscera are costly to handle and dispose of in an environmentally responsible way. These expenses may be offset by

Utilization of Food Waste and Byproducts. Mudasir Ahmad Malik and Kshirod Kumar Dash (Eds.)

DOI: 10.1201/9781779641151-9

innovation that provides products with more value, thereby increasing the firm's profitability. Intriguing vaporization strategies involve the hydrolysis of waste to get value-added products such as bioactive peptides with physiological effects such as antihypertensive, antioxidant, anti-diabetic, and so forth, which have promising food and non-food applications. This chapter thus covers the various waste products generated by the meat industry and their potential use as byproducts. In addition, the most recent developments and trends in the utilization and value addition of byproducts and wastes from the meat industry are also covered in this chapter.

9.1 INTRODUCTION

A significant amount of food waste, such as blood, bones, skin, fatty tissues, feet, viscera, and many more, is produced by the meat processing industries. The handling and disposal of these wastes in an environmentally friendly manner is an expensive process (Ryder et al., 2015). Through innovation and waste utilization, these expenses can be compensated to produce products with added value and raise profitability. However, improper handling or treatment of these byproducts in the past led to pertinent crises like the emergence of spongiform encephalopathies. Acceptance of meat-based byproducts varies among different localities, traditions, and religions. Many countries around the world include byproducts (such as brain, kidney, liver, and blood) with high nutritional value in their diets and dishes. If properly managed to produce substances or products with additional value, meat-based byproducts may represent a valued reserve (Zhang et al., 2010).

Although there are many different types of meat byproducts, most of them, on average, include appropriate levels of nutrients, providing the meat industry with an excellent chance for value-adding (Valta et al., 2015). Figure 9.1 shows the various applications of meat-based byproducts for food and non-food uses. Meat wastes can be used as raw material to produce biomolecules of interest, for example, protein hydrolysates (Lasekan et al., 2013), and bioactive peptides (Martínez-Alvarez et al., 2015). Other usages focus on non-food items such as fertilizers, raw materials for chemical or pharmaceutical usages, or energy production (Figure 9.1). Recently, the production of biodiesel from used animal fats has become a popular subject (Adewale et al., 2016), similar to the second generation of biogas oil, also called bioderived diesel fuel (Baladincz and Hancsók, 2015).

The usage and valorization of meat industry wastes for the development of useful byproducts, related recent advancements, and trends are covered in this chapter.

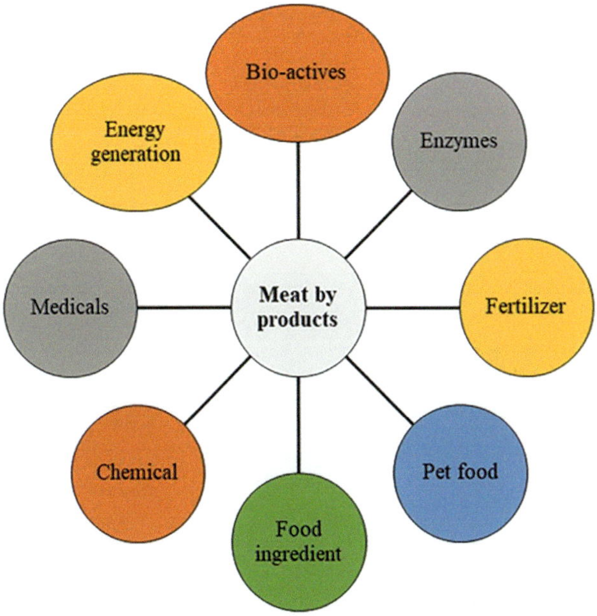

FIGURE 9.1 Main routes of applications for meat byproducts.

9.2 UTILIZATION OF MEAT WASTE

9.2.1 AS FUNCTIONAL INGREDIENTS

According to market needs and to avoid wastage of food products, byproducts are taken into consideration (Figure 9.2). Therefore, some non-carcass goods which are discarded in one nation may be regarded as priceless commodities in another. Meat-based byproducts possess various nutrients; for example, the liver contains a high number of vitamins, and the liver and kidney have a significant amount of minerals despite their high cholesterol level (García-Llatas et al., 2011). The majority of the calorie intake comes from other foods that are high in fat tissue, like lard or tallow. It is important to remember that the use and function of certain organs vary among the animal species from which they are derived because of variations in the nutrient content. For instance, vitamin A varies respectively as 50,000

RE μg, 20,000 RE μg, and 18,000 RE μg for the heart organ of lamb, pork, and chicken (Honikel, 2011).

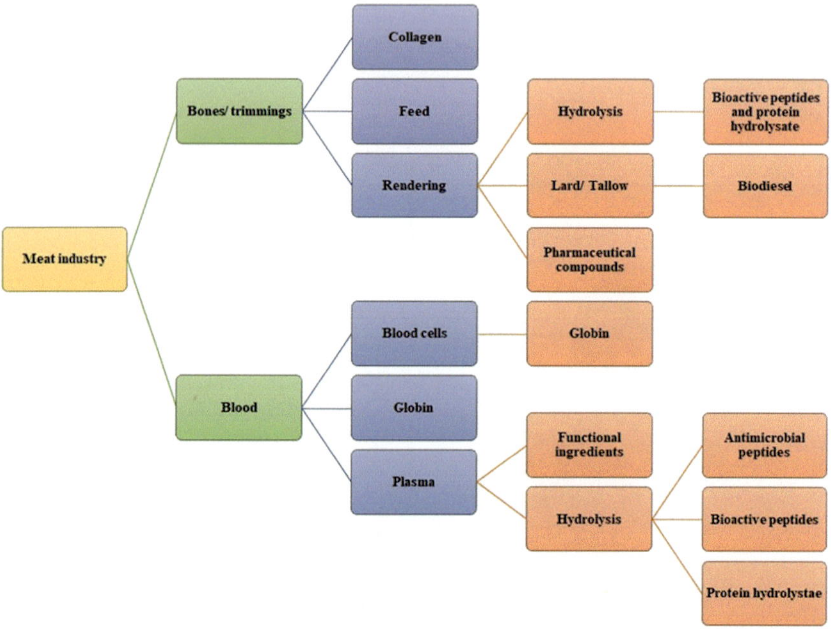

FIGURE 9.2 Primary paths for adding value to meat byproducts.

9.2.2 PROTEINS: PRODUCTION OF BIOACTIVE PEPTIDES

Production of free amino acids and smaller peptide units as a result of proteolytic actions occurring during the processing of meat is of paramount importance. These proteolysis actions are caused by endogenous meat muscle enzymes such as calpains and cathepsins in the earlier stages, followed by peptidyl peptidases (Arihara, 2006; Sentandreu and Toldrá, 2007). Moreover, the peptides can also be produced industrially by hydrolyzing waste meat protein (organs, collagen, hemoglobin) using specialized proteases such as bromelain, papain, ficin, thermolysin, and pronase. A reactor (either continuous or batch) with an ultrafiltration membrane can be used to carry out the enzymatic process. The product is subsequently subjected to partial purification and fractionation using chromatographic or filtering methods.

Numerous peptides are produced by these hydrolysis reactions, but only bioactive peptides are of great interest as they have the potential to improve the consumer's health through a variety of bioactivities, such as antioxidant activity, antihypertensive action, or opioid activity. The most significant bioactivity is probably the suppression of the angiotensin I-converting enzyme (ACE). This enzyme takes part in the renin-angiotensin system, which causes angiotensin I to become angiotensin II, which narrows the arteries and raises blood pressure. Through ACE inhibition, blood pressure can be decreased in a practical method (Ahmed and Mugurama, 2010). Peptides must pass to the circulatory system in an undamaged state for this reason.

Peptides with hypocholesterolemic, antioxidant, and antithrombotic properties influence the functioning of the cardiovascular system, peptides with mineral binding affect the gastrointestinal functioning, while those with immunomodulatory properties influence the immune systems. However, some peptide plays a role in more than one activity. Bioactive peptides must be released from their origin protein to perform appropriate biological functions (Vercruysse et al., 2005). Production of some bioactive peptides also occurs during the fermentation or curing stages of food production, while some are produced upon gastrointestinal digestion. However, because of the simultaneous action of numerous endogenous enzymes and the diverse profile of peptides produced, the fundamental issue with these naturally occurring biopeptides is the difficulty in controlling and manipulating their hydrolysis parameters and conditions (Mora et al., 2014).

To manage the bioactive peptides produced and to produce more uniform batches, protein extracts are digested under carefully controlled conditions of hydrolysis settings using recognized enzymes, for instance, trypsin, alcalase, pepsin, and thermolysin. There has been a lot of research on the utilization of meat-based byproducts as bioactive peptides. The utmost commonly tested meat wastes in this regard are blood and collagen, which are significant byproducts of the meat industry and slaughterhouses (Ryder et al., 2016).

9.2.3 PROTEINS: PRODUCTION OF ANTIMICROBIAL PEPTIDES

Proteins have been reported to exhibit antimicrobial properties. Fragment 33–61 of the bovine hemoglobin chain was the first to be identified as a peptide in bovine hemoglobin with antibacterial action (Toldrá et al., 2012).

This peptide was effective against *Boophilus microplus*, a tick parasite that infests cattle, as well as other Gram-positive bacteria and fungi. From peptic-digested bovine hemoglobin, four additional antimicrobial peptides with activity against *M. luteus* A270, *L. innocua*, *E. coli*, and *S. enteritidis* were discovered (Nedjar-Arroume et al., 2006). The most prevalent protein complex in blood, hemoglobin, which contains iron, provides a rich supply of proteins (Ofori and Hsieh, 2011). It may be found throughout the world, and although Europe, Asia, and Africa use it as a culinary ingredient, its production is excessive.

Both the cellular and plasma fractions and their hydrolysates have shown ACE-inhibitory, antibacterial, antioxidant, and opioid properties (Chang et al., 2007). However, most research has been done on antimicrobial peptides made from hemoglobin hydrolysates (Adje et al., 2011). The type of enzymes, substrate, and digesting or hydrolyzing conditions (such as temperature, pH, and time) plays a significant impact on the production of bioactive peptides (Yu et al., 2006).

These antimicrobial biopeptides are primarily hydrophobic because a higher hydrophobicity is required for their affinity with microbials' outer membranes. Also, some opioid peptides obtained from animal blood have the potential to affect the nervous and gastrointestinal systems (Kapel et al., 2003). Studies on the antioxidant capacity of peptides generated from hemoglobin are few, nevertheless. The protein collagen is the most prevalent in numerous byproducts of the meat industry (primarily from bones, cartilage, skin, and hide) and is an excellent source for the production of bioactive peptides.

9.2.4 BLOOD AND RELATED PRODUCTS

Bovine blood typically has 80.9% of its weight in water, 17.3% protein, and the rest in other nutrients (Duarte et al., 1999). Blood-derived products are used as value-added additives and nutritional supplements. Moreover, blood proteins offer relevant technological qualities such as gelation, foaming agents, and emulsification (Ofori and Hsieh, 2011) and thus may have found application in food products (Cofrades et al., 2000).

Besides these properties, some plasma proteins have demonstrated good major protein cross-linking capacity and protease inhibitory activity (Yousif et al., 2003). Other precipitated blood plasma components, such as fibrinogen and the enzyme thrombin, act as binding agents and are useful in creating restructured meat-based food items. This is because when thrombin is applied on the meat surface, it transforms the soluble fibrinogen to form an insoluble fibrin polymer, that is, protofibril, which results in a 3-D network of fibrin clot (Lennon et al., 2010). The pH, temperature, moisture, size of the meat cuts, as well as direction of the meat fibers, all affect the strength of the adhesion (Chen and Lin, 2002).

Despite taking hygienic care, blood may still be significantly contaminated by microbes, necessitating the use of preservation techniques (Toldrá et al., 2007). The production of globin, sometimes referred to as decolorized blood, is a superior substitute to avoid sensory side effects. Bovine hemoglobin creates heme iron polypeptide by digestion with enzymes and is marketed to boost the absorption of iron (Nissenson et al., 2003). The degree of hydrolysis was found to be high, between 30–40%, upon two-stage enzymatic hydrolysis, resulting in a heme-iron enriched peptide product (In et al., 2002). Moreover, the generated heme-enriched peptides had an iron content that was higher than that of hemoglobin. This suggests an intriguing method for preparing heme-iron that could be employed for the fortification of everyday and widely consumed food products, thereby preventing iron deficiency.

9.2.5 PREPARATION AND USE OF MEAT BYPRODUCTS

The nervous system, including the brain and spinal cord, is processed only for human consumption. Before the final cooking, these organs are blanched to preserve their textural characteristics. Before cooking kidneys, the ureter and blood veins are often removed. The most common edible organ is the liver. It is used to make liver paste, numerous processed meat items, and liver sausages (Devatkal et al., 2004). Braising, grilling, or boiling are used to prepare livers for consumption.

Normally, intestines are utilized in fertilizers, animal feed, and meat meals. Additionally, they are widely utilized in sausages (Bhaskar et al., 2007). Animal intestines are separated from the carcasses, the excess fat is removed, and the feces are peeled off to prepare the sausage casings. The manufacture of livestock and poultry feed has frequently employed slaughter leftovers in poorer nations. Wastes contain important amino acids that can be used to make cattle protein feed (Ristic and Kormanjos, 1992). Nutrients such as protein and minerals such as phosphorus and calcium can be used as a source of nitrogen and important minerals in animal feed. Table 9.1 provides a quick overview of some usual uses for various parts and organs of harvested animals. Also, meat-based byproducts have been the subject of inventive applications, which are discussed in subsections.

TABLE 9.1 Potential Applications and Methods for Preparing Edible Meat Byproducts

Source	Storage/ Preparation	Method of Usage
Blood	Frozen and refrigerated	Sausages, puddings
Skin	Frozen and refrigerated	Gelatin
Fat	Frozen, fresh, refrigerated	Shortening, Lard
Ear	Frozen, fresh, refrigerated	Stewed, smoked, and salted
Brain	Piece, whole	Braised, broiled, cooked, scrambled
Liver	Frozen, fresh, refrigerated	Braised, broiled, cooked, scrambled
Spleen	Frozen, fresh, refrigerated, pre-cooked	Fried, In blood sausages
Intestine	Frozen, fresh, refrigerated	As sausage casing

9.3 FEED AND PET FOOD APPLICATIONS

Traditional feeds and pet meals have used raw or rendered animal byproducts as a component. Animal byproducts contain vital amino acids, fatty acids, minerals, and vitamins. Blood-derived amino acid solutions, protein or blood meals (Pérez-Galvez et al., 2011), or meat and bone meal ashes are a few examples (Coutand et al., 2008). Meat and bone meal are good sources of group B vitamins and important amino acids and can be employed for feeding animals (Jayathilakan et al., 2012). It has been found that protein hydrolysates are effective in aquaculture (Gilbert et al., 2008).

Due to the lack of anti-nutritional components and abundance of essential amino acids, these protein hydrolysates offer an intriguing substitute for soybean meal. Additionally, byproducts containing keratin including hair, nails, and the epidermis of the skin might be profitable after being hydrolyzed by the enzyme keratinase (Lasekan et al., 2013) and can thus be utilized and commercialized as a source of keratin protein.

9.4 ENERGY GENERATION

Biodiesel has recently been accepted as a substitute for diesel fuel owing to its advantages such as biodegradability, non-toxicity, and favorable combustion emission profile with lower levels of carbon monoxide, carbon dioxide, and particulate matter (Gerpen, 2005; Moreira et al., 2010). Additionally, the usage of biodiesel does not require major engine changes. A combination of glycerol and fatty acid methyl esters is produced and used as a byproduct of the transesterification of inexpensive animal fat byproducts (Bhatti et al., 2008; Moreira et al., 2010). It has been suggested that tubular reactors can boost the profitability of biodiesel through hydro-oxygenation, hydroisomerization, and supercritical transesterification (Marulanda et al., 2010). Other recent studies concentrate on enhancing biodiesel production by employing advanced techniques such as the use of ultrasounds (Adewale et al., 2016).

Owing to their high amount of saturated fatty acids, animal fats must be decreased by the winterization process or inclusion of chemicals (Banković-Ilić et al., 2014). The advancements have continued, and a new second generation, so-called biogas oil, is already facing prompt applications (Baladincz and Hancsók, 2015).

9.5 MEDICAL AND PHARMACEUTICAL APPLICATIONS

Pork skin can be used to treat human skin ulcers and burns. Glands and organs are high-nutrient-content edible beef byproducts that are consumed worldwide. In reality, several of them have been used for medicinal purposes or as a source of certain pharmaceutical chemicals. Bile, melatonin, heparin, progesterone, and insulin from the gallbladder, pineal gland, liver, ovaries, and pancreas, respectively, are some examples (Jayathilakan et al., 2012).

Protein hydrolysates, particularly collagen hydrolysates, can create peptides that can be deposited in joint cartilage to treat osteoarthritis (Bello and Oesser, 2006). Bones and joints benefit from the effects of hydrolyzed collagen. Hyaluronic acid-infused hydrolysates are being sold for improved joint function and pain reduction in people. Laboratory guinea pigs were used to test ultrafiltrates (b30 kDa) from pig aortas, and it was discovered that these extracts significantly reduced atherogenic index, atherogenic lipoproteins, and total and residual cholesterol.

9.6 FERTILIZER APPLICATIONS

All countries produce significant amounts of meat and bone meal, and one intriguing method involved in their production is thermochemical processing, which includes pyrolysis, combustion, and gasification. Pyrolysis and coal combustion are the most extensively studied. The generated ashes show a high phosphorus concentration, making them useful as fertilizers. Additionally, the gas emissions meet contain combustibles that can be used to produce energy (Cascarosa et al., 2012). Animal waste is burned to produce high-quality mineral fertilizers. Furthermore, the application of heat recovery enables effective energy recoveries (Bujak, 2015).

9.7 CHEMICAL APPLICATIONS

In the cosmetics sector, rendered fats are employed in various products, including body lotions, creams, and bath treatments. In the chemical industry, fatty acids are used to make plasticizers, softeners, and lubricants. The chemical industry also uses collagen and gelatin as components in surfactants, antifreeze, paints, cleaners, adhesives, and polishes. There have been reports of new uses for rendered fats, such as in the manufacturing of polyhydroxyalkanoates using a recombinant strain of *Ralstonia eutropha* (Riedel et al., 2015). This polymer offers an appealing replacement for plastics made from petroleum since it is biodegradable. The hides that are usually used to make leather goods like clothing, shoes, belts, handbags, and purses can be utilized for a variety of other things.

9.8 CONCLUSION

In the modern world, where there is a high demand for nutrient-dense foods, there is also a lot of concern about environmental degradation, which has resulted in rising costs and a frequently decreasing supply of raw resources. Recycling and garbage recovery have consequently gained significant importance. To reduce costs, it is essential to apply new or enhanced processing methods where waste effluents and byproducts can be recovered and commonly converted to higher value useable products in addition to environmental and safety concerns.

The applications for meat byproducts are incredibly varied. In addition to the rendered fat for cosmetics, and hides for leather, recent innovative techniques also include the use of proteins for better technological or nutritional properties, the creation of bioactive and antimicrobial peptides, or the use of animal fats for the production of biodiesel. The most advanced and effective tools must still be used to analyze byproducts for nutritional properties and to look for key active molecules in food and nutrition. Thus the exploitation of meat industry wastes in particular is attracting growing attention because these wastes constitute a potentially utilizable resource for conversion.

KEYWORDS

- **antimicrobial peptides**
- **biodiesel production**
- **bovine hemoglobin**
- **meat byproducts**
- **protein hydrolysates**
- **value-added applications**

REFERENCES

Adewale, P., Dumont, M. J., & Ngadi, M. (2016). Enzyme-catalyzed synthesis and kinetics of ultrasonic-assisted methanolysis of waste lard for biodiesel production. *Chemical Engineering Journal, 284*, 158–165.

Adje, E. Y., Balti, R., Kouach, M., Dhulster, P., & Guillochon, D. (2011). Obtaining antimicrobial peptides by controlled peptic hydrolysis of bovine hemoglobin. *International Journal of Biological Macromolecules, 49*, 143–153.

Ahmed, A. M., & Mugurama, M. (2010). A review of meat protein hydrolysates and hypertension. *Meat Science, 86*, 110–118.

Arihara, K. (2006). Strategies for designing novel functional meat products. *Meat Science, 74*, 219–229.

Baladincz, P., & Hancsók, J. (2015). Fuel from waste animal fats. *Chemical Engineering Journal, 282*, 152–160.

Banković-Ilić, I. B., Stojković, I. J., Stamenković, O. S., Veljkovic, V. B., & Hung, Y. T. (2014). Waste animal fats as feed stocks for biodiesel production. *Renewable and Sustainable Energy Reviews, 32*, 238–254.

Bello, A. E., & Oesser, S. (2006). Collagen hydrolysate for the treatment of osteoarthritis and other joint disorders: A review of the literature. *Current Medical Research and Opinion, 22*, 2221–2232.

Bhaskar, N., Modi, V. K., Govindaraju, K., Radha, C., & Lalitha, R. G. (2007). Utilization of meat industry byproducts: Protein hydrolysate from sheep visceral mass. *Bioresource Technology, 98*, 388–394.

Bhatti, H. N., Hanif, M. A., Qasim, M., & Rheman, A. U. (2008). Biodiesel production from waste tallow. *Fuel, 87*, 2961–2966.

Bujak, J. W. (2015). New insights into waste management: Meat industry. *Renewable Energy, 83*, 1174–1186.

Cascarosa, E., Gea, G., & Arauzo, J. (2012). Thermochemical processing of meat and bone meal: A review. *Renewable and Sustainable Energy Reviews, 16*, 942–957.

Chang, C. Y., Wu, K. C., & Chiang, S. H. (2007). Antioxidant properties and protein compositions of porcine hemoglobin hydrolysates. *Food Chemistry, 100*, 1537–1543.

Chen, M. J., & Lin, C. W. (2002). Factors affecting the water-holding capacity of fibrinogen/plasma protein gels optimized by response surface methodology. *Journal of Food Science, 67*, 2579–2582.

Cofrades, S., Guerra, N. I. A., Carballo, J., Fernández-Martin, F., & Jiménez-Colmenero, F. (2000). Plasma protein and soy fiber content effect on bologna sausage properties as influenced by fat levels. *Journal of Food Science, 65*, 281–287.

Coutand, M., Cyr, M., Deydier, D., Guilet, R., & Clastres, P. (2000). Characteristics of industrial and laboratory meat and bone meal ashes and their potential applications. *Journal of Hazardous Materials, 150*, 522–532.

Darine, S., Christophe, V., & Gholamreza, D. (2010). Production and functional properties of beef lung protein concentrates. *Meat Science, 84*, 315–322.

Devatkal, S., Mendiratta, S. K., Kondaiah, N., Sharma, M. C., & Anjaneyulu, A. S. R. (2004). Physicochemical, functional and microbiological quality of buffalo liver. *Meat Science, 68*, 79–86.

Duarte, R. T., Carvalho Simoes, M. C., & Sgarbieri, V. C. (1999). Bovine blood components: Fractionation, composition, and nutritive value. *Journal of Agricultural and Food Chemistry, 47*, 231–236.

García-Llatas, G., Alegría, A., Barberá, R., & Farré, R. (2011). Minerals and trace elements. In L. M. L. Nollet & F. Toldrá (Eds.), *Handbook of Analysis of Edible Animal Byproducts* (p. 183). CRC Press: USA.

Gerpen, J. V. (2005). Biodiesel processing and production. *Fuel Processing Technology, 86*, 1097–1107.

Gilbert, E. R., Wong, E. A., & Webb, K. E. (2008). Board-invited review: Peptide absorption and utilization: Implications for animal nutrition and health. *Journal of Animal Science, 86*, 2135–2155.

Honikel, K. O. (2011). Composition and calories. In L. M. L. Nollet & F. Toldrá (Eds.), *Handbook of Analysis of Edible Animal Byproducts* (p. 183). CRC Press: USA.

In, M. J., Chae, H. J., & Oh, N. S. (2002). Process development for heme-enriched peptide by enzymatic hydrolysis of hemoglobin. *Bioresource Technology, 84*, 63–68.

Jayathilakan, K., Sultana, K., Radhakrishna, K., & Bawa, A. S. (2012). Utilization of byproducts and waste materials from meat, poultry and fish processing industries: A review. *Journal of Food Science and Technology, 49*, 278–293.

Kapel, R., Froidevaux, R., Nedjar-Arroume, N., Fertin-Bazus, A., Dhulster, P., & Guillochon, D. (2003). Continuous production of a peptidic fraction containing the intermediate opioid peptide LVV-haemorphin-7 (LVVh-7) by peptic hydrolysis of bovine hemoglobin in a continuous membrane reactor. *Biotechnology and Applied Biochemistry, 37*, 317–324.

Lasekan, A., Abu Bakar, F., & Hashim, D. (2013). Potential of chicken byproducts as sources of useful biological resources. *Waste Management, 33*, 552–565.

Lennon, A. M., McDonald, K., Moon, S. S., Ward, P., & Kenny, T. A. (2010). Performance of cold-set binding agents in re-formed beef steaks. *Meat Science, 85*, 620–624.

Martínez-Alvarez, O., Chamorro, S., & Brenes, A. (2015). Protein hydrolysates from animal processing byproducts as a source of bioactive molecules with interest in animal feeding: A review. *Food Research International, 73*, 204–212.

Marulanda, V. F., Anitescu, G., & Tavlarides, L. L. (2010). Investigations on supercritical transesterification of chicken fat for biodiesel production from low-cost lipid feedstocks. *The Journal of Supercritical Fluids, 54*, 53–60.

Mora, L., Reig, M., & Toldrá, F. (2014). Bioactive peptides generated from meat industry byproducts. *Food Research International, 65*, 344–349.

Mora, L., Sentandreu, M. A., Koistinen, K. M., Fraser, P. D., Toldrá, F., & Bramley, P. M. (2009). Naturally generated small peptides derived from myofibrillar proteins in Serrano dry-cured ham. *Journal of Agricultural and Food Chemistry, 57*, 3228–3234.

Moreira, A. L., Dias, J. M., Almeida, M. F., & Alvim-Ferraz, C. M. (2010). Biodiesel production through transesterification of poultry fat at 30 °C. *Energy & Fuels, 24*, 5717–5721.

Nedjar-Arroume, N., Dubois-Delval, V., Miloudi, K., Daoud, R., Krier, F., Kouach, M., Briand, G., & Guillochon, D. (2006). Isolation and characterization of four antibacterial peptides from bovine hemoglobin. *Peptides, 27*, 2082–2089.

Nissenson, A. R., Berns, J. S., Sakiewicz, P., Ghaddar, S., Moore, G. M., Schleicher, R. B., & Seligman, P. A. (2003). Clinical evaluation of heme iron polypeptide: Sustaining a response to rHuEPO in hemodialysis patients. *American Journal of Kidney Diseases, 42*, 325–330.

Pérez-Galvez, R., Almecija, M. C., Espejo, F. J., Guadix, E. M., & Guadix, A. (2011). Biobjective optimization of the enzymatic hydrolysis of porcine blood protein. *Biochemical Engineering Journal, 53*, 305–310.

Riedel, S. L., Jahnsa, S., Koeniga, S., Bocka, M. C. E., Brighamc, C. J., Baderd, J., & Stahla, U. (2015). Polyhydroxyalkanoates production with *Ralstonia eutropha* from low-quality waste animal fats. *Journal of Biotechnology, 214*, 119–127.

Ristic, M., & Kormanjos, S. (1992). Non-edible poultry slaughter byproducts as raw material for producing feed for cattle. *Veterinarski Glasnik, 46*, 233–238.

Ryder, K., El-Din Bekhit, A., McConnell, M., & Carne, A. (2016). Towards generation of bioactive peptides from meat industry waste proteins: Generation of peptides using commercial microbial proteases. *Food Chemistry, 208*, 42–50.

Ryder, K., Ha, M., El-Din Bekhit, A., & Carne, A. (2015). Characterization of novel fungal and bacterial protease preparations and evaluation of their ability to hydrolyze meat myofibrillar and connective tissue proteins. *Food Chemistry, 172*, 197–206.

Sentandreu, M. A., & Toldrá, F. (2007). Evaluation of ACE inhibitory activity of dipeptides generated by the action of porcine muscle dipeptidyl peptidases. *Food Chemistry, 102*, 511–515.

Toldrá, F., Aristoy, M. C., Mora, L., & Reig, M. (2012). Innovations in value-addition of edible meat byproducts. *Meat Science, 92*, 290–296.

Toldrá, M., Dávila, E., Saguer, E., Fort, N., Salvador, P., Parés, D., & Carretero, C. (2007). Functional and quality characteristics of the red blood cell fraction from bio preserved porcine blood as influenced by high pressure processing. *Meat Science, 80*, 380–388.

Valta, K., Damala, P., Orli, E., Papadaskalopoulou, C., Moustakas, K., Malamis, D., & Loizidou, M. (2015). Valorization opportunities related to wastewater and animal byproducts exploitation by the Greek slaughtering industry: Current status and future potentials. *Waste and Biomass Valorization, 6*, 927–945.

Vercruysse, L., Van Camp, J., & Smagghe, G. (2005). ACE inhibitory peptides derived from enzymatic hydrolysate of animal protein: A review. *Journal of Agricultural and Food Chemistry, 53*, 8106–8115.

Yousif, A. M., Cranston, P., & Deeth, H. C. (2003). Incorporation of bovine dry blood plasma into biscuit flour for the production of pasta. *LWT–Food Science and Technology, 36*, 295–302.

Yu, Y., Hu, J., Miyaguchi, Y., Bai, X., Du, Y., & Lin, B. (2006). Isolation and characterization of angiotensin I-converting enzyme inhibitory peptides derived from porcine hemoglobin. *Peptides, 27*, 2950–2956.

Zhang, W., Xiao, S., Samaraweera, H., Lee, E. J., & Ahn, D. U. (2010). Improving functional value of meat products. *Meat Science, 86*, 15–31.

CHAPTER 10

Poultry: Wastes, Byproducts, and Utilization

DARAKHSHAN TAHSEEN SIDDIQUI,[1] SWETA JOSHI,[1,*]
NAZIA TABASSUM,[2] and GAZIA NASIR[3]

[1]*Department of Food Technology, School of Interdisciplinary Sciences, Jamia Hamdard, New Delhi, India*

[2]*Department of Post-Harvest Engineering and Technology, Faculty of Agricultural Sciences, Aligarh Muslim University, Aligarh, Uttar Pradesh, India*

[3]*Department of Bioengineering, Faculty of Engineering and Technology, Integral University, Lucknow, Uttar Pradesh, India*

*Corresponding author

ABSTRACT

India is known for its vast livestock wealth. Poultry slaughtering and different post-processing operations in abattoirs generate perishable wastes and byproducts. Effective management of waste and the utilization of byproducts has always been a task for both food industries and the government. The waste utilization has a huge impact on the country's economy and in combating environmental pollution. The ineffective utilization of byproducts and management of waste from poultry can raise aesthetic as well as appalling health concerns. Therefore, there is a shift of paradigm for the waste to be seen as a valuable resource for producing value-added products. Both waste and byproducts from the poultry industry are excellent sources of protein. The poultry manure can be used as fertilizers and

Utilization of Food Waste and Byproducts. Mudasir Ahmad Malik and Kshirod Kumar Dash (Eds.)

DOI: 10.1201/9781779641151-10

to produce biogas and electricity. The poultry byproducts are rendered to produce the byproduct meals containing around 64 to 68% protein, which are used as animal feed. The feathers as a byproduct, due to their abundance in keratin proteins and amino acids, can be used to make important goods including feather meal, biodiesel, biodegradable plastic, and fertilizer. This chapter focuses on summarizing such proper management practices, innovative processing methods, and effective alternatives for recycling of poultry waste and byproducts for sustainable development.

10.1 INTRODUCTION

Global living standards have improved as a result of technological advancements. The shift of paradigm to view wastes and byproducts as valuable resources has led to the development of sustainable measures for reducing environmental pollution and revenue generation. While developing nations likewise aspire to better and more sustainable living, they do not always have access to better and more modern technologies. Providing the right amount of food and managing their waste would be the first step towards the advancements. Waste is mostly produced by the agricultural, industrial, and communal sectors of the population; the chain begins with wholesalers and concludes with consumers. The challenges associated with disposing of these wastes and utilizing the byproducts produced are getting worse. The byproduct utilization and waste management task are difficult owing to their biological instability, high pathogenic nature, high moisture content, susceptibility to rapid auto-oxidation and enzymatic activity, and regulatory requirements for food safety and quality. Therefore, proper techniques must be established in order to limit pollution and make use of the byproducts produced; proper systems for recycling and reducing waste and byproducts must also be devised. To achieve these goals, researchers should concentrate on waste management and the application of effective and economical methods (Jayathilakan et al., 2012).

Aside from agricultural waste, some are produced on a ground level by farms, such as animal and poultry byproducts used as manure, and by livestock processing facilities, such as the byproducts of slaughtering and tanneries. The primary consideration towards the waste is to convert them into feed for livestock because of the high protein content they hold. Other uses of these waste could be as fuel from manure and fertilizers.

As per the Department of Animal Husbandry and Dairying, and MoFPI, Govt. of India, the meat production in the country is estimated to be about 5.3 million metric tons annually, with the largest proportion for poultry that approximates to around 36% of total meat produced (https://www.mofpi.gov.in/sites/default/files/OpportunityinMeat&PoultrysectorinIndia.pdf). The poultry industry generates a variety of wastes, such as solid waste and waste water. The byproducts produced can be divided into primary byproducts (hide and skins, feathers, and bones), as well as hatchery byproducts (such as infertile eggs and egg shells), and secondary byproducts (such as blood meal, bone meal, blood albumen, and egg albumen). Utilizing waste and byproducts from egg production and poultry processing is crucial for the development of the poultry industry.

Sun drying, steam heating, and roasting are just a few of the straightforward techniques for drying cage layer manure that were described by Kese and Donkoh (1982). According to a 2019 study by Saim and Kaan, high-quality wastes require effective processing methods to discover prospective feed additives as well as alternative waste treatment technologies that provide the opportunity for energy recovery and the exploitation of high-quality nutrients for crop production. The soft and trimmed bones are rendered to produce animal fat, meat meal, and can be further processed as feed for animals, insects, aquafeed, or renewable energy sources (Del Nery et al., 2007; Kantarli et al., 2016).

In this chapter, we will study the proper management practices, innovative processing methods, and effective alternatives for recycling of poultry waste and byproducts for sustainable development.

10.2 TYPES OF POULTRY WASTES AND BYPRODUCTS

The broiler meat industry produces a variety of organic byproducts and related solid wastes during different processing stages. The main stages of the broiler slaughtering procedures are depicted in Figure 10.1. The accumulation of waste begins at the reception stage and continues throughout the slaughtering stages with the accumulation of poultry excrement and mortalities, blood from the bleeding stage, feathers from the defeathering process, offal and intestinal deposits from the evisceration process, and bone and inedible meat deposits from the trimming and deboning stage (Yoon et al., 2014). After the sellable carcass has been removed, the concentration of inedible byproduct, computed on the basis of live weight, accounts for 30% of the live bird weight, thus representing the very need to utilize them (Salminen and Rintala, 2002). As illustrated in Figure 10.1, depending on the properties of waste, chicken abattoir wastes can be categorized into soft solid wastes (such as offal organs, intestines, and bones), blood, feathers, and waste water. The wastewater produced during the washing, scalding, cleaning, and washing processes contains a high content of suspended solids and biodegradable organic matter. Therefore, inappropriate disposal and improper management of these wastes can cause environmental pollution.

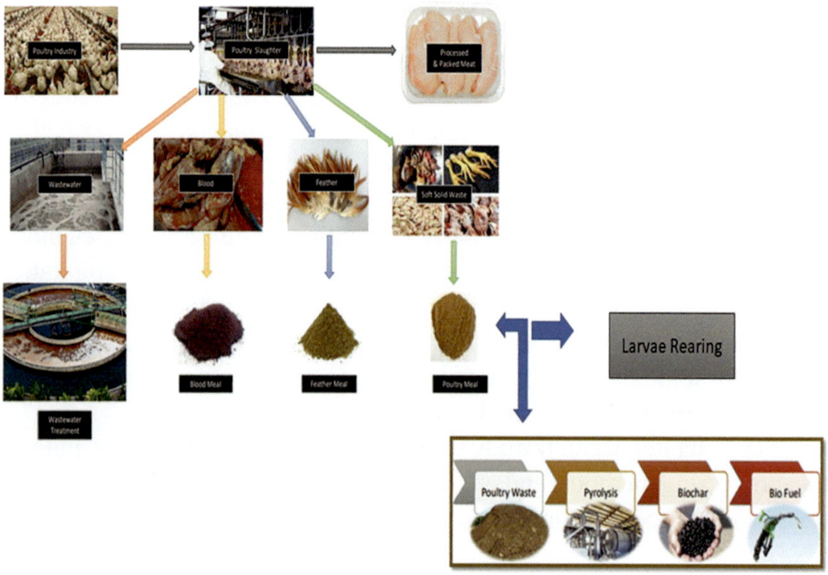

FIGURE 10.1 Slaughtering and poultry processing stages.

10.2.1 SOFT SOLID WASTE AND BYPRODUCTS

The major classes of the poultry abattoir waste and byproducts portrayed in Figure 10.2 include the head, poultry skin, viscera, feet, intestines, bones, fat trimmings, blood, and feathers, which are not fit for immediate human consumption (Ferreira et al., 2018). These wastes and byproducts are primarily made up of chicken proteins and lipids. When these wastes are used improperly, critical biological resources including protein, lipids, and enzymes are lost. Previous research states that around 22–23% is classified as broiler chicken soft waste from the poultry abattoir (Sari et al., 2016).

The earlier study exemplified that the poultry waste and byproducts can be utilized to produce sustainable and value-added products. It is reported that blood (3.2–3.7% of poultry live weight) is used to produce blood meal, heads (2.5–3.0% of poultry live weight) for poultry meal, glands and intestines (8.5–9.0% of poultry live weight) for meat meal, poultry grease, feet (3.5–4.0% of poultry live weight) for soup and fat/poultry grease, and litter and manure for recycled feed and surface dressing of agricultural land (Jayathilakan et al., 2012).

Post-slaughtering process, meat tissues tend to lose their antioxidant potential, thereby accelerating the oxidation of lipids, myoglobin, and proteins in mishandled byproducts, resulting in deterioration of raw materials (Papuc et al., 2017). Therefore, the effective management of poultry solid wastes pursuant to regulatory requirements is of utmost importance for processing new products with added value. The presence of excessive levels of organic matter, nutrients, sources of disease transmission, and odor emissions, due to ineffective management and utilization of waste and byproducts, can pose a serious threat to the water, soil, and even public health (Franke and Insam, 2013).

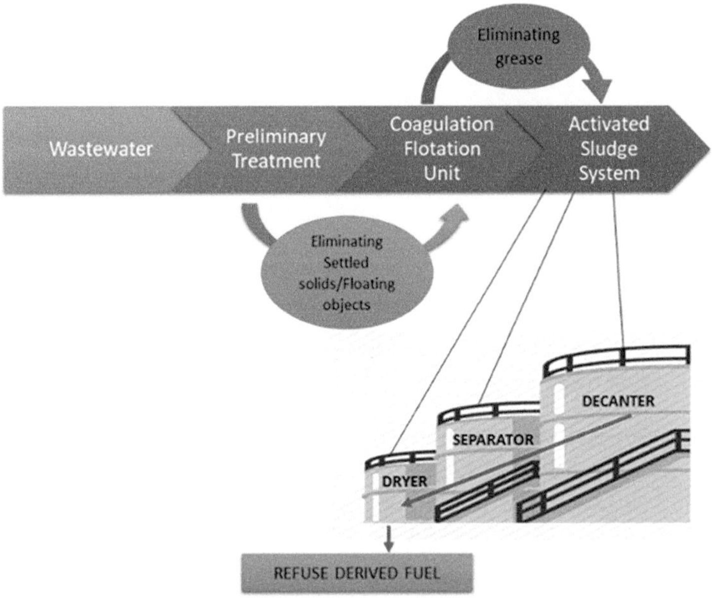

FIGURE 10.2 Types of soft solid waste.

10.2.2 BLOOD

The blood, an important edible byproduct, consists of 95% protein, is produced during the slaughtering of poultry, and makes up to 2% of the chicken's live weight (Salminen and Rintala, 2002). The dry poultry blood, as processed blood meal, is of huge nutritive and functional quality, possesses excellent amino acid balance, and significant quantities of micronutrients like iron (Marquez et al., 2005). Blood is often removed from the other solid wastes during slaughter and managed with the utmost hygiene to produce high-quality blood meal. However, in some small slaughterhouses, blood is discarded along with feather, wastewater, and intestinal content. High-quality blood meal (Figure 10.3) can be utilized as a nutrient source for industrial crops and in the preparation of feed for many animal species. Owing to its nutritional and functional attributes, blood finds its application in manufacturing of non-food items like feedstuffs, binders, and fertilizer. But, the use of blood in the meat industry results in products with dark color and less palatability. However, the portion of blood, the plasma, due to its lack of color, is suitable to be used in the meat processing industry.

FIGURE 10.3 . Poultry blood meal.

10.2.3 FEATHER

Another significant and protein-rich waste produced by the poultry business is feather, which accounts for approximately 7–8% of the whole weight of chickens and more than 50 g of waste per chicken. Feathers are frequently processed into feather meal, agricultural fertilizers, bedding material, and decorative items (Figure 10.4) (Brandelli et al., 2015).

The poultry industry produces huge amounts of feathers as byproducts. Feathers are mainly constituted of non-digestible keratin proteins. The utilization of poultry byproducts as a nutrient source for aquatic life sustenance is one of the economically sustainable ways for producing value-added products (Klinger and Naylor, 2012; Naylor et al., 2000, 2009). The physical, chemical, or biological pre-treatments facilitate solubilization of keratin into a conveniently digested protein (Jayathilakan et al., 2012). The most popular technique is rendering with high-pressure autoclaving at a temperature above 133°C. Feather meal is typically created by the rendering process after the hydrolyzed feathers have been dried and homogenized into a powder. Another popular technique to turn feathers into a digestible protein source is biological treatment utilizing the feather-decomposing bacteria, *Bacillus licheniformis*, or keratin hydrolyzing enzymes (Mezes et al., 2015).

According to Bureau et al. (2000, 2008), Klemesrud et al. (2000), hydrolyzed feather meal (HF), a potent protein source with 74 to 91% crude protein (CP), has a high cystine content but very little lysine and methionine. In addition to being a rich source of protein, HF also has polyunsaturated fatty acids (PUFA), antioxidant peptides, and is low in phosphorus and ash, making it suitable for use in formulations that reduce water pollution (Sugiura et al., 2000). The impact of using the feather protein hydrolysate created by the keratinolytic microorganism *Vibrio* sp. kr2 on Wistar rats was demonstrated, and the research reported its usefulness as protein feed formulations (Grazziotin et al., 2008).

FIGURE 10.4 Feather utilized as Feather Meal and as pillow stuffing (Bedding material).

The feather is preferred for yarn production and technical textile applications due to its distinctive physicochemical structure, which includes low density, a high slenderness ratio, moderate strength, spinnability, good pliability, fineness and length, microstructure, thermal stability, and durability properties (Tesfaye et al., 2018). Utilizing waste made from keratin to make yarn and technical textile goods adds value and integrates beneficiation on an economic and environmental level. Another study depicted the use of hydrolysates from chicken byproducts as a source of peptone in microbiological applications due to its functional and bioactive properties (Lasekan et al., 2013).

10.2.4 WASTE-WATER

The continuous and increasing industrialization has increased the demands of goods and services. To meet this demand, the poultry industry has increased its production in the past decade, thus increasing water consumption for poultry processing industries. This increased water consumption, if not effectively managed, poses a threat to the environment due to increasing water scarcity, as by 2050 global water demand is expected to be 20–30% higher than the current scenario (Beck et al., 2021). The potable water consumption in poultry abattoirs is around 26L per bird for different cleaning operations (Barbut, 2015).

An enormous amount of wastewater is produced at the poultry abattoir facility during various slaughtering and poultry meat processing stages, including the holding and receiving of the birds, bird washing, stunning, slaughtering, scalding, defeathering, and evisceration, as well as equipment and facility sanitization and cooling of mechanical equipment like compressors and pumps. Other significant sources of waste include further processing and rendering operations involving uncollected blood, feathers, viscera, soft tissue removed during trimming and cutting, bone, and soil from feathers. The water in question is a high-strength wastewater with a high level of organic matter, chemical oxygen demand (COD), biochemical oxygen demand (BOD), total suspended solids (TSS), total organic carbon (TOC), and nitrogen and phosphorous source. A slaughter house with a daily slaughter capacity of 100,000 birds uses 1050 m^3 of process water and generates 2.2 mg of sludge cake that is discharged from the wastewater treatment facility. Additionally, it was stated that 1000 heads of poultry manure could yield 22.4 kg of nitrogen (Yoon et al., 2014), which could serve as a nitrogen source for plant nutrition. The sludge from poultry slaughterhouses is frequently a source of plant nutrients (Ozdemir and Yetilmezsoy, 2019).

Along with other materials like cooking oils, the processing and rendering processes serve as additional sources of animal fat and other soft tissue production. The main elements of poultry processing wastewater are various biodegradable organic compounds, primarily fats and proteins, present in both particulate and dissolved forms. In order to reduce particulate matter concentration before treatment, the wastewater from poultry processing is screened, which increases the collection of rendered byproducts (Jayathilakan et al., 2012).

Hence, the wastewater undergoes preliminary treatment to get rid of settled solids and floating objects, then goes through a coagulation-floatation unit to get rid of grease, and finally gets its final treatment in an activated sludge system (Yetilmezsoy et al., 2011). The intensity of treatment required is estimated by the discharge limitations specified by the industrial wastewater treatment guideline. Wastewater sludge is commonly dewatered in three ways to produce sludge with a moisture content under 40%: a decanter, separator, and drying, respectively. Finally, dried sludge is given to a waste processing industry in order to be processed into RDF (refuse derived fuel) for combustion in cement clinker (Figure 10.5).

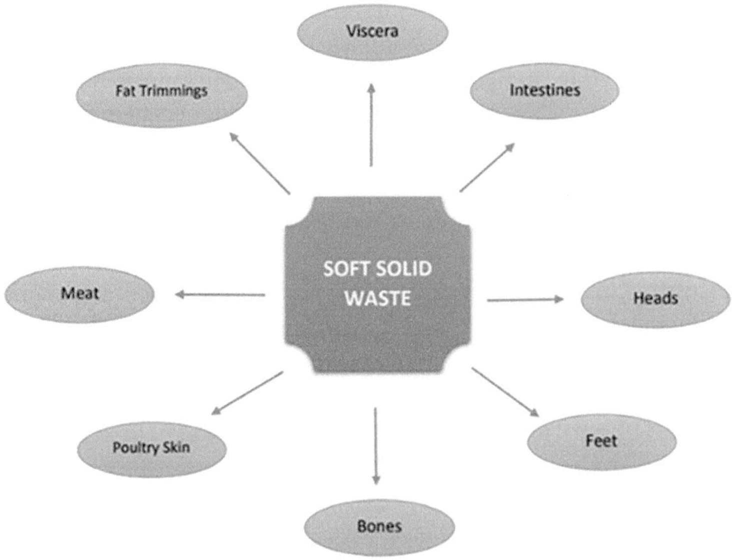

FIGURE 10.5 Flow process for wastewater treatment into RDF (refuse derived fuel).

One of the studies reviewed that the activated sludge and anaerobic digestion (AD) are the most cost-effective, eco-friendly, low-maintenance-required methods of wastewater treatment. Multi-integrated systems can be designed and added as pre-treatments prior to the anaerobic digester, as well as post-treatment after the digester for recovery of nutrients and removal of pathogens from wastewater (Gutu et al., 2021).

10.2.5 HATCHERY WASTE

The poultry industry generates a lot of solid waste and wastewater that is used in hatcheries. Empty eggshells, infertile eggs, dead embryos, late hatchlings, dead birds, and a viscous liquid made of eggs and decomposing flesh make up the solid hatchery waste. The water used to clean incubators, hatcheries, and areas where chicks are handled produces the effluent. Hatcheries and the poultry industry as a whole are in danger of going out of business. One of the reasons that the concentration of nitrates and phosphates in drinking water is rising is the seepage of livestock industry effluent into ground water or runoff into streams and rivers. The expansion of the agricultural industry, which is essential for boosting the economy and creating jobs, depends on the main industries' consistent expansion. The primary industry's ability to expand is expected to have a substantial impact on the region's landscape and local communities, and it is crucial that it can continue to access affordable, high-quality land and water. Land filling, composting, rendering, and incineration are among the conventional techniques for disposing of solid hatchery waste (Das et al., 2002). The majority of hatchery waste is disposed of in landfills or composting facilities, incurring annual disposal expenses of millions of dollars for the chicken meat business (Das et al., 2002). Some of the waste from the hatcheries is rendered.

According to Hamm and Whitehead (1982) and Vanderpopuliere et al. (1977), hatchery waste has a high protein content and a moisture content of 43–71%. According to Sharara et al. (1992a), dried hatchery waste contains 21.5% ash, 12.1% crude fiber, 33.1% CP, 29.0% ether extract, and 28.8 MJ/kg of gross energy. The hatchery waste byproduct meal's apparent metabolizable energy (AME) and apparent amino acid availability are 23.9 MJ/kg and 73.5% (Sharara et al., 1992a, b), respectively. Hatchery waste can be made into meals high in protein, other goods with added value, or, after proper processing, an organic fertilizer.

Alternative management and processing techniques for hatchery waste are as follows:

- Centrifuging can be used to separate hatchery waste into solid waste and liquid waste (Cawthon, 1998);
- Inclined screens, belts, and filter presses can also be used to separate the components of the waste (Van Slyke et al., 2005);
- Flexible multi-layer filters can be utilized for the separation of liquid wastes from solid wastes (Schilling and Mintz, 1998);
- A conveyor with upper and lower conveyor rollers can also be used to separate liquid and solid waste. Separated liquid and solid wastes are put in collectors next to the upper and lower rollers (Van Slyke et al., 2007).

The following methods can be used to separate shells from hatchery waste: (World Intellectual Property Organization-WO/2001/074491, Eggshell waste processing method and device).

- A strong suction vacuum is utilized to only collect the dry, very light shells from the waste, leaving the heavier sterile eggs behind.
- Eggshell waste can be further separated from heavier components in hatchery waste using a cyclone forced-air separator and a vibrating or shaking device.
- Instead, live chicks, unhatched eggs, or clear eggs from the hatching tray are placed on a moving belt with fixed gaps that only allow chicks to slide through, while shells and unhatched eggs are kept on the belt and dead embryos are disposed of into a different container.

Products developed from shell helps in:

- Increasing the mineral content of compost;
- Deterring slugs and snails around plants;
- Fertilizing when mixed with garden soil;
- Feeding aviary birds when mixed with seeds;
- Increasing the strength of cement when added;
- Making mosaics by artists;
- Making textured paint for 3D effects in artwork;
- Producing collagen from the egg shell membrane.

The following techniques can be used to treat the solid waste:

- Heating the waste in a furnace to create steam that powers a turbine generator and creates electricity (Kennedy, 1986).
- To manufacture pet or animal feed, it may be rendered (Swan, 1992), autoclaved (Verma and Rao, 1974), extruded (Miller, 1984), boiled (Kirkpinar et al., 2004), ensiled (Kompiang, 1994), enzyme-treated (Kim and Patterson, 2000), or composted to make fertilizer (Cambardella et al., 2003; Imbeah, 1998).
- Onsite product stabilization using an in-vessel composter.

The installation of an AD system is the most efficient way to handle hatchery waste locally. In all other organic waste industries, it is by far the most widely utilized method of treating organic wastes. The method benefits from being highly efficient and creates biogas that may be utilized for heating or generating electricity. After the digestion process, the biosolids that are left over can be used to make high-quality fertilizer (Lin et al., 1986).

10.3 POULTRY RENDERING

Processing of poultry solid waste produced from slaughtering is an abattoir's responsibility. Waste from poultry slaughterhouses can quickly deteriorate if kept at room temperature for a longer period and treated improperly prior to preliminary processing. A range of tools and procedures are used in the physical and chemical transformation of waste animal tissue to make it usable: a process known as rendering. Rendering is an important emerging method frequently used to preserve and enhance the nutritional quality of poultry abattoir waste using currently available advanced technologies, thereby converting them into high-grade hygienic marketable products (Farmanesh et al., 2019). The rendering processes vary according to the raw material composition. Organic poultry wastes, like viscera and entrails, bones and feathers, blood, offal, and trimming, have been treated to create a feed fit for animals via the rendering process. Soft wastes, feathers, bones, and other pieces of the sacrificed poultry can all be broken down completely and indefeasibly by autoclaving at high temperatures and pressure (Kazemi et al., 2017). Each rendering product is a great resource for a particular usage and typically offers an affordable source of animal protein, edible fats, and inedible fats for industrial usage as fuel blending agents (Kamalak et al., 2005).

The gathering and sanitary transportation of raw materials to a facility, where they are ground into uniform particle sizes and transported to a cooking vessel, either in continuous-flow or batch mode, is a requirement of all rendering system technologies. Depending on the method and materials, cooking is typically performed with steam at temperatures between 240°F and 290°F (approximately 115°C–145°C) for 40–90 minutes (David et al., 2006). Regardless of the method of cooking, a significant amount of moisture is removed, and the melted fat is separated from the protein and bone particles. Cooking, most crucially, renders bacteria, viruses, protozoa, and parasites inactive. Alternative methods of disposing of raw materials, like burial, composting, or landfill usage, rarely result in the inactivation of microbes (David et al., 2006).

Rendering procedures and technology have evolved over time and are always getting better. The rendering process largely comprises heat application, moisture extraction, and fat separation. The steps in the rendering operation involve raw material handling, sizing, heat treatment, pressing, protein and fat separation, grinding, and storage (Hamilton, 2004). The handling of raw materials is segregated from the processing and storage sections in modern rendering plants. Computer technology is used for process control and monitoring so that time/temperature measurements for the right thermal kill values for microorganisms are made. Temperatures that are far higher than what is necessary to reach thermal kill are unnecessary and should be avoided since they can reduce nutritional value and digestibility. Upon heating, the maximum amount of fat is removed by draining the excess water, where a screw press inside the closed container separates the remaining fat from the boiled raw materials. After heating and fat separation, the "cracklings" or "crax," which contain protein, minerals, and some leftover fat, undergo additional moisture removal and grinding before being moved for storage or transportation. Either feed bin structures or covered buildings are used to store the protein, whereas the fat is kept and delivered in tanks (David et al., 2006).

There is an increase in global demand for meals high in protein proposed to be used as feed in the fishery and pig industry. For example, it has been stated that the annual consumption of animal feed in Finland approximates 370 million kg, the majority of which is made from the byproducts of the meat and fish industries (Salminen and Rintala, 2002).

10.3.1 RENDERED PROTEIN PRODUCTS

All animal body tissues contain proteins, which are crucial components of all living creatures. Protein concentrations are higher in organ and muscle tissue. The main sources of protein and other nutrients in the diets of humans are meals with an animal origin. Similarly, various protein meals used in animal feeds are made from animal waste and processed tissues that are not used in human food (Figure 10.6). Following are some poultry rendered protein products which are used as animal feed.

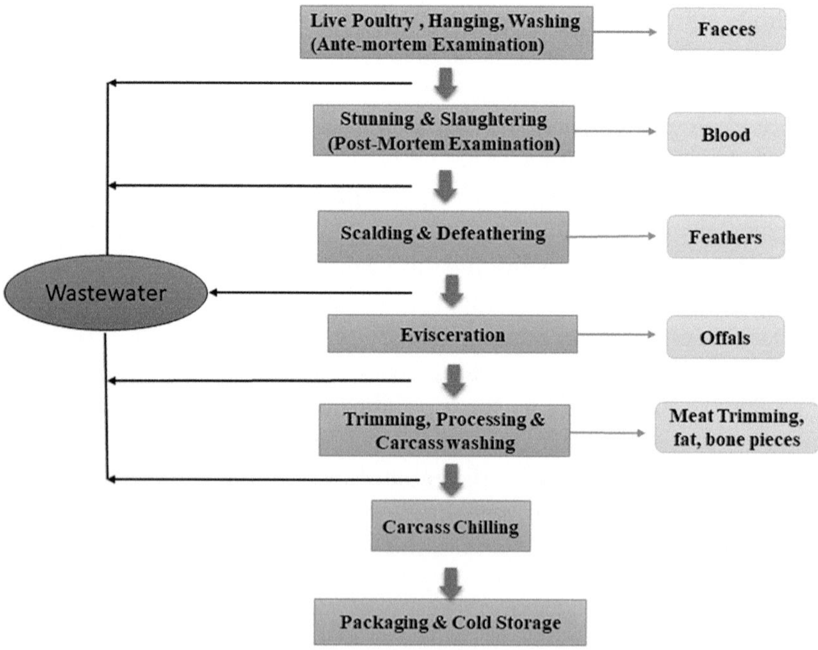

FIGURE 10.6 Poultry abattoir's waste and byproduct utilization.

Source: Adapted from Ozdemir and Yetilmezsoy (2019).

10.3.1.1 MEAT AND BONE MEAL (MBM)

The pepsin digestibility of food is normally considered for the prediction of amino acid yields. No more than 12% of the CP in MBM may be pepsin-indigestible residue, and no more than 9% of the CP may be pepsin-indigestible. An enzyme called pepsin, which hydrolyses proteins into polypeptides and oligopeptides, is secreted by the stomach. Animals might not be able to digest a protein if it is pepsin indigestible. All livestock, poultry, and aquaculture species may consume MBM, although ruminant feed may only contain non-ruminant source material (per FDA regulations).

10.3.1.2 POULTRY BYPRODUCT MEAL (PBM)

Poultry Byproduct Meal is made from ground, rendered, clean parts of slaughtered chicken carcasses, such as the necks, feet, unhatched eggs, and intestines, but not the feathers, unless they are present in amounts that are unavoidably present during good processing procedures. Solid organic raw materials have a strong capacity for digestion and contain 32% protein and 54% lipid (Salminen and Rintala, 2002). A great supply of balanced amino acids as well as mineral elements like phosphorus is provided by cooked poultry meal (Wang et al., 2006). Poultry meal has a high energy value due to its rich fat and protein content that includes crucial amino acids, necessary fatty acids, vitamins, and minerals, and relatively low ash level, thereby making it excellent for a variety of poultry-derived food, feed, and intermediate energy products (David et al., 2006).

10.3.1.3 POULTRY FEATHER MEAL (FEM)

Poultry Feather Meal is made from clean, undecomposed feathers from slaughtered poultry that have not been added with any chemicals or accelerators. By using the pepsin digestibility method, at least 75% of the CP content must be digestible. The protein in the feathers is partially hydrolyzed, and the keratinaceous linkages that give feather fibers their distinctive structure, are broken by modern processing techniques that roast the feathers under pressure with live steam. The resulting feather meal is a freely flowing, delectable product that all types of animals may digest with ease (Brandelli et al., 2015).

10.3.1.4 BLOOD MEAL

Blood Meal is made from clean, fresh animal blood, free of unwanted substances like hair, stomach gurgles, and urine, except for those that may inevitably arise through good manufacturing procedures. By condensing during heating to a semi-solid condition or by mechanically dewatering, a significant amount of moisture (water) is often eliminated. The more securely bonded water is then quickly removed from the semi-solid blood mass before it is transported to a rapid drying plant. The biological activity of lysine must be at least 80%. The amino acid isoleucine is naturally low in blood meal, and the vat-drying techniques utilized at the time to treat raw blood were severe enough to reduce lysine's bioavailability. Blood meals made using more recent processing techniques (ring or flash-drying) have amino acid digestibility of at least 90% (David et al., 2006).

10.4 RENDERED FAT

Most of the poultry fat is recovered from poultry tissue during the commercial rendering or extraction process. This finished product serves as an energy source for animal and aquaculture feed. Rendering facilities that use waste from poultry create fat, which may be edible or inedible, with low cost and so may be used as blending agents in fuel, non-livestock animal meal, and chemical sector goods. Soft solid waste from poultry (like heads, bone, feet, trims, and intestines) are both crushed and processed distinctly, or are rendered with feathers (Farmanesh et al., 2019). It has been reported that the proportion of fat that can be extracted from poultry wastes ranges from 2–30%, depending on the raw material and the rendering techniques used (Jayathilakan et al., 2012). Because rendered fat typically costs less than vegetable oils, it is a major component of feed (Lewis et al., 2019). Fat is added to aquafeed to increase the amount of metabolizable energy. Notably, energy-rich meal encourages growth and enhances feed conversion ratio.

In fat storage, moisture hastens the deterioration of equipment used to handle them and may encourage the growth of rust, a potent catalyst for oxidation and rancidity, thereby making sampling challenging. The FFA (free fatty acids) content is one way to gauge the quality of a fat. Three fatty acids joined to glycerol by ester bonds typically make up fats. When those fatty acids are released through hydrolysis, FFA is created. High quantities of FFA therefore show that the fat was exposed to water, acids, and/or enzymes. To avoid hydrolysis during storage, fats should be treated to have the lowest possible moisture content as it also does not offer energy, lubricity, or other benefits to feed.

10.5 AQUACULTURE FEED

The fish species raised for aquaculture have been fed high-protein, high-energy feeds. As rendered products have a high amount of protein, they can be employed as a significant source for aquaculture (Al-Souti et al., 2019). However, the nutritional value, digestibility, and amino acid profile vary depending on the rendering method utilized as well as the type and fineness of the raw materials used to make the meal (Lewis et al., 2019). In order to help the aquaculture industry, which heavily relies on high-quality fish meal, and to maintain a sustainable, ecologically friendly business model, poultry byproduct meal may be a highly good solution. With a typical use of 5–10%, poultry meal can be integrated into animal feed due to its well-balanced amino acid profile. For fish production, protein digestibility for chicken byproduct meal is found to be more than 80% (Tacon et al., 2009). When specified proportions of poultry meal or feather meal are replaced with fish meal, it has been shown that dietary supplementation of 5 to 25% has a positive impact on the growth and feed utilization ratio of fish species (Lewis et al., 2019). The poultry byproducts (raw material) are processed via a conditioner to facilitate curing, and then they are sent to an extruder to convert them into pellets, which are then graded, packed, and stored or sent to the facilities to be used as aquaculture feed (Figure 10.7).

In one study, the effects of substituting HF for fish meal on the muscle fatty acid composition and growth performance of European seabass

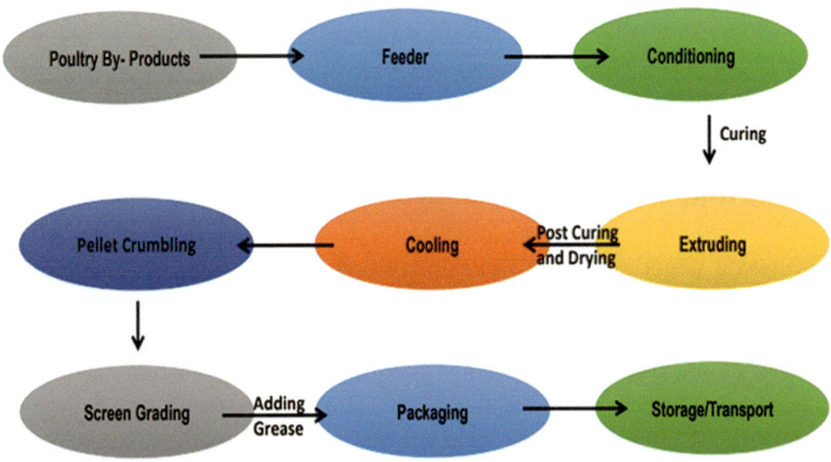

FIGURE 10.7 Methodology to produce aquaculture feed.

(*Dicentrarchus labrax*) were examined. According to the findings of Campos et al. (2017), European seabass fed 76% fishmeal substituted with 12.5% HF showed no evidence of impaired growth, immune response, or EPA and DHA levels in the muscle.

10.6 LARVAE REARING

The rearing of detritivore insect larvae is an interesting emerging technology which can be of much need for the poultry processing industries. The insect larvae, fed on highly nutritive poultry abattoir wastewater sludge, are potent sources to be utilized as animal/aqua feed alternatives and bio-fertilizers (Figure 10.6). The study showed that *Musca domestica,* house fly larvae when fed on poultry waste contained high protein (63.1%) and fat (15.5%) with well-balanced and better amino acid profile than soyabean meal could be used as chicken feed (Ozdemir and Yetilmezsoy, 2019).

Another similar study on maggots, which are produced when house flies biodegrade chicken droppings, found that using them as a feed supplement improved broiler performance. Broiler chicken growth performance and carcass quality were both enhanced by the maggots' high protein content (63.99%), essential amino acid profile (29.46%), and high protein digestibility (98.50%) (Hwangbo et al., 2009). According to estimates, a 1200 m^2 poultry shed that generates about 113 tons of manure per year could produce about 65 tons of bio-fertilizer and 47 tons of insect larvae to produce protein if it were used for insect rearing (Allegretti et al., 2018).

10.7 SLAUGHTERHOUSE WASTE AS RENEWABLE BIOMASS FUEL SOURCE

The waste generated from the poultry has proved to be an exceptional source of renewable biomass fuel (Xu et al., 2018). The frequently used technologies for the conversion of poultry waste to biomass fuel are AD and direct combustion. This is also helpful in reducing carbon emissions, it is cost efficient and environmental and health friendly like it has no or minimum odor with less disease transmission as well (Ozdemir and Sezer, 2013).

At various volatile solids concentrations, poultry abattoir wastes and their feed, which are heavy in proteins and lipids, have a tremendous potential to create large amounts of methane (Sakar et al., 2009, Yamashiro et al., 2013). Longer retention time of slaughterhouse wastes such as 50 days (Lalit et al., 2019), are required for the biochemical methane reaction chains to complete. This is most likely because long-chain fatty acids hinder methanogen processes. Moreover, the duration is influenced by the source, different inoculum concentrations, and incubation temperature (Thyagarajan et al., 2013).

One of the newest technologies, flash pyrolysis produces solid, liquid, and gaseous fuel that is appropriate for storage and transportation while operating in the absence of air. The main benefits of this method are the quick AD process and energy production from waste resources at lower operating temperatures, typically between 250 and 500°C as opposed to 800 to 900°C for gasification (Sari et al., 2016). Figure 10.1 depicts the stages involved in producing biofuel from poultry slaughterhouse waste.

Up to 12% of the fat in slaughterhouse wastes including feathers, blood, and viscera is inedible. Biodiesel is made from wastes with a high oil-grease content using environmentally acceptable procedures. In order to make biodiesel, mostly fat from poultry slaughterhouse wastes is extracted in water that is boiling (70°C) and then trans esterified with nitrogen, potassium, and methane to produce 7–11% biodiesel (Emiroglu et al., 2018). The biodiesel created from poultry waste had superior physico-chemical and energetic fuel characteristics in comparison to biodiesel produced from classic crop-based feedstocks.

10.8 CONCLUSION

In India, the poultry production among all livestock is the largest; subsequently, the waste produced from poultry would also be the largest. Therefore, a huge amount of waste is produced from poultry that needs to be managed effectively. The waste produced by poultry slaughterhouses is enormous and needs to be efficiently managed. The natural decomposing mechanism of these organic wastes affects both the ambience and quality of life, posing a majority of health risks. One of the promising alternatives is food waste valorization or rendering, which involves increasing its commercial value. Prior to legal limits, solid wastes from poultry slaughterhouses were processed by conventional rendering, which produces useful byproducts and hygienic feed components for ruminants and birds alike. The need for feed and food for non-livestock still outweighs other commercial materials due to their specific nutritional value obtained from protein and fat-rich poultry wastes. Therefore, rendering is the utmost preferred process for poultry wastes that reduces the environmental and health effects, satisfies the regulation and cost, and produces a range of valorized products. Also, the waste generated from the poultry has proved to be an outstanding source of renewable biomass fuel.

KEYWORDS

- **aqua feed**
- **byproducts**
- **hatchery waste**
- **poultry waste**
- **rendering**
- **renewable biofuel**

REFERENCES

Allegretti, G., Talamini, E., Schmidt, V., Bogorni, P. C., & Ortega, E. (2018). Insects as feed: An energy assessment of insect meal as a sustainable protein source for the Brazilian poultry industry. *Journal of Cleaner Production, 171*, 403–412.

Al-Souti, A., Gallardo, W., Claereboudt, M., & Mahgoub, O. (2019). Attractability and palatability of formulated diets incorporated with chicken feather and algal meals for juvenile gilthead seabream, *Sparus aurata*. *Aquaculture Reports, 14*, 100199.

Barbut, S. (2015). *The Science of Poultry and Meat Processing.*

Beck, S. E., Suwan, P., Rathnayeke, T., Nguyen, T. M. H., Huanambal-Sovero, V. A., Boonyapalanant, B., & Koottatep, T. (2021). Woven-fiber microfiltration (WFMF) and ultraviolet light emitting diodes (UV LEDs) for treating wastewater and septic tank effluent. *Water, 13*(11), 1564.

Brandelli, A., Sala, L., & Kalil, S. J. (2015). Microbial enzymes for bioconversion of poultry waste into added-value products. *Food Research International, 73*, 3–12.

Cambardella, C. A., Richard, T. L., & Russell, A. (2003). Compost mineralization in soil as a function of composting process conditions. *European Journal of Soil Biology, 39*(3), 117–127.

Campos, I., Matos, E., Marques, A., & Valente, L. M. (2017). Hydrolyzed feather meal as a partial fishmeal replacement in diets for European seabass (*Dicentrarchus labrax*) juveniles. *Aquaculture, 476*, 152–159.

Cawthon, D. (1998). *In-Vessel Composting of Un-Separated Hatchery Waste*. Department of Agricultural Sciences, Texas A&M University-Commerce: Commerce, TX, USA.

Das, K. C., Minkara, M. Y., Melear, N. D., & Tollner, E. W. (2002). Effect of poultry litter amendment on hatchery waste composting. *Journal of Applied Poultry Research, 11*(3), 282–290.

Del Nery, V., De Nardi, I. R., Damianovic, M. H. R. Z., Pozzi, E., Amorim, A. K. B., & Zaiat, M. (2007). Long-term operating performance of a poultry slaughterhouse wastewater treatment plant. *Resources, Conservation and Recycling, 50*(1), 102–114.

Emiroğlu, A. O., Keskin, A., & Şen, M. (2018). Experimental investigation of the effects of turkey rendering fat biodiesel on combustion, performance and exhaust emissions of a diesel engine. *Fuel, 216*, 266–273.

Farmanesh, A., Mohtasebi, S. S., & Omid, M. (2019). Optimization of rendering process of poultry byproducts with batch cooker model monitored by electronic nose. *Journal of Environmental Management, 235*, 194–201.

Ferreira, A., Kunh, S. S., Cremonez, P. A., Dieter, J., Teleken, J. G., Sampaio, S. C., & Kunh, P. D. (2018). Brazilian poultry activity waste: Destinations and energetic potential. *Renewable and Sustainable Energy Reviews, 81*, 3081–3089.

Franke-Whittle, I. H., & Insam, H. (2013). Treatment alternatives of slaughterhouse wastes, and their effect on the inactivation of different pathogens: A review. *Critical Reviews in Microbiology, 39*(2), 139–151.

Gutu, L., Basitere, M., Harding, T., Ikumi, D., Njoya, M., & Gaszynski, C. (2021). Multi-integrated systems for treatment of abattoir wastewater: A review. *Water, 13*(18), 2462.

Hamilton, C. R. (2004). Real and perceived issues involving animal proteins. In *Protein Sources for the Animal Feed Industry: FAO Expert Consultation and Workshop*, Bangkok, Thailand, 29 April–3 May 2002 (pp. 255–276). Food and Agriculture Organization of the United Nations (FAO).

Hamm, D., & Whitehead, W. K. (1982). Holding techniques for hatchery wastes. *Poultry Science, 61*(6), 1025–1028.

Hwangbo, J., Hong, E. C., Jang, A., Kang, H. K., Oh, J. S., Kim, B. W., & Park, B. S. (2009). Utilization of house fly-maggots, a feed supplement in the production of broiler chickens. *Journal of Environmental Biology, 30*(4).

Imbeah, M. (1998). Composting piggery waste: A review. *Bioresource Technology, 63*(3), 197–203.

Jayathilakan, K., Sultana, K., Radhakrishna, K., & Bawa, A. S. (2012). Utilization of byproducts and waste materials from meat, poultry and fish processing industries: A review. *Journal of Food Science and Technology, 49*(3), 278–293.

Kamalak, A., Canbolat, O., Gurbuz, Y., & Ozay, O. (2005). *In situ* ruminal dry matter and crude protein degradability of plant- and animal-derived protein sources in Southern Turkey. *Small Ruminant Research, 58*(2), 135–141.

Kantarli, I. C., Kabadayi, A., Ucar, S., & Yanik, J. (2016). Conversion of poultry wastes into energy feedstocks. *Waste Management, 56*, 530–539.

Kazemi-Bonchenari, M., Alizadeh, A., Javadi, L., Zohrevand, M., Odongo, N. E., & Salem, A. Z. (2017). Use of poultry pre-cooked slaughterhouse waste as ruminant feed to prevent environmental pollution. *Journal of Cleaner Production, 145*, 151–156.

Kennedy, L. D. (1986). Solid and liquid waste incinerator system. *U.S. Patent No. 4,619,210*. http://www.freepatentsonline.com/4619210.html (accessed on 1 January 2023).

Kese, A. G., & Donkoh, A. (1982). Evaluation of methods of processing dried poultry waste in terms of performance and carcass quality of broiler chickens. *Poultry Science, 61*(12), 2500–2502.

Kim, W. K., & Patterson, P. H. (2000). Recycling dead hens by enzyme or sodium hydroxide pretreatment and fermentation. *Poultry Science, 79*(6), 879–885.

Kirkpinar, F., Açikgöz, Z., Bozkurt, M., & Ayhan, V. (2004). Effects of inclusion of poultry byproduct meal and enzyme-prebiotic supplementation in grower diets on performance and feed digestibility of broilers. *British Poultry Science, 45*(2), 273–279.

Kompiang, S. (1994). Ensiling of hatchery waste using a mixture of formic and propionic acids. In *Proceedings of the 7th AAAP Animal Science Congress: Sustainable Animal Production and the Environment* (Vol. 2, pp. 179–180).

Latifi, P., Karrabi, M., & Danesh, S. (2019). Anaerobic co-digestion of poultry slaughterhouse wastes with sewage sludge in batch-mode bioreactors: Effect of inoculum-substrate ratio and total solids. *Renewable and Sustainable Energy Reviews, 107*, 288–296.

Lewis, M. J., Francis, D. S., Blyth, D., Moyano, F. J., Smullen, R. P., Turchini, G. M., & Booth, M. A. (2019). A comparison of *in-vivo* and *in-vitro* methods for assessing the digestibility of poultry byproduct meals using barramundi (*Lates calcarifer*): Impacts of cooking temperature and raw material freshness. *Aquaculture, 498*, 187–200.

Lin, C. Y., Sato, K., Noike, T., & Matsumoto, J. (1986). Methanogenic digestion using mixed substrate of acetic, propionic and butyric acids. *Water Research, 20*(3), 385–394.

Márquez, E., Bracho, M., Archile, A., Rangel, L., & Benítez, B. (2005). Proteins, isoleucine, lysine and methionine content of bovine, porcine and poultry blood and their fractions. *Food Chemistry, 93*(3), 503–505.

Meeker, D. L., & Hamilton, C. R. (2006). An overview of the rendering industry. In *Essential Rendering* (pp. 1–16).

Mézes, L., Nagy, A., Gálya, B., & János, T. (2015). Poultry feather wastes recycling possibility as soil nutrient. *Eurasian Journal of Soil Science, 4*(4), 244–252.

Miller, B. F. (1984). Extruding hatchery waste. *Poultry Science, 63*(6), 1284–1286.

Ozdemir, S., & Yetilmezsoy, K. (2019). A mini literature review on sustainable management of poultry abattoir wastes. *Journal of Material Cycles and Waste Management*, 1–11.

Ozdemir, S., Ozdemir, S., & Yetilmezsoy, K. (2019). Agro-economic and ecological assessment of poultry abattoir sludge as bio-nutrient source for walnut plantation in low-fertility soil. *Environmental Progress & Sustainable Energy, 38*(6), 13225.

Papuc, C., Goran, G. V., Predescu, C. N., & Nicorescu, V. (2017). Mechanisms of oxidative processes in meat and toxicity induced by postprandial degradation products: A review. *Comprehensive Reviews in Food Science and Food Safety, 16*(1), 96–123.

Sakar, S., Yetilmezsoy, K., & Kocak, E. (2009). Anaerobic digestion technology in poultry and livestock waste treatment: A literature review. *Waste Management & Research, 27*(1), 3–18.

Salminen, E., & Rintala, J. (2002). Anaerobic digestion of organic solid poultry slaughterhouse waste: A review. *Bioresource Technology, 83*(1), 13–26.

Sari, O. F., Ozdemir, S., & Celebi, A. (2016). Utilization and management of poultry slaughterhouse wastes with new methods. In *Eurasia 2016 Waste Management Symposium* (pp. 02–04).

Schilling, M. R., & Mintz, M. F. (1998). *U.S. Patent No. 5,776,567*. Washington, DC: U.S. Patent and Trademark Office.

Sharara, H. H., El-Hammady, H. Y., & Abdel-Fattah, H. (1993). The nutritive value of some non-conventional byproducts as poultry feed ingredients. III. Apparent and true availability of amino acids. *Assiut Journal of Agricultural Sciences (Egypt), 24*, 311–323.

Sharara, H. H., El-Hammady, H. Y., & El-Fattah, H. A. (1992a). Nutritive value of some non-conventional byproducts as poultry feed ingredients. 1. Chemical composition. *Assiut Journal of Agricultural Sciences (Egypt), 23*, 333–349.

Sharara, H. H., El-Hammady, H. Y., & El-Fattah, H. A. (1992b). Nutritive value of some non-conventional byproducts as poultry feed ingredients. 2. Biological determination of metabolizable energy. *Assiut Journal of Agricultural Sciences (Egypt), 23*, 349–363.

Swan, J. E. (2000). Animal byproduct processing. *Encyclopedia of Food Science and Technology, 2*, 35–43.

Tacon, A. G., Metian, M., & Hasan, M. R. (2009). *Feed Ingredients and Fertilizers for Farmed Aquatic Animals: Sources and Composition* (No. 540). Food and Agriculture Organization of the United Nations (FAO).

Tesfaye, T., Sithole, B., & Ramjugernath, D. (2018). Valorization of chicken feather barbs: Utilization in yarn production and technical textile applications. *Sustainable Chemistry and Pharmacy, 8*, 38–49.

Thyagarajan, D., Barathi, M., & Sakthivadivu, R. (2013). Scope of poultry waste utilization. *IOSR Journal of Agriculture and Veterinary Science, 6*(5), 29–35.

Van Slyke, J. V., Van Kempen, M. H., & Van Kempen, T. A. (2007). *U.S. Patent No. 7,232,036*. Washington, DC: U.S. Patent and Trademark Office.

Van Slyke, V., Timmenga, H., Helle, S., Watkinson, P., & Bi, X. (2005). *U.S. Patent No. 6,916,426*. Washington, DC: U.S. Patent and Trademark Office.

Vandepopuliere, J. M., Kanungo, H. K., Walton, H. V., & Cotterill, O. J. (1977). Broiler and egg type chick hatchery byproduct meal evaluated as laying hen feedstuffs. *Poultry Science, 56*(4), 1140–1144.

Verma, A. K., & Rao, P. V. (1974). Hatchery byproduct meal—a source of protein. *Indian Poultry Gazette, 59*, 138–139.

Wang, Y., Guo, J. L., Bureau, D. P., & Cui, Z. H. (2006). Replacement of fish meal by rendered animal protein ingredients in feeds for cuneate drum (*Nibea miichthioides*). *Aquaculture, 252*(2–4), 476–483.

Xu, X., Ma, Z., Chen, Y., Gu, X., Liu, Q., Wang, Y., & Chang, D. (2018). Circular economy pattern of livestock manure management in Longyou, China. *Journal of Material Cycles and Waste Management, 20*(2), 1050–1062.

Yamashiro, T., Lateef, S. A., Ying, C., Beneragama, N., Lukic, M., Masahiro, I., Ihara, I., Nishida, T., & Umetsu, K. (2013). Anaerobic co-digestion of dairy cow manure and high concentrated food processing waste. *Journal of Material Cycles and Waste Management, 15*(4), 539–547.

Yetilmezsoy, K., Turkdogan-Aydinol, I., Gunay, A., & Ozis, I. (2011). Post-treatment of poultry slaughterhouse wastewater and appraisal of the economic outcome. *Environmental Engineering & Management Journal, 10*(11).

Yoon, Y. M., Kim, S. H., Oh, S. Y., & Kim, C. H. (2014). Potential of anaerobic digestion for material recovery and energy production in waste biomass from a poultry slaughterhouse. *Waste Management, 34*(1), 204–209.

CHAPTER 11

Coffee: Wastes, Byproducts, and Utilization

PIR MOHAMMAD JUNAID,[1,*] M. KAMALISH,[2] P. SHRUTHI,[3] and SADAF ZAIDI[1]

[1]*Department of Post-Harvest Engineering and Technology, Aligarh Muslim University, Aligarh, Uttar Pradesh, India*

[2]*Department of Clinical Nutrition, Sri Ramachandra Institute of Higher Education and Research, Chennai, Tamil Nadu, India*

[3]*Department of Community Science, College of Agriculture, Vellayani, Kerala Agricultural University, Thiruvananthapuram, Kerala, India*

*Corresponding author

ABSTRACT

Coffee, a valuable product, is widely cultivated, traded, and consumed across the globe and generates a huge number of byproducts from the farm level to the consumption end. The coffee cherry husks, coffee pulp, silver skin, spent grounds, and pulp are the main residues of coffee production. The mismanagement of such waste/residues of coffee results decisively to the environment and biodiversity. Mostly, coffee residues are rich in valuable phytochemical compounds and can be utilized in food, pharmaceutical, and biochemical areas. The phytochemicals in coffee byproducts can help in novel functional food formulations. The biofertilizer and biodiesel production from the coffee husk and pulp are diversified sources to meet the upcoming challenge for a healthy and safe environment. This chapter will focus on the types of coffee waste and their application in different areas of life.

Utilization of Food Waste and Byproducts. Mudasir Ahmad Malik and Kshirod Kumar Dash (Eds.)

DOI: 10.1201/9781779641151-11

11.1 INTRODUCTION

Coffee is one of the most valued agricultural products with strong commerce and cultural affiliations. It is the typical aromatic, stimulating, and flavoring global commodity from traditional times. The seeds are obtained from the green coffee beans, roasted, and formulated into the powder. The various drinks are then prepared from these powdered coffee seeds. The use of coffee dates back to old times and is believed to have originated from the Arabian subcontinent (AL-Asmari et al., 2020). Nowadays, coffee production and consumption have gained worldwide attention due to its organoleptic characteristics. Generally, the two varieties of *C. arabica* and *C. robusta* of coffee are grown widely in over 70 countries, mainly in America, Africa, India, and Southeast Asia (Volk and Byrne, 2020). It is believed that a 1000 acres of land in Brazil alone are used for the mass cultivation of coffee. Therefore, Brazil is one of the leading countries, producing 35% of the total world's coffee production (Ibrahim and Zailani, 2010). The commercial utility of coffee has made it the leading cash crop in many countries of Africa as well as in some parts of Asia (Achterbosch et al., 2014). Around 175,647,000 60 kg bags of coffee were produced in 2020, with 39% production only from Brazil (Djufry and Wulandari, 2021). The South Indian states of Karnataka, Kerala, and Tamil Nadu produce the majority of India's 8,200 tons of coffee (Amaravathi, 2014). Karnataka accounts for 53% of the coffee production in India, which is followed by 20% of production from Kerala and 11% from Tamil Nadu (Ayyasamy and Regupathy, 2014). Despite the billion-dollar tradable commodity, ecological and environmental concerns are always the topic of discussion. One square inch of land is required to produce one cup of coffee (Rueda and Lambin, 2013). The usage of land habitat and incorporation of chemicals for coffee production ultimately disrupts the ecological balance. The shrinkage of rainforest land mass for the cultivation of coffee and other crops is prominently affecting the environment (Smith et al., 2018; Somarriba and López, 2018). The environmental threats are not only in respect to land shrinkage or land erosion but also the enormous amount of waste production. Coffee, which produces an alarming amount of waste, is simply discarded during the processing methods (Cerino-Córdova et al., 2020; Saldaña et al., 2022). Gunter Pauli, a sustainability expert, estimates that the world's coffee business generates more than 23 million tons of waste each year (Passadis et al., 2020). The separation of coffee from the surrounding pulp and the process of fermentation results in the production of waste in various forms (Campos et al., 2021). Bottausci et al. (2022) suggests an estimated 1.1 million tons of pulp production during six months of time in central America only. The production of the waste directly or indirectly accounts for the massive water and soil pollution (Kim et al., 2014).

The coffee consists of the outer layers and the bean, which is soaked and fermented to weaken the bean coating and make it easier to remove the inner bean, leaving a slimy, pulpy residue behind (Morris, 2018). The remaining residual material gets dumped into the nearby water bodies, where it gets deposited and decomposed steadily. The outcome of such deposition leads to decreased oxygen within the water system and proves to be a viable threat to aquatic life (Tekleet al., 2015). The ineffectiveness of the government bodies towards the proper utilization of waste results in climatic problems. The disturbance and disruption in the aquatic life are predominantly affected by the mass disposal of waste from the coffee processing.

11.2 COFFEE PROCESSING

Coffee processing is the biggest challenge as it is accompanied by different process operations. Coffee can undergo processing through one of three fundamental methods, namely dry, wet, or semi-dry processing, depending on the region and the resources accessible (Figure 11.1) (Rodriguez et al., 2020). All methods attempt to remove the coffee cherry's fruit flesh, but they each go about it in a somewhat different way.

FIGURE 11.1 Coffee processing.

The plantation, harvesting till the consumption of coffee, and the different processing steps are required to end up with a perfect drink (Haile and Kang, 2019). The shaded nurseries are required to plant the coffee seeds in the bed layout (Megersa, 2022). The growth of seedlings requires care and protection until they are transferred permanently for plantation. A new plant takes up to 4 years to grow as a fully matured fruit-bearing plant. The fruit (coffee cherry) is harvested according to its ripening stage, which is to be detected by the change to deep red (Hu et al., 2020). The processing of coffee fruit should be done as quickly as possible to avoid the spoilage of the coffee bean (Vincent, 1987). The method of processing, such as the wet method or dry method, depends upon the location and the resources available (Gemechu, 2020). The traditional dry method employs sunlight for drying. This is a lengthy process and will continue for a few weeks till the moisture content is reduced to nearly 11%. The wet method is a complex processing method that involves the removal of pulp from the coffee cherry. The beans are then separated on a weight basis in a running channel to get the lighter and heavier beans. Also, the size-based separation is carried out by passing the beans through the rotating drums. After separation, the beans are subjected to fermentation in their tanks for 12–48 hrs to remove the mucilage (Krajangsang et al., 2022). The beans, after fermentation, are rough in texture and are rinsed with water, followed by the drying process. The dried beans are subjected to hulling and polishing process after proper drying. The beans are graded and sorted based upon the size represented on the scale of 10 to 20 hole diameter of 1/64 of an inch to remove imperfect beans (Umroh et al., 2022). After the coffee beans are milled, they are referred to as green coffee and can be transported either in plastic-lined containers or in jute or sisal bags loaded into shipping containers on board ships. The coffee is then checked for its overall vision as well as the other sensory properties by the specialist called a cupper (Nadaleti et al., 2022). The process of roasting is then followed to intensify the bean aroma, and the color is changed from green to brown (Grzelczyk et al., 2022). This process facilitates the leakage of caffeoyl from the bean and accounts for the flavor of the coffee (Rojas-González et al., 2022). The coffee beans are ground to get an acceptable amount of flavor; an ideal grind is determined by the contact time of the grounds with water (Angeloni et al., 2023; Rao et al., 2023). The coffee is brewed by pouring hot water on the coffee ground by different methods (drip-brewed coffee, pour-over coffee, filtered coffee, immersion-brewed coffee) to get the brewed coffee drink (Chen and Fang, 2023).

11.3 COFFEE WASTE

The coffee cherry usually comprises two coffee beans, positioned with their flat sides adjacent to each other. The silver skin refers to the tight-fitting tegument that covers each bean. Each bean is loosely covered by the parchment, with a second yellowish skin. The mucilaginous pulp envelops the parchment-coated coffee beans, with the cherry flesh situated in between and covered by the outer skin or peel (Aryal, 2023). Green coffee beans only make up 50–55% of the dry matter of ripe cherries. Depending on how it is processed, the leftover material is diverted to different forms of garbage. The rise in global coffee production and consumption is leading to a corresponding increase in the generation of waste. These coffee wastes are generated as its byproduct form in respect of pulp, skin, silverskin, mucilage, and the rejected coffee beans (Montoya et al., 2019; Pattanayak, 2023). During the coffee preparation process, around 55–60% of coffee beans and 40–45% of coffee bean skins are produced. These skins comprise of mesocarp (pulp), exocarp (skin), mucilage, and endocarp (parchment) (Durán-Aranguren et al., 2021). Different byproducts are generated at each stage of industrial coffee processing methods, such as wet or dry processing and brewing. The amateur or defective coffee beans alone account for 50% of rejection of the total production (Subba, 2020). Therefore, coffee byproducts and coffee wastes are a significant source of pollution and an adverse environment across all coffee-producing nations. The aim of the postharvest processing of coffee cherry is to separate the seed from the remaining components and ensure good product processing and preservation. These operations ensure the quality and quantity effectiveness of coffee production. The processing of coffee needs the separation of five outer protective coverings of the bean (Aryaet al., 2021). The outermost covering of the coffee fruit is known as the exocarp, which is sometimes referred to as the peel, skin, or epicarp (Ferreira et al., 2019). It is the monolayer waxy covering having different colors depending upon the variety of coffee. This skin is formed from compact parenchyma cells having thin walls containing chloroplast. In the amateur stage, the color is of green tint and changes to red or yellow upon maturation.

In the dry method of processing coffee cherry, the prime byproduct is coffee cherry husk comprising of pulp, parchment, and dried skin, which accounts for 12% of the coffee cherry (Haile and Kang, 2019). It is estimated that from one metric ton of coffee fruit, approximately 0.18 metric tons of husk are obtained (Salazaret al., 2020). The flesh of the coffee berry is known as the mesocarp, sometimes known as the mucilage. While the term "pulp" might sometimes only refer to the mesocarp, it usually refers to a mixture of the exocarp and a portion of the mesocarp that is removed during pulping. In the case of a semi-drying process or wet process, the coffee pulp is the foremost byproduct generated and comprises 29% weight of coffee fruit (Sangta et al., 2021). The ratio of coffee production and pulp generation is 2:1, respectively means for 2 kg of coffee, 1 kg of the pulp is produced (Serna et al., 2022). The pulp obtained during coffee processing is directly disposed of in water streams or deposited as such. This enormous amount of coffee pulp waste accounts for water and air pollution (odor pollution). The seed's topmost layer is referred to as the silverskin, also known as the perisperm or spermoderm. It forms an outer layer around the coffee bean representing 4.2% of the total weight of coffee fruit. The coffee silverskin (CS), also known as coffee chaff, is an industrial processing reside produced during the roasting process. Although the silverskin of the bean is polished off, it is generally agreed that this reduces the flavor of the coffee (Aryal, 2023). Fine particle waste produced after brewing roasted coffee grounds is known as spent coffee. According to data, for 1 kg of brewed coffee about 0.65 kg of dried spent coffee or 2 kg of wet spent coffee are produced (Carta et al., 2022). This huge amount of waste produced at each stage of coffee processing results in a serious environmental threat. Wet and semi-wet processing has a considerable impact on the environment. Problems occur when massive amounts of wastewater are thrown into watercourses that are overflowing with organic materials, not because of its inherent toxicity (Bisekwa, 2022). The higher aquatic life is put in danger if the watercourse's capacity for self-purification is exceeded because microbiological deterioration causes the oxygen concentration to drop to anaerobic levels. In the modern or the traditional approach of mucilage removal, wastewater generation problems are always a point of concern. The water usage must be employed as per the capacity and affordably of the treatment procedure. The principal cause of pollution is the release of organic material during pulping, which is accompanied by the quick fermenting of carbohydrates and proteins. Also, the biochemical modifications of these organic materials result in froth development on the water surfaces von (Endenet al., 2002). This in turn result in the more effluent generation and increased water acidity. Therefore, such contamination of water bodies, air pollution, and soil pollution, ultimately result in health and ecological problems (Saravanan et al., 2022).

11.4 TYPES OF COFFEE WASTE

According to sustainability researcher Gunter Pauli, the pulp from fresh coffee fruits contributes more than 23 million tons of garbage annually (Duncan Pike, 2018). Used coffee grounds, which are discarded after each fresh brew, are the most obvious example of this waste at the user end in the supply chain. The dry coffee beans, separated from the coffee fruit parts such as parchment, outer skin, and pulp, are traded and marketed globally. Coffee cherries are processed through various steps before becoming coffee beans. Coffee cherries are typically processed using one of two methods, the dry process and the wet process, which differ in their approach. The primary distinguishing factor between the two methods is the use of water in the removal of the coffee skin and subsequent fermentation (Ijanu et al., 2020). Therefore, the two different processing methods produce waste of different forms. Red coffee cherries are dried to a moisture level of between 10% and 12% during the dry coffee processing process, either in the sun or with a hot air machine (Silva, 2014). Initially, the coffee fruit is cleaned, dried, and the dry peel and pulp are removed. Wet processing, on the other hand, involves a series of complex procedures, including mechanical removal of the coffee skin and pulp, fermentation of the mucilage layer, sun drying, and hulling to eliminate the parchment skin (Haile and Kang, 2019). The steps of the dry and wet coffee processing procedures are combined in semi-dry (semi-wet) processing, which involves mechanically processing the coffee cherries and drying the beans without removing the mucilage (Farah and dos Santos, 2015). During such processing of coffee, enormous quantities of byproducts are produced. In nations where this crop is cultivated and processed, these byproducts have adverse environmental consequences. Depending on the method employed for the process of coffee, several byproducts, that include pulp, husk, silverskin, and used coffee grinds, are created.

11.4.1 HUSK

The solid byproducts formed from dehulling coffee cherries during the dry method are referred to as coffee husks. The main byproduct of dry coffee cherry processing is a mixture of the outer skin, pulp, and parchment, which are all combined together in a single portion (Oliveir and

Franca, 2015). Coffee husks of around 0.18 tons are produced for every tons of harvested coffee fruit (Murthy and Naidu, 2012). The coffee husk is the major processing byproduct that has no marketable uses and whose disposal poses a serious environmental issue. Most of the Arabica and Robusta coffee varieties harvested in Brazil undergo dry processing. During handling, after harvesting coffee cherries, the moisture is reduced to 10–11% by drying. A dehulling machine is used to separate the coffee beans from their covering, which includes the outer skin, pulp, and parchment. The coffee beans are usually dried with the natural sun-drying process on farms (Siagian et al., 2017). The beans are spread evenly throughout the large surface to facilitate uniform drying and avoid fermentation (Murthy and Naidu, 2012). It is a very slow process requiring the time of around 3–4 weeks. In contrast to wet processing, there is no drying requirement for this method. The mechanical removal of the outer skin and pulp results in the production of coffee pulp, which comprises both the skin and pulp in a single fraction. This pulp accounts for 40% to 50% of the fresh weight of coffee berries. Fermentation is employed for the removal of further pulp substance adhering to it to get the pulped coffee (Girma and Sualeh, 2022). The beans are then dried to obtain the de-hulled coffee. In both of these processes, the beans undergo dehulling to eliminate the parchment. Approximately 0.18 metric tons of husk are generated from every metric ton of coffee fruit.

11.4.2 COFFEE PULP

Around 35% of the coffee fruit is made up of coffee pulp, which is a byproduct of the wet method of coffee hulling. A significant amount of coffee pulp is produced during coffee processing; typically, about 1 ton of pulp is obtained from every 2 tons of coffee produced. Compared to coffee husk, the amount of phenolic acids in coffee pulp is slightly higher (Pandey et al., 2000). The worldwide production of coffee pulp is approximately 9.4 million metric tons annually and is a primary source of environmental concerns in coffee-producing nations. The coffee pulp is responsible for the pollution of rivers and lakes close to coffee processing facilities, as well as agricultural lands. It is responsible for the decreased level of oxygen percentage in water bodies and blockage of sunlight due to suspended solid presence. Environmentalists are particularly concerned about the production of methane (CH_4) and nitrous oxide (N_2O) gases resulting from the fermentation of coffee pulp (Bisekwa, 2022). The huge collection of pulp causing unpleasing odors with a growing chance of infections is also a major concern for health. The chemical toxicity, such as the presence of tannins and caffeine in coffee husk and coffee pulp, makes it a threat to the environment (Skorupa et al., 2023). Moreover, the presence of tannin compounds in coffee pulp has an antinutritional effect that limits its utilization as animal feed beyond a certain threshold (10%).

11.4.3 COFFEE SILVERSKIN (CS)

CS is a thin layer that is separated from the outer layer of the coffee bean during the roasting process. It is the most abundant solid byproduct generated during roasting and is composed primarily of cellulose, hemicellulose, and lignin (Saldaña et al., 2022). CS is a significant byproduct resulting from the coffee roasting process, and if not disposed of adequately, it can lead to environmental pollution. CS contains high levels of phenolic compounds, which can have a negative environmental impact if released into the environment (Lorbeer et al., 2022). Additionally, CS can release volatile organic compounds into the air, which can contribute to air pollution (Skorupa et al., 2023). Proper disposal and recycling of CS is crucial to minimize its environmental impact. Due to its high cellulose, hemicellulose, and lignin content, CS can have an adverse impact on the environment. These components are not easily biodegradable, and when discarded in landfills, they can contribute to the emission of greenhouse gases. CS can also contribute to soil acidification, which can also have an adverse effect on the surrounding environment (Malara et al., 2018). Additionally, CS can cause water pollution if it is not disposed of properly.

11.4.4 SPENT COFFEE GROUNDS (SCFS)

Spent coffee grounds (SCFs) are the leftover coffee grounds that have been used for brewing coffee, usually in a filter or French press. They are typically dark, moist, and aromatic and contain a significant amount of caffeine (Sermyagina et al., 2021). This coffee brewing leftover is from a domestic source or from an industrial beverage. The coffee brew is prepared using a blend of coffee varieties such as arabica and robusta. Also, the coffee brew can be made from a single coffee of different geographical origins, whatever is available to the consumers, like whole or ground, roasted beans, or instant soluble coffee. The term "spent coffee grounds" can encompass both the grounds produced after brewing coffee at home or in cafes, as well as the grounds obtained from the soluble coffee industry (Kovalcik et al., 2018). For every gram of ground coffee, approximately 0.91 grams of SCFs are generated. Additionally, for every one kilogram of instant soluble coffee, around two kilograms of wet wasted coffee grounds are produced (Cho et al., 2022). Although used coffee grounds contain a number of extremely useful chemical components, they are typically viewed as waste and are either thrown away or composted. It is estimated that more than 15 million metric tons of used coffee grounds are generated each year (Cervera-Mata et al., 2021). The efficiency of extracting coffee from spent grounds depends on several factors, including the coffee type, the grind size, the coffee-water ratio, the water quality, the degree of roasting, the temperature, and the percolation time. Consequently, different extraction methods can yield distinct sensory and chemical characteristics in the brew, resulting in various types of SCG. In domestic/cafeteria extracted spent coffee after brewing, the spent is more chemically active than the one obtained from the industrial source. Industrial sources make sure the spent coffee is free from chemical constituents, and thus the brewing methods are effectively employed (Al-Hamamre et al., 2012). Usually, used coffee grounds are thrown in with other trash and taken to a landfill, where they decompose and release methane. The gas released from the deposit accounts for global warming, and it is said that methane is 25 times stronger than carbon dioxide. They also contain high levels of caffeine and phenolic compounds, both of which can have negative environmental impacts if released into the environment. Additionally, coffee spent grounds can release volatile organic compounds into the air, which can contribute to air pollution (Limousy et al., 2013). Proper disposal and recycling of coffee spent grounds is crucial to minimize their environmental impact.

11.4.5 COFFEE WASTEWATER

Wet and semi-wet processing methods used in coffee production can have notable environmental impacts. Despite not being intrinsically toxic, the processes can lead to issues such as the discharge of substantial amounts of wastewater into waterways that already contain organic matter. Coffee processing leaves behind wastewater, also referred to as coffee effluent (Chanakya et al., 2004). The presence of a high content of organic material that gets dumped in the water streams results in environmental toxicity. When the watercourse's natural capacity to cleanse itself is surpassed, the microbial decay process can cause a drop in oxygen levels to the point where it becomes anaerobic (Butnariu, 2022). Depending upon the process utilized for processing, the amount of wastewater produced also varies. In contrast to the conventional completely washed approach without recycling, modern mechanical mucilage removal machines require only roughly 1 m^3 per tons of fresh cherry for producing semi-washed coffee without finishing fermentation and washing (Blinová et al., 2017). The amounts of wastewater must be kept to a minimum in order to treat it properly and affordably. The organic matter released during pulping, when the coffee pulp is removed and the mucilage around the parchment is partially broken down, is what causes the majority of the pollution in the coffee wastewater. Sugars from pulp and mucilage components both quickly ferment to produce pulping water (Dias et al., 2014). The majority of the ingredients in pulp and mucilage are proteins, sugars, and pectin's, which are polysaccharide carbohydrates. Additional wastewater develops from fermentation and washing in the form of hydrolyzed pectin's, depending on the processing technique used (Chanakya and De Alwis, 2004; Rattan et al., 2015). During fermentation by pectinase and pectate enzymes, long-chain pectin's are converted into short-chain pectin oligosaccharides. These oligosaccharides are soluble in neutral and alkaline solutions, but in acidic environments, they precipitate and form pectic acid particles. In the presence of calcium or other multivalent ions, these pectic acid particles are cross-bonded and form a non-soluble gel called calcium pectate. After pulping, the wastewater generated by mechanical mucilage removers contains some disaccharide carbohydrates but appears gel-like due to undigested mucilage and pectic materials that have been mechanically removed from the parchment. To biodegrade these solid materials, they need to undergo natural fermentation, acidification, and hydrolysis at a later stage. Wastewater generated during the manufacturing of coffee is a type of industrial water pollution, hence its treatment and disposal are significant environmental considerations.

Also, in the process of harvesting and roasting, the defective and immature beans come up as byproducts (Alves et al., 2017). In order to retain the quality of the coffee, such defective beans are to be separated from the perfect ones. These amateur beans are high in amino acid content and could have potential applications in food (Brzezicha et al., 2021).

11.5 APPLICATION OF COFFEE WASTE

The chemical composition, presence of extractable chemicals, and range of functional groups in the biomass are significant factors that impact the revalorization of coffee waste and other biomasses. To boost the sustainability of the process, it is crucial that the coffee industry makes an effort to value the byproducts of the processing of coffee. The utilization of coffee waste and byproducts can present a chance to boost economic output and generate new employment. There is a lot of social and political prominence nowadays to lessen the pollution caused by industrial activity. Almost all developed and developing nations are making changes to their operations to allow for the recycling of their leftovers in an effort to adapt to this reality. Nowadays, most large corporations no longer consider their leftovers as waste. Instead, they see them as valuable raw materials that can be utilized in other processes. Similarly, various methods are available for utilizing different types of coffee waste, such as coffee cherry husks, coffee pulp, CS, and wasted grounds. These wastes can be utilized as a sorbent for removing heavy metals (Blinová et al., 2017) and dyes from aqueous solutions (Lafi et al., 2014), as a substrate for cultivating edible mushrooms (Doroški et la., 2022), as a source of naturally occurring antioxidants (Andrade et al., 2022), for manufacturing reusable cups (Stylianou et al., 2018), and as a substrate for producing biogas, alcohol, and biodiesel (Battista et al., 2021; Blinová et al., 2017). Additionally, they can be used for the production of fuel pellets or briquettes and preparing ion-exchange materials (Ke et al., 2022). As already discussed, if these leftovers are kept as such, they pose a hazardous risk to the surrounding environment. Therefore, finding new strategies/alternatives for their use is of utmost importance. Coffee waste is rich in organic materials that can be converted into high-value bioproducts, such as lipids, pectin's, proteins, and carbs. The presence of caffeine, free phenols, and tannins (polyphenols) in coffee waste can pose significant challenges to the large-scale management and utilization of coffee waste globally, as these compounds can be harmful to various life processes. However, various approaches –

such as hot water pre-treatment, microbial biodegradation, and aerobic fermentation – have been demonstrated to diminish the levels of these hazardous compounds (Mazzeo et al., 2020). As a result, only small-scale production of bio-products has been accomplished, such as silage, biogas, worms, animal feed, ethanol, vinegar, single-cell protein, enzymes, bio-pesticides, and probiotics (Chatzipavlidis et al., 2013; Rattan et al., 2015).

Coffee waste is mainly composed of lignocellulosic biomass, which is primarily comprised of the essential living elements (C, H, O, and N). The primary components of this biomass are cellulose (59.2–62.94 weight percent), hemicellulose (5–10 weight percent), and lignin (19.8–26.5 weight percent) (Cerino-Córdova et al., 2020). Moreover, these substances are present in the form of recoverable chemicals, such as flavonoids (Abdeltaif et al., 2018) and essential oils (Cerino-Córdova et al., 2020). However, as this material has already undergone a hydrothermal extraction process, the amount of these compounds is typically lower (around 10 weight percent) compared to lignocellulosic components. Coffee waste also often contains inorganic micronutrients, such as calcium, magnesium, or sodium, but their amounts are typically less than 5.0% by dry weight. Gurram et al. (2016) observed the presence of xylose, arabinose, galactose, and mannose in coffee residues. The use of coffee waste for these purposes has the potential to have social effects and enhance housing conditions for vulnerable groups, particularly coffee farmers. This would allow for better living conditions and production facilities, which would increase the final product's added value. However, the majority of the outcomes reported in the literature are from exploratory tests that rarely see actual pilot or full-scale deployment. The main aspects in which the coffee pulp has seen the scope of utilization are soil recovery (Cruz et al., 2012), biological control (Moreira et al., 2018), and organic fertilizer (Padmapriya et al., 2013). Coffee wastewater and coffee pulp obtained from the processing of coffee are suggested to be good substrates for ethanol production by fermentation technique (Bonilla-Hermosa et al., 2014). The other study suggests that coffee mucilage is a novel substrate for the production of hydrogen gas (Hernández et al., 2014). The coffee pulp has the potential to be used in agriculture as fertilizer, food substrate for microorganisms, feed for livestock, and in the production of some beverages as well (Carmen et al., 2020; Hoseini et al., 2021; Kadian et al., 2022). Also, the use of coffee pulp in the area, such as phytoremediation (Gómez-Aguilar et al., 2021), inoculating mycorrhizal (Mendoza-Villarreal et al., 2022), soil compositing

(Flores-Solórzano et al., 2022), and management of soil treatment cycles (Carmen et al., 2020), has seen a growing scope. The improvement in the distribution of the *Trichoderma* spp. (significant agent to control plant pathogens in soil) has also been reported by the incorporation of coffee pup in soil (Carmen et al., 2020). The control of rust disease that is caused by *Hemileia vastatrix* and is controlled by the *proanthocynins*. Ramirez-Coronel (2004) reported the presence of *proanthocyanins* in the coffee pulp. Thus, coffee pulp can therefore be used as a cheap source of biopesticide (Das et al., 2021). Fermented coffee pulp has been used for making animal feed for improved nutrition; the recommended quantity is 25g/kg of feed for the birds (Serna-Jiménez et al., 2022). The studies suggest the presence of tannins, caffeine, and other antinutritional factors should be reduced so as to make the supplement fit for animal feed formulations. Foods for human consumption can also be produced by the employment of coffee pulp. The cultivation of edible fungi *Pleurotus* spp., *Auricularia* spp., *Volvariella vivace* was done by using coffee pulp as the substrate (Díaz and Díaz-Godínez, 2022). Agustin (2022) studies the development of the drink "cascara" from the coffee pulp. Researchers have also developed methods for producing toffee-like sweets using coffee pulp and honey (Sutanto et al., 2020). Additionally, some researchers have investigated the production of vinegar from coffee pulp (Padmapriya et al., 2013). The potential of coffee pulp to reduce cardio vascular diseases is also suggested by Corro et al. (2013) keeping in view the presence of antioxidants and high polyphenolic compounds. The presence of high iron content in offer pulp can be exploited to tackle the malnutrition problem and anemia by value addition of the foods. The high presence of the dietary fire 19% in the coffee pulp can compete with the fiber-rich high-consumption cereals such as maize and quinoa. The use of a controlled release membrane system can be developed from the extracted pectin from the coffee pulp. Some researchers suggest the potential benefits of coffee pulp as a substrate for making biopolymer packaging (Garcia and Kim, 2021).

Coffee husk, the outer layer of the coffee cherry fruit, is among the various byproducts generated during coffee production. The coffee husk has been the subject of numerous studies due to its significant abundance and potential for utilization. It is a sustainable material that can be used in a market that has great demand. Coffee husk and parchment can also be utilized to make agglomerates for insulation in buildings. Coffee husk is a versatile byproduct that can be utilized in various industrial processes, such as the production of biofuels, as a sorbent for contaminants, and for the extraction of enzymes. It can also be used in agriculture, such as feeding animals or for making compost, silage, biochar, or mushroom substrates (Ayalew and Aragaw, 2020). Furthermore, coffee husk is a rich source of both macro and micronutrients, which make it a potential dietary fiber and source of bioactive compounds. Given its potential as a valuable resource, further exploration and development of coffee husk utilization could be crucial for the coffee industry and the environment (Chali and Wakgari, 2021; Hoseini el. al., 2021). Particle boards can be made by mixing coffee husk and parchment with polymers (Bekalo and Reinhardt, 2010). A possible alternative valorization technique involves using coffee husks and parchment to create building boards or coatings (Serna-Jiménez et al., 2022). Using coffee husk ash and K_2O as a precursor, glass-ceramic materials can be made (Eixenberger et al., 2022). Producing glass ceramics using coffee peel ash as a substitute for commercial potassium carbonate is an environmentally friendly method. Coffee husk contains potassium, phosphorus, nitrogen, calcium, magnesium, iron, and other minerals, which makes it a valuable byproduct. The protein content of the coffee husk is around 10%, with a profound content of glutamic acid and aspartic acid. The highly valuable amino acids in coffee husk extend their applications for the use of dietary phytochemicals. The fortification of the coffee husk in food supplements is a great approach for the beneficial utilization of humans. Secondary metabolites like phenolic compounds, which can be exploited as a source of raw materials to create bioactive molecules of interest in food, are present in coffee husks (*Coffea arabica* L.). The potential of coffee husk to be used as a bioactive material having antioxidant, antifungal, and antibacterial properties can be evaluated on the basis of its oil content (Hoseini et al., 2021). The coffee husk is said to be rich in butylated hydroxy, 1,2-benzene dicarboxylic acid, 2,3-isopropylidene-6-decoxyhexo (Andrade et al., 2012). Coffee husk can also be used for biorefineries due to the presence of lignin, which can further be utilized in biochemical production (Rambo et al., 2015). Skorupa et al. (2023) suggests coffee husk as an effective agent for the removal of heavy metals from water. The use of coffee husk can be employed for the generation of alternative sources of energy by the conversion of it to bio pellets. The coffee husk can be used as a nutritional supplement for cattle, swine, fish, sheep, fowl, and horses due to low-cost availability during the harvesting and processing season. The utilization of coffee husks presents a promising way to increase the value of this waste and reduce the amount of agro-industrial waste. Coffee husk ash, for example, contains alkaline and alkaline-earth metals and can replace the expensive and scarce feldspars used in ceramic formulations as a fluxing component (Hoseini et al., 2021). Another potential application of coffee husk is in the production of structural and nonstructural panel products, which showed excellent flexural and internal bond qualities (Malik and Rawat, 2021). Additionally, after steam-treatment to remove caffeine and chlorogenic acids, coffee husk can be used for scent synthesis using fungi from the genus *Ceratocystis* (Malik and Rawat, 2021). Furthermore, coffee husk can serve as an inexpensive and readily available substrate for the production of organic acids like citric acid and gibberellin through fermentation (Arya and Kumar, 2022). The viability of using coffee husk as a substrate for creating gibberellic acid has been explored in both solid-state fermentation and submerged fermentation studies (Machado et al., 2000). Coffee husk is a suitable substrate for the growth of mushrooms since it is readily available, inexpensive, and fragmented, so it does not require grinding before use. The fungus *Pleurotus ostreatus* fermented coffee husk, increasing the protein and cellulose content while reducing the amount of lignin, tannins, and caffeine. As the fermentation of coffee husk progresses, there is a reduction in both the volatile fatty acid and digestible dry matter. Coffee husk also demonstrates an excellent precursor for biochar production. The chemical characteristics of soil were improved by coffee husk biochar by raising pH, electrical conductivity, cation exchange capacity, organic matter, total N, and accessible Phosphorus (Heredia et al., 2021). Coffee husk can be recycled in various ways, with direct or indirect use as fuel being one of the most common methods. However, due to the presence of nutrients, it is also an excellent resource for use in agriculture, particularly as a soil compost for acidic soils. On the other hand, SCFs are the primary and only byproduct generated during coffee consumption.

SCGs are typically disposed of in landfills, resulting in significant environmental concerns. In addition to taking up valuable landfill space, the decomposition of SCGs produces methane, a potent greenhouse gas that contributes to climate change. Moreover, the disposal of SCGs also represents a loss of potentially valuable resources, such as their rich content of organic matter, nutrients, and bioactive compounds like polyphenols and caffeine. Therefore, researchers have been exploring various sustainable approaches for utilizing SCGs, such as the production of biofuels, animal feed, and compost. The utilization of SCGs can not only reduce environmental impacts but also generate economic benefits and contribute to the development of a circular economy. SCGs are a possible green substitute for conventional organic and inorganic phase change materials (PCMs) since they include a significant proportion of fatty acids (Lauberts et al., 2022). The spent coffee is considered rich in organic acids such as palmitic and linoleic acids, and also a source of vitamin E (Bijla et al., 2022). SCGs are a useful resource even though they are typically discarded after brewing because they are a source of recoverable sugars and have high concentrations of oils, antioxidants, and other high-value ingredients. The SCFs can therefore be utilized as an effective source of antioxidants. Studies on the oil extraction from spent coffee suggest that it includes between 60 and 80% of fatty acids, which is suggestive of a strong source of organic non-paraffin PCM (Acevedo et al., 2013). SCGs can be recycled through a biorefinery-based process to produce biofuels and other value-added products, thereby preventing the environmental issues, hazardous pollutants, and costs associated with their disposal. Some studies suggest that SCGs have potential as a low-cost source for biogas production. Additionally, many researchers have investigated the potential of SCGs as a feedstock for biotechnological processes and their ability to produce various value-added products, including aromatics and enzymes. By utilizing SCGs in such processes, it is possible to convert this waste material into valuable resources, thereby contributing to the development of a more sustainable and circular economy. One of the researchers exploited the usage of SCG for the bio chelates for food fortification. SCG biomass can be directly employed as the feedstock for animals and also for biogas and bio-crude oil production by direct transesterification and anaerobic digestion (AD), respectively. Also, the magnetically modified SCG can be used for the removal of water–soluble pigments (Safarik et al., 2012), biosorbents for xenobiotics removal (Safarik et al., 2011). The spent coffee extracts can be obtained by the US extraction methods, pressurized liquid extraction (PLE) for the production of caffeine, Phenolics, cafetal, and kahweol to be used in the food industry. The presence of different phytochemicals (Caffeic acid, Chlorogenic acid, Catechin, Gallic acid, and Protocatechuic acid) in SCG extracts can extend their utility as strong antioxidants and free radical scavengers (Andrade et al., 2022). In addition, the presence of sugars, proteins, and cellulose in SCG opens the options for usage in different fields of life. The various application of coffee byproducts is enlisted in Table 11.1.

TABLE 11.1 List of the Coffee Byproducts and Their Utilization

Coffee Byproduct	**Utilization**	**References**
Coffee Husk	Production of biofuels	Oliveira and Franca (2015)
	Animal feed	Mazzafera (2002)
	Biocomposites	Ilangovan et al. (2021)
	Antioxidants, functional food ingredients, enzymes	Avila et al. (2021) Hoseini et al. (2021)
	Organic acids	Akacha and Gargouri (2015)
	Scent synthesis	
Spent Coffee Grounds	Production of biofuels	Haile (2014)
	Animal feed	Campos-Vega et al. (2015)
	Compost	Liu and Price (2011)
	Dietary supplements	Panusa et al. (2013)
	Nutraceuticals and food additives	Salazar-López et al. (2020)
	Value-added foods	Kourmentza et al. (2018)
Coffee Pulp	Production of toffee-like sweets	Sutanto et al. (2020)
	Vinegar	Carmen et al. (2020)
	Organic fertilizer	Mazzafera (2002)
	Animal feed	Carmen et al. (2020)
	Bioactive compounds	Torres-Valenzuela et al. (2020)
	Concentrate in dairy cattle feeds	Pedraza-Beltrán et al. (2012)
	Substrate for growing mushrooms	Martínez-Carrera et al. (2000)
	Compost production	Kassa et al. (2011)
	Development of soft and alcoholic drinks	Lopes et al. (2019)
	Utilization as functional drinks	Muzaifa and Rahmi (2021)
Coffee Cherry Husks	Production of biogas	Chala et al. (2018)
	Animal feed	Oliveira and Franca (2015)
	Organic fertilizer	Dzung et al. (2013)
	Clay-based ceramic formulations	Munirwan et al. (2022)
	Production of food flavoring compounds	Arya et al. (2021)
	Production of Particle boards	Bekalo and Reinhardt (2010)
	Solid-fuel gasification	Orosco et al. (2018)
	Ethanol production	Blinová et al. (2017)
	Adsorption of contaminants	Skorupa et al. (2023)
	Producing animal feed	Oliveira et al. (2009)
	Used as mushroom beds	Oliveira and Franca (2015)
	Biochar	Blinová et al. (2017)
	Silage and composting	Hoseini et al. (2021)

TABLE 11.1 *(Continued)*

Coffee Byproduct	Utilization	References
Coffee Grounds	Sorbent for the removal of heavy metals Source of antioxidants Production of reusable cups Substrate for biogas, biodiesel, and alcohol production Cultivation of edible mushrooms	Roh et al. (2012) Mussatto et al. (2021) Blinová et al. (2017) Battista et al. (2020) Obruca et al. (2015)
Coffee silver skin	Production of biofuel Animal feed Used in food supplements, food products, antioxidants, and functional food ingredients	Procentese et al. (2019) Arya et al. (2021) Gocmen et al. (2019) Borrelli et al. (2004)
Fermented coffee husk	Production of organic acids (citric acid and gibberellic acid) Enzyme production (pectinase, tannase, and caffeinase) Source of bioactive compounds (anthocyanins and dietary fiber)	Singhania et al. (2008) Pandey et al. (2000) Oliveira and Franca (2015)

Coffee silver skin is also a byproduct that is much more valuable due to its phytochemical composition and low moisture content. Coffee silver skin is a byproduct that is obtained after the roasting process and comprises approximately 4.2% (w/w) of the total weight of the coffee seed (Martuscelli et al., 2021). It is the thin, papery layer that covers the coffee bean and is usually removed during the roasting process. Due to its high content of dietary fiber, protein, and antioxidants, coffee silver skin is considered a potential source of valuable bioactive compounds that can be utilized in various industrial applications. Researchers have explored several approaches for utilizing coffee silver skin, such as the production of biofuels, animal feed, and dietary supplements. Additionally, it has been proposed as a potential source of antioxidants and functional food ingredients, as well as a substrate for the production of enzymes and other biochemicals. Given its significant potential as a valuable resource, further exploration and development of CS utilization could be crucial for the coffee industry and the environment. CS is high in total proteins (10.9–19%) and insoluble dietary fiber (45–66%), low in reducing sugar (0.2%) and fat (2.2%), and active in antioxidants (Castaldo et al., 2020). Although CSS is a less common byproduct, its availability supports interest in its recovery and use in various applications in several industries. The presence of fibers, sugars, and polyphenols is exploited for utilization in the food and pharmaceutical sector. The CS has already been employed in the bakery industry as fat-replacing and moisturizing ingredients (Ateş and Elmacı, 2018). The presence of high levels of protein and fiber uplifts the scope of silverskin as a food ingredient, such as partial flour (Lorbeer et al., 2022). The richness of the flavoring compounds in the silver skin makes it a suitable food flavoring agent.

11.6 CONCLUSION

The richness of the coffee drink makes it one of the top consumable stimulating beverages across the globe. The consumption of coffee is increasing, and so is the demand for production and cultivation on the rise. The increased production and processing of coffee subsequently pose a great threat to the environment. The enormous number of byproducts as residues from every step of coffee processing is directly responsible for environmental problems. The residues in the form of liquid, semisolid mass, or solid fractions deplete the surrounding ecosystem. Despite the emphasis on waste utilization, the coffee processing sector needs to implement the necessary steps to tackle this problem. The effective utilization of the waste from coffee production in various fields is an approach to waste utilization. The utilization of such waste in the sectors of biotechnology, pharmacology, food, agriculture, and other areas will boost waste management and waste utilization. The chemical composition of the coffee byproducts makes it a cheap and promising material for biochemical processes. The introduction of new methods and strategies with the latest research techniques can serve as a milestone for the best utilization of potential coffee byproducts.

KEYWORDS

- **animal byproducts**
- **bioactive compounds**
- **biopolymers**
- **enzymatic hydrolysis**
- **protein hydrolysates**
- **value-added products**

REFERENCES

Abdeltaif, S. A., SirElkhatim, K. A., & Hassan, A. B. (2018). Estimation of phenolic and flavonoid compounds and antioxidant activity of spent coffee and black tea (processing) waste for potential recovery and reuse in Sudan. *Recycling, 3*(2), 27.

Acevedo, F., Rubilar, M., Scheuermann, E., Cancino, B., Uquiche, E., Garcés, M., & Shene, C. (2013). Spent coffee grounds as a renewable source of bioactive compounds. *Journal of Biobased Materials and Bioenergy, 7*(3), 420–428.

Achterbosch, T. J., van Berkum, S., Meijerink, G. W., Asbreuk, H., & Oudendag, D. A. (2014). *Cash Crops and Food Security: Contributions to Income, Livelihood Risk and Agricultural Innovation* (No. 2014-15). LEI Wageningen UR.

Agustin, S., Wiliana, A., Kumalaputri, A. J., & Abduh, M. Y. (2022). Production and characterization of cascara powder from coffee pulp. *Biological and Natural Resources Engineering Journal, 6*(1).

Akacha, N. B., & Gargouri, M. (2015). Microbial and enzymatic technologies used for the production of natural aroma compounds: Synthesis, recovery modeling, and bioprocesses. *Food and Bioproducts Processing, 94*, 675–706.

Al-Asmari, K. M., Zeid, I. M. A., & Al-Attar, A. M. (2020). Coffee Arabica in Saudi Arabia: An overview. *International Journal of Pharmaceutical and Phytopharmacological Research (eIJPPR), 10*(4), 71–78.

Al-Hamamre, Z., Foerster, S., Hartmann, F., Kröger, M., & Kaltschmitt, M. (2012). Oil extracted from spent coffee grounds as a renewable source for fatty acid methyl ester manufacturing. *Fuel, 96*, 70–76.

Alves, R. C., Rodrigues, F., Nunes, M. A., Vinha, A. F., & Oliveira, M. B. P. (2017). State of the art in coffee processing byproducts. In *Handbook of Coffee Processing Byproducts* (pp. 1–26).

Amaravathi, M. (2014). Indian coffee production and export destinations: An overview. *International Journal of Scientific Research, 3*(10), 505–508.

Andrade, C., Perestrelo, R., & Câmara, J. S. (2022). Bioactive compounds and antioxidant activity from spent coffee grounds as a powerful approach for its valorization. *Molecules, 27*(21), 7504.

Andrade, K. S., Gonçalvez, R. T., Maraschin, M., Ribeiro-do-Valle, R. M., Martínez, J., & Ferreira, S. R. (2012). Supercritical fluid extraction from spent coffee grounds and coffee husks: Antioxidant activity and effect of operational variables on extract composition. *Talanta, 88*, 544–552.

Angeloni, G., Masella, P., Spadi, A., Guerrini, L., Corti, F., Bellumori, M., & Parenti, A. (2023). Using ground coffee particle size and distribution to remodel beverage properties. *European Food Research and Technology*, 1–10.

Arya, P., & Kumar, P. (2022). Processing of coffee and tea waste. In *Food Processing Waste and Utilization* (pp. 155–173). CRC Press.

Arya, S. S., Venkatram, R., More, P. R., & Vijayan, P. (2021). The wastes of coffee bean processing for utilization in food: A review. *Journal of Food Science and Technology*, 1–16.

Aryal, B. (2023). *Study of Physicochemical Properties of Coffee Beans from Different Processing Methods* (Doctoral dissertation, Department of Food Technology, Central Campus of Technology, Institute of Science and Technology, Tribhuvan University, Nepal, 2022).

Ateş, G., & Elmacı, Y. (2018). Coffee silverskin as fat replacer in cake formulations and its effect on physical, chemical and sensory attributes of cakes. *LWT, 90*, 519–525.

Avila, P. F., Martins, M., & Goldbeck, R. (2021). Enzymatic production of xylooligosaccharides from alkali-solubilized arabinoxylan from sugarcane straw and coffee husk. *BioEnergy Research, 14*, 739–751.

Ayalew, A. A., & Aragaw, T. A. (2020). Utilization of treated coffee husk as low-cost bio-sorbent for adsorption of methylene blue. *Adsorption Science & Technology, 38*(5, 6), 205–222.

Ayyasamy, R., & Regupathy, A. (2014). Outbreak of short horned grasshopper, *Diabolocatantops pinguis* (Stål, 1861) (Orthoptera: Acrididae) in Yercaud coffee plantation, Tamil Nadu.

Battista, F., Barampouti, E. M., Mai, S., Bolzonella, D., Malamis, D., Moustakas, K., & Loizidou, M. (2020). Added-value molecules recovery and biofuels production from spent coffee grounds. *Renewable and Sustainable Energy Reviews, 131*, 110007.

Battista, F., Zuliani, L., Rizzioli, F., Fusco, S., & Bolzonella, D. (2021). Biodiesel, biogas and fermentable sugars production from spent coffee grounds: A cascade biorefinery approach. *Bioresource Technology, 342*, 125952.

Bekalo, S. A., & Reinhardt, H. W. (2010). Fibers of coffee husk and hulls for the production of particleboard. *Materials and Structures, 43*(8), 1049–1060.

Bijla, L., Ibourki, M., Bouzid, H. A., Sakar, E. H., Aissa, R., Laknifli, A., & Gharby, S. (2022). Proximate composition, antioxidant activity, mineral and lipid profiling of spent coffee grounds collected in Morocco reveal a great potential of valorization. *Waste and Biomass Valorization, 13*(11), 4495–4510.

Bisekwa, E. (2022). *Wastewater in Burundi: Environmental Pollution and Management of Wet Coffee Processing* (Doctoral dissertation, JKUAT–IEET).

Blinová, L., Sirotiak, M., Bartošová, A., & Soldán, M. (2017). Utilization of waste from coffee production. *Research Papers Faculty of Materials Science and Technology Slovak University of Technology, 25*(40), 91–101.

Bonilla-Hermosa, V. A., Duarte, W. F., & Schwan, R. F. (2014). Utilization of coffee byproducts obtained from semi-washed process for production of value-added compounds. *Bioresource Technology, 166*, 142–150.

Borrelli, R. C., Esposito, F., Napolitano, A., Ritieni, A., & Fogliano, V. (2004). Characterization of a new potential functional ingredient: Coffee silverskin. *Journal of Agricultural and Food Chemistry, 52*(5), 1338–1343.

Bottausci, S., Midence, R., Serrano-Bernardo, F., & Bonoli, A. (2022). Organic waste management and circular bioeconomy: A literature review comparison between Latin America and the European Union. *Sustainability, 14*(3), 1661.

Brzezicha, J., Błażejewicz, D., Brzezińska, J., & Grembecka, M. (2021). Green coffee vs dietary supplements: A comparative analysis of bioactive compounds and antioxidant activity. *Food and Chemical Toxicology, 155*, 112377.

Butnariu, M. (2022). Aquatic pollution and marine ecosystems. In *Bacterial Fish Diseases* (pp. 1–29). Academic Press.

Campos, R. C., Pinto, V. R. A., Melo, L. F., da Rocha, S. J. S. S., & Coimbra, J. S. (2021). New sustainable perspectives for "coffee wastewater" and other byproducts: A critical review. *Future Foods, 4*, 100058.

Campos-Vega, R., Loarca-Pina, G., Vergara-Castaneda, H. A., & Oomah, B. D. (2015). Spent coffee grounds: A review on current research and future prospects. *Trends in Food Science & Technology, 45*(1), 24–36.
Carmen, M. T., Lorena, Z. C., Alexander, V. A., Amandio, V., & Raúl, S. (2020). Coffee pulp: An industrial byproduct with uses in agriculture, nutrition and biotechnology. *Reviews in Agricultural Science, 8*, 323–342.
Carta, S., Tsiplakou, E., Nicolussi, P., Pulina, G., & Nudda, A. (2022). Effects of spent coffee grounds on production traits, hematological parameters, and antioxidant activity of blood and milk in dairy goats. *Animal, 16*(4), 100501.
Castaldo, L., Narváez, A., Izzo, L., Graziani, G., & Ritieni, A. (2020). *In vitro* bio accessibility and antioxidant activity of coffee silverskin polyphenolic extract and characterization of bioactive compounds using UHPLC-Q-Orbitrap HRMS. *Molecules, 25*(9), 2132.
Cerino-Córdova, F. J., Dávila-Guzmán, N. E., León, A. M. G., Salazar-Rabago, J. J., & Soto-Regalado, E. (2020). Revalorization of coffee waste. In *Coffee: Production and Research* (p. 133).
Cervera-Mata, A., Fernández-Arteaga, A., Navarro-Alarcón, M., Hinojosa, D., Pastoriza, S., Delgado, G., & Rufián-Henares, J. Á. (2021). Spent coffee grounds as a source of smart biochelates to increase Fe and Zn levels in lettuces. *Journal of Cleaner Production, 328*, 129548.
Chala, B., Oechsner, H., Latif, S., & Müller, J. (2018). Biogas potential of coffee processing waste in Ethiopia. *Sustainability, 10*(8), 2678.
Chali, G., Abera, T., & Wakgari, T. (2021). Effect of coffee husk compost and NPS fertilizer rates on growth and yield of coffee (*Coffea arabica* L.) at Haru Research Sub-center, Western Ethiopia. *American Journal of Bioscience and Bioengineering, 9*(3), 81–87.
Chanakya, H. N., & De Alwis, A. A. P. (2004). Environmental issues and management in primary coffee processing. *Process Safety and Environmental Protection, 82*(4), 291–300.
Chatzipavlidis, I., Kefalogianni, I., Venieraki, A., & Holzapfel, W. (2013). Commission on genetic resources for food and agriculture: Status and trends of the conservation and sustainable use of microorganisms in agro-industrial processes. FAO.
Chen, C. L., Kuo, Y. H., & Fang, M. (2023). Study on coffee flavor of various pour-over brews. *Flavor and Fragrance Journal, 38*(1), 3–12.
Cho, E. J., Lee, Y. G., Song, Y., Nguyen, D. T., & Bae, H. J. (2022). An integrated process for conversion of spent coffee grounds into value-added materials. *Bioresource Technology, 346*, 126618.
Corro, G., Paniagua, L., Pal, U., Bañuelos, F., & Rosas, M. (2013). Generation of biogas from coffee pulp and cow dung co-digestion: Infrared studies of post combustion emissions. *Energy Conversion and Management, 74*, 471–481.
Cruz, R., Baptista, P., Cunha, S., Pereira, J. A., & Casal, S. (2012). Carotenoids of lettuce (*Lactuca sativa* L.) grown on soil enriched with spent coffee grounds. *Molecules, 17*(2), 1535–1547.
Das, M. M., Aguilar, C. N., Haridas, M., & Sabu, A. (2021). Production of bio-fungicide, *Trichoderma harzianum* CH1 under solid-state fermentation using coffee husk. *Bioresource Technology Reports, 15*, 100708.

Dias, D. R., Valencia, N. R., Franco, D. A. Z., & López-Núñez, J. C. (2014). Management and utilization of wastes from coffee processing. In *Cocoa and Coffee Fermentations* (pp. 376–382).

Díaz, R., & Díaz-Godínez, G. (2022). Substrates for mushroom, enzyme and metabolites production: A review. *Journal of Environmental Biology, 43*(3), 350–359.

Djufry, F., & Wulandari, S. (2021). Climate-smart agriculture implementation facing climate variability and uncertainty in the coffee farming system. In *IOP Conference Series: Earth and Environmental Science* (Vol. 653, No. 1, p. 012116). IOP Publishing.

Doroški, A., Klaus, A., Režek Jambrak, A., & Djekic, I. (2022). Food waste originated material as an alternative substrate used for the cultivation of oyster mushroom (*Pleurotus ostreatus*): A review. *Sustainability, 14*(19), 12509.

Durán-Aranguren, D. D., Robledo, S., Gomez-Restrepo, E., Arboleda Valencia, J. W., & Tarazona, N. A. (2021). Scient metric overview of coffee byproducts and their applications. *Molecules, 26*(24), 7605.

Dzung, N. A., Dzung, T. T., & Khanh, V. T. P. (2013). Evaluation of coffee husk compost for improving soil fertility and sustainable coffee production in rural central highland of Vietnam. *Resources and Environment, 3*(4), 77–82.

Eixenberger, D., Carballo-Arce, A. F., Vega-Baudrit, J. R., Trimino-Vazquez, H., Villegas-Peñaranda, L. R., Stöbener, A., & Liese, A. (2022). Tropical agro-industrial biowaste revalorization through integrative biorefineries—Review part II: Pineapple, sugarcane and banana byproducts in Costa Rica. *Biomass Conversion and Biorefinery*, 1–28.

Farah, A., & dos Santos, T. F. (2015). The coffee plant and beans: An introduction. In *Coffee in Health and Disease Prevention* (pp. 5–10). Academic Press.

Farias, M. B., Araújo, A. J., Medeiros, V. S., Macedo, D. A., De Miranda, A., Paskocimas, C. A., & Nascimento, R. M. (2022). Use of coffee husk ash for preparation of glass-ceramics as potential sealants for solid oxide cells. *Ceramics International, 48*(16), 23145–23150.

Ferreira, T., Shuler, J., Guimarães, R., & Farah, A. (2019). Introduction to coffee plant and genetics.

Flores-Solórzano, S. B., Huerta-Lwanga, E., Cuevas-González, R., & Guillén-Navarro, K. (2022). Optimal conditions to produce extracts of compost and vermicompost from oil palm and coffee pulp wastes. *Journal of Material Cycles and Waste Management, 24*(2), 801–810.

Garcia, C. V., & Kim, Y. T. (2021). Spent coffee grounds and coffee silverskin as potential materials for packaging: A review. *Journal of Polymers and the Environment, 29*, 2372–2384.

Gemechu, F. G. (2020). Embracing nutritional qualities, biological activities and technological properties of coffee byproducts in functional food formulation. *Trends in Food Science & Technology, 104*, 235–261.

Girma, B., & Sualeh, A. (2022). A review of coffee processing methods and their influence on aroma. *International Journal of Food Engineering and Technology, 6*(7).

Gocmen, D., Sahan, Y., Yildiz, E., Coskun, M., & Aroufai, İ. A. (2019). Use of coffee silverskin to improve the functional properties of cookies. *Journal of Food Science and Technology, 56*, 2979–2988.

Gómez-Aguilar, D. L., Esteban-Muñoz, J. A., Rodríguez-Miranda, J. P., Baracaldo-Guzmán, D., & Salcedo-Parra, O. J. (2022). Desorption of coffee pulp used as an adsorbent material for Cr (III and VI) ions in synthetic wastewater: A preliminary study. *Molecules, 27*(7), 2170.

Grzelczyk, J., Fiurasek, P., Kakkar, A., & Budryn, G. (2022). Evaluation of the thermal stability of bioactive compounds in coffee beans and their fractions modified in the roasting process. *Food Chemistry, 387*, 132888.

Gurram, R., Al-Shannag, M., Knapp, S., Das, T., Singsaas, E., & Alkasrawi, M. (2016). Technical possibilities of bioethanol production from coffee pulp: A renewable feedstock. *Clean Technologies and Environmental Policy, 18*, 269–278.

Haile, M. (2014). Integrated valorization of spent coffee grounds to biofuels. *Biofuel Research Journal, 1*(2), 65–69.

Haile, M., & Kang, W. H. (2019). The harvest and post-harvest management practices' impact on coffee quality. In *Coffee: Production and Research* (pp. 1–18).

Heredia Salgado, M. A., Säumel, I., Cianferoni, A., & Tarelho, L. A. (2021). Potential for farmers' cooperatives to convert coffee husks into biochar and promote the bioeconomy in the north Ecuadorian Amazon. *Applied Sciences, 11*(11), 4747.

Hernández, M. A., Susa, M. R., & Andres, Y. (2014). Use of coffee mucilage as a new substrate for hydrogen production in anaerobic co-digestion with swine manure. *Bioresource Technology, 168*, 112–118.

Hoseini, M., Cocco, S., Casucci, C., Cardelli, V., & Corti, G. (2021). Coffee byproducts derived resources: A review. *Biomass and Bioenergy, 148*, 106009.

Hu, G., Peng, X., Wang, X., Li, X., Li, X., & Qiu, M. (2020). Excavation of coffee maturity markers and further research on their changes in coffee cherries of different maturity. *Food Research International, 132*, 109121.

Ibrahim, H. W., & Zailani, S. (2010). A review on the competitiveness of global supply chain in a coffee industry in Indonesia. *International Business Management, 4*(3), 105–115.

Ijanu, E. M., Kamaruddin, M. A., & Norashiddin, F. A. (2020). Coffee processing wastewater treatment: A critical review on current treatment technologies with a proposed alternative. *Applied Water Science, 10*, 1–11.

Ilangovan, M., Guna, V., Hu, C., Takemura, A., Leman, Z., & Reddy, N. (2021). Dehulled coffee husk-based bio composites for green building materials. *Journal of Thermoplastic Composite Materials, 34*(12), 1623–1638.

Kadian, S. S., Sarao, L. K., & Mohan, C. (2022). Utilization of food industry wastes. In *Food Processing Waste and Utilization* (pp. 1–21). CRC Press.

Kassa, H., Suliman, H., & Workayew, T. (2011). Evaluation of composting process and quality of compost from coffee byproducts (coffee husk & pulp). *Ethiopian Journal of Environmental Studies and Management, 4*(4).

Ke, Y., Cui, S., Fu, Q., Hough, R., Zhang, Z., & Li, Y. F. (2022). Effects of pyrolysis temperature and aging treatment on the adsorption of Cd^{2+} and Zn^{2+} by coffee grounds biochar. *Chemosphere, 296*, 134051.

Kim, M. S., Min, H. G., Koo, N., Park, J., Lee, S. H., Bak, G. I., & Kim, J. G. (2014). The effectiveness of spent coffee grounds and its biochar on the amelioration of heavy metals-contaminated water and soil using chemical and biological assessments. *Journal of Environmental Management, 146*, 124–130.

Kourmentza, C., Economou, C. N., Tsafrakidou, P., & Kornaros, M. (2018). Spent coffee grounds make much more than waste: Exploring recent advances and future exploitation strategies for the valorization of an emerging food waste stream. *Journal of Cleaner Production, 172*, 980–992.

Kovalcik, A., Obruca, S., & Marova, I. (2018). Valorization of spent coffee grounds: A review. *Food and Bioproducts Processing, 110*, 104–119.

Krajangsang, S., Mahingsapun, R., Tantayotai, P., Panyachanakul, T., Samosorn, S., Dolsophon, K., & Srisuk, N. (2022). Enhancement of Arabica coffee quality with selected potential microbial seed culture under controlled fermentation in wet process. *Available at SSRN 4008103*.

Lafi, R., Ben Fradj, A., Hafiane, A., & Hameed, B. H. (2014). Coffee waste as potential adsorbent for the removal of basic dyes from aqueous solution. *Korean Journal of Chemical Engineering, 31*, 2198–2206.

Lauberts, M., Mierina, I., Pals, M., Latheef, M. A. A., & Shishkin, A. (2022). Spent coffee grounds valorization in biorefinery context to obtain valuable products using different extraction approaches and solvents. *Plants, 12*(1), 30.

Limousy, L., Jeguirim, M., Dutournié, P., Kraiem, N., Lajili, M., & Said, R. (2013). Gaseous products and particulate matter emissions of biomass residential boiler fired with spent coffee grounds pellets. *Fuel, 107*, 323–329.

Liu, K., & Price, G. W. (2011). Evaluation of three composting systems for the management of spent coffee grounds. *Bioresource Technology, 102*(17), 7966–7974.

Lopes, A. C. A., Eda, S. H., Andrade, R. P., Amorim, J. C., & Duarte, W. F. (2019). New alcoholic fermented beverages—Potentials and challenges. In *Fermented Beverages* (pp. 577–603).

Lorbeer, L., Schwarz, S., Franke, H., & Lachenmeier, D. W. (2022). Toxicological assessment of roasted coffee silver skin (*testa* of *Coffea* sp.) as novel food ingredient. *Molecules, 27*(20), 6839.

Machado, C. M. M., Oliveira, B. H., Pandey, A., & Soccol, C. R. (2000). Coffee husk as substrate for the production of gibberellic acid by fermentation. In *Coffee Biotechnology and Quality: Proceedings of the 3rd International Seminar on Biotechnology in the Coffee Agro-Industry, Londrina, Brazil* (pp. 401–408). Springer Netherlands.

Malara, A., Paone, E., Frontera, P., Bonaccorsi, L., Panzera, G., & Mauriello, F. (2018). Sustainable exploitation of coffee silverskin in water remediation. *Sustainability, 10*(10), 3547.

Malik, T., & Rawat, S. (2021). Biotechnological interventions for production of flavor and fragrance compounds. In *Sustainable Bioeconomy: Pathways to Sustainable Development Goals* (pp. 131–170).

Martínez-Carrera, D., Aguilar, A., Martínez, W., Bonilla, M., Morales, P., & Sobal, M. (2000). Commercial production and marketing of edible mushrooms cultivated on coffee pulp in Mexico. In *Coffee Biotechnology and Quality: Proceedings of the 3rd International Seminar on Biotechnology in the Coffee Agro-Industry, Londrina, Brazil* (pp. 471–488). Springer Netherlands.

Martinez-Saez, N., García, A. T., Pérez, I. D., Rebollo-Hernanz, M., Mesías, M., Morales, F. J., & Del Castillo, M. D. (2017). Use of spent coffee grounds as food ingredient in bakery products. *Food Chemistry, 216*, 114–122.

Martuscelli, M., Esposito, L., Di Mattia, C. D., Ricci, A., & Mastrocola, D. (2021). Characterization of coffee silver skin as potential food-safe ingredient. *Foods, 10*(6), 1367.

Mazzafera, P. (2002). Degradation of caffeine by microorganisms and potential use of decaffeinated coffee husk and pulp in animal feeding. *Scientia Agricola, 59*, 815–821.

Mazzeo, D. E. C., Misovic, A., Oliveira, F. A., Levy, C. E., Oehlmann, J., & de Marchi, M. R. R. (2020). Effects of bio stimulation by sugarcane bagasse and coffee grounds on sewage sludges, focusing agricultural use: Microbial characterization, respirometric assessment and toxicity reduction. *Waste Management, 118*, 110–121.

Megersa, H. G. (2022). Coffee (*Coffea arabica* L.) field establishment and management practices in Ethiopia. *American Journal of Engineering and Technology Management, 7*, 48–58.

Mendoza-Villarreal, R., González-Fuentes, J. A., Hernández-Montiel, L. G., Chiquito-Contreras, R. G., Paredes-Jácome, J. R., & Robledo-Torres, V. (2022). Arbuscular mycorrhizal fungus and organic substrates effect on bean plant morphology and minerals. *Terra Latinoamericana, 40*.

Montoya, A. C. V., da Silva Mazareli, R. C., Delforno, T. P., Centurion, V. B., Sakamoto, I. K., de Oliveira, V. M., & Varesche, M. B. A. (2019). Hydrogen, alcohols and volatile fatty acids from the co-digestion of coffee waste (coffee pulp, husk, and processing wastewater) by applying autochthonous microorganisms. *International Journal of Hydrogen Energy, 44*(39), 21434–21450.

Moreira, M. D., Melo, M. M., Coimbra, J. M., Dos Reis, K. C., Schwan, R. F., & Silva, C. F. (2018). Solid coffee waste as alternative to produce carotenoids with antioxidant and antimicrobial activities. *Waste Management, 82*, 93–99.

Morris, J. (2018). *Coffee: A Global History*. Reaktion Books.

Munirwan, R. P., Taib, A. M., Taha, M. R., Abd Rahman, N., & Munirwansyah, M. (2022). Utilization of coffee husk ash for soil stabilization: A systematic review. *Physics and Chemistry of the Earth, Parts A/B/C*, 103252.

Murthy, P. S., & Naidu, M. M. (2012). Sustainable management of coffee industry byproducts and value addition—A review. *Resources, Conservation and Recycling, 66*, 45–58.

Mussatto, S. I., Ballesteros, L. F., Martins, S., & Teixeira, J. A. (2011). Extraction of antioxidant phenolic compounds from spent coffee grounds. *Separation and Purification Technology, 83*, 173–179.

Muzaifa, M., & Rahmi, F. (2021). Utilization of coffee byproducts as profitable foods—A mini review. In *IOP Conference Series: Earth and Environmental Science* (Vol. 672, No. 1, p. 012077). IOP Publishing.

Nadaleti, D. H. S., de Rezende Abrahão, J. C., Malta, M. R., Dos Santos, C. S., Pereira, A. A., & Carvalho, G. R. (2022). Influence of postharvest processing on the quality and sensory profile of groups of arabica coffee genotypes. *Journal of the Science of Food and Agriculture, 102*(15), 6899–6906.

Obruca, S., Benesova, P., Kucera, D., Petrik, S., & Marova, I. (2015). Biotechnological conversion of spent coffee grounds into polyhydroxyalkanoates and carotenoids. *New Biotechnology, 32*(6), 569–574.

Oliveira, L. C., Pereira, E., Guimaraes, I. R., Vallone, A., Pereira, M., Mesquita, J. P., & Sapag, K. (2009). Preparation of activated carbons from coffee husks utilizing $FeCl_3$ and $ZnCl_2$ as activating agents. *Journal of Hazardous Materials, 165*(1–3), 87–94.

Oliveira, L. S., & Franca, A. S. (2015). An overview of the potential uses for coffee husks. In *Coffee in Health and Disease Prevention* (pp. 283–291). Academic Press.

Orosco, J., Patiño, F. J., Quintero, M. J., & Rodríguez, L. (2018). Residual biomass gasification on a small scale and its thermal utilization for coffee drying. *Livestock Research for Rural Development, 30*(1).

Padmapriya, R., Tharian, J. A., & Thirunalasundari, T. (2013). Coffee waste management—An overview. *International Journal of Current Science, 9*, 83–91.

Pandey, A., Soccol, C. R., Nigam, P., Brand, D., Mohan, R., & Roussos, S. (2000). Biotechnological potential of coffee pulp and coffee husk for bioprocesses. *Biochemical Engineering Journal, 6*(2), 153–162.

Panusa, A., Zuorro, A., Lavecchia, R., Marrosu, G., & Petrucci, R. (2013). Recovery of natural antioxidants from spent coffee grounds. *Journal of Agricultural and Food Chemistry, 61*(17), 4162–4168.

Passadis, K., Fragoulis, V., Stoumpou, V., Novakovic, J., Barampouti, E. M., Mai, S., & Loizidou, M. (2020). Study of valorization routes of spent coffee grounds. *Waste and Biomass Valorization, 11*, 5295–5306.

Pattanayak, A. K. (2023). Coffee fibres from coffee waste. In *Sustainable fibers for Fashion and Textile Manufacturing* (pp. 287–307). Woodhead Publishing.

Pedraza-Beltrán, P., Estrada-Flores, J. G., Martínez-Campos, A. R., Estrada-López, I., Rayas-Amor, A. A., Yong-Angel, G., & Castelán-Ortega, O. A. (2012). On-farm evaluation of the effect of coffee pulp supplementation on milk yield and dry matter intake of dairy cows grazing tropical grasses in central Mexico. *Tropical Animal Health and Production, 44*, 329–336.

Pike, D. (2018, October 9). What goes around: How coffee waste is fueling a circular economy. *Roast Magazine*.

Procentese, A., Raganati, F., Olivieri, G., Russo, M. E., & Marzocchella, A. (2019). Combined antioxidant–biofuel production from coffee silverskin. *Applied Microbiology and Biotechnology, 103*, 1021–1029.

Rambo, M. K. D., Schmidt, F. L., & Ferreira, M. M. C. (2015). Analysis of the lignocellulosic components of biomass residues for biorefinery opportunities. *Talanta, 144*, 696–703.

Ramirez-Coronel, M. A., Marnet, N., Kolli, V. K., Roussos, S., Guyot, S., & Augur, C. (2004). Characterization and estimation of proanthocyanidins and other phenolics in coffee pulp (*Coffea arabica*) by thiolysis–high-performance liquid chromatography. *Journal of Agricultural and Food Chemistry, 52*(5), 1344–1349.

Rao, N. Z., Wilkinson, F. H., & Yust, B. G. (2023). The effects of contact time, brewing method, and bean roast on the chemistry of cold brew coffee. In *Coffee Science* (pp. 165–174). CRC Press.

Rattan, S., Parande, A. K., Nagaraju, V. D., & Ghiwari, G. K. (2015). A comprehensive review on utilization of wastewater from coffee processing. *Environmental Science and Pollution Research, 22*, 6461–6472.

Rodriguez, Y. F., Guzman, N. G., & Hernandez, J. G. (2020). Effect of the postharvest processing method on the biochemical composition and sensory analysis of arabica coffee. *Engenharia Agrícola, 40*, 177–183.

Roh, J., Umh, H. N., Yoo, C. M., Rengaraj, S., Lee, B., & Kim, Y. (2012). Waste coffee-grounds as potential biosorbents for removal of acid dye 44 from aqueous solution. *Korean Journal of Chemical Engineering, 29*, 903–907.

Rojas-González, A., Figueroa-Hernández, C. Y., González-Rios, O., Suárez-Quiroz, M. L., González-Amaro, R. M., Hernández-Estrada, Z. J., & Rayas-Duarte, P. (2022).

Coffee chlorogenic acids incorporation for bioactivity enhancement of foods: A review. *Molecules, 27*(11), 3400.

Rueda, X., & Lambin, E. F. (2013). Linking globalization to local land uses: How eco-consumers and gourmands are changing the Colombian coffee landscapes. *World Development, 41*, 286–301.

Safarik, I., Horska, K., Svobodova, B., & Safarikova, M. (2012). Magnetically modified spent coffee grounds for dyes removal. *European Food Research and Technology, 234*, 345–350.

Salazar-López, N. J., López-Rodríguez, C. V., Hernández-Montoya, D. A., & Campos-Vega, R. (2020). Health benefits of spent coffee grounds. In *Food Wastes and By-Products: Nutraceutical and Health Potential* (pp. 327–351). Wiley.

Saldaña-Mendoza, S. A., Martínez-Hernandez, J. L., Rodríguez-Herrera, R., Palacios-Ponce, A. S., Sugathan, S., & Aguilar, C. N. (2022). Use of coffee waste for the production of biofuels. *Environmental Quality Management, 32*(1), 463–471.

Sangta, J., Wongkaew, M., Tangpao, T., Withee, P., Haituk, S., Arjin, C., & Cheewangkoon, R. (2021). Recovery of polyphenolic fraction from arabica coffee pulp and its antifungal applications. *Plants, 10*(7), 1422.

Saravanan, A., Kumar, P. S., Hemavathy, R. V., Jeevanantham, S., Harikumar, P., Priyanka, G., & Devakirubai, D. R. A. (2022). A comprehensive review on sources, analysis and toxicity of environmental pollutants and its removal methods from water environment. *Science of the Total Environment, 812*, 152456.

Sermyagina, E., Martinez, C. L. M., Nikku, M., & Vakkilainen, E. (2021). Spent coffee grounds and tea leaf residues: Characterization, evaluation of thermal reactivity and recovery of high-value compounds. *Biomass and Bioenergy, 150*, 106141.

Serna-Jiménez, J. A., Siles, J. A., de los Ángeles Martín, M., & Chica, A. F. (2022). A review on the applications of coffee waste derived from primary processing: Strategies for revalorization. *Processes, 10*(11), 2436.

Setter, C., Borges, F. A., Cardoso, C. R., Mendes, R. F., & Oliveira, T. J. P. (2020). Energy quality of pellets produced from coffee residue: Characterization of the products obtained via slow pyrolysis. *Industrial Crops and Products, 154*, 112731.

Siagian, P., Setyawan, E. Y., Gultom, T., Napitupulu, F. H., & Ambarita, H. (2017). A field survey on coffee beans drying methods of Indonesian small holder farmers. In *IOP Conference Series: Materials Science and Engineering* (Vol. 237, No. 1, p. 012037). IOP Publishing.

Silva, C. F. (2014). Microbial activity during coffee fermentation. In *Cocoa and Coffee Fermentations* (pp. 368–423). CRC Press.

Singhania, R. R., Soccol, C. R., & Pandey, A. (2008). Application of tropical agro-industrial residues as substrate for solid-state fermentation processes. In *Current Developments in Solid-State Fermentation* (pp. 412–442). Springer.

Skorupa, A., Worwąg, M., & Kowalczyk, M. (2023). Coffee industry and ways of using byproducts as bio adsorbents for removal of pollutants. *Water, 15*(1), 112.

Smith, N. J., Williams, J. T., Plucknett, D. L., & Talbot, J. P. (2018). *Tropical Forests and Their Crops*. Cornell University Press.

Somarriba, E., & López Sampson, A. (2018). Coffee and cocoa agroforestry systems: Pathways to deforestation, reforestation, and tree cover change. In *Agroforestry and Biodiversity Conservation in Tropical Landscapes* (pp. 149–173). Island Press.

Stylianou, M., Agapiou, A., Omirou, M., Vyrides, I., Ioannides, I. M., Maratheftis, G., & Fasoula, D. (2018). Converting environmental risks to benefits by using spent coffee grounds (SCG) as a valuable resource. *Environmental Science and Pollution Research, 25*, 35776–35790.

Subba, R. (2020). *Byproduct Utilization of Coffee Cherry as an Alcoholic Beverage* (Doctoral dissertation, Department of Food Technology, Central Campus of Technology, Dharan, Institute of Science and Technology, Tribhuvan University, Nepal).

Sutanto, A., Achyani, A., Noor, R., Subandowo, D., & Thresia, F. (2020). Coffee pulp fermentation techniques for the sustainable organic agriculture. In *Journal of Physics: Conference Series* (Vol. 1469, No. 1, p. 012019). IOP Publishing.

Tekle, D. Y., Hailu, A. B., Wassie, T. A., & Tesema, A. G. (2015). Effect of coffee processing plant effluent on the physicochemical properties of receiving water bodies, Jimma Zone, Ethiopia. *American Journal of Environmental Protection, 4*(2), 83–90.

Torres-Valenzuela, L. S., Serna-Jiménez, J. A., & Martínez, K. (2020). Coffee byproducts: Nowadays and perspectives. In *Coffee—Production and Research* (pp. 1–18). IntechOpen.

Umroh, S. B., Nur, I. A., & Rastuti, U. (2022). Continuous improvement of total quality management in Arabica coffee quality improvement and control in Brebes Regency. *Sustainable Competitive Advantage (SCA), 11*(1).

Vincent, J. C. (1987). Green coffee processing. In *Coffee: Volume 2: Technology* (pp. 1–33). Springer.

Volk, G., & Byrne, P. (2020). Case study: Coffee wild species and cultivars. In *Crop Wild Relatives and Their Use in Plant Breeding* (pp. 1–23). CRC Press.

Von Enden, J. C., Calvert, K. C., Sanh, K., Hoa, H., Tri, Q., Vietnam, S. R., & Consulting, C. E. O. R. (2002). Review of coffee waste water characteristics and approaches to treatment. *PPP Project: Improvement of Coffee Quality and Sustainability of Coffee Production in Vietnam* (pp. 1–10). German Technical Cooperation Agency (GTZ).

CHAPTER 12

Chili: Wastes, Byproducts, and Utilization

SHAKIRA, AROOJ MATTOO, TAJAMUL ROUF SHAH, SAJAD AHMAD SOFI, KHALID MUZZAFER, HILAL AHMAD MAKROO, B. N. DAR, and DARAKSHAN MAJID*

Department of Food Technology, Islamic University of Science and Technology, Awantipora, Jammu and Kashmir, India

*Corresponding author

ABSTRACT

People in Jammu and Kashmir regularly consume red chili, a horticulture product. It is one of the regions in India where the reddest chilies are grown. Among all the Indian states, Andhra Pradesh produced the most chilies during the fiscal year 2022. In that year, there were 700,000 metric tons of output. Telangana and Madhya Pradesh came after this. During that year, India produced roughly two million metric tons of chili. In Jammu and Kashmir, the area planted with green chilies had a 2018–19 output of 15.76 thousand metric tons, while the area planted with dried chilies had a production of 0.35 thousand metric tons. Red chilies are a vital source of nutrients that are consumed as culinary additions to improve flavor and aroma. Traditional remedies and cosmetics both employ red chilies. When included in a daily diet, red peppers show energizing potential for reducing chronic degenerative diseases including cardiomyopathy and disease. Additionally, spices have anti-inflammatory, antiviral, antibacterial, and antioxidant properties. The main chemical substances generated from spices are oleoresins and spice oils. In the flavor oil and oleoresin

Utilization of Food Waste and Byproducts. Mudasir Ahmad Malik and Kshirod Kumar Dash (Eds.)

DOI: 10.1201/9781779641151-12

industries, accumulation makes approximately 80% to 90% of the waste produced. This accumulation contains minerals such as iron, magnesium, calcium, and zinc that are necessary for the majority of the body's metabolic cycles, as well as proteins, nutrients, and polyphenols. Flavors also have a significantly higher content of dietary fiber (44%–62%) than many leafy foods. The idea that the spent buildup from flavors like chili after their necessary handling could give another source of affordable and regular sources of cell reinforcement accumulates, colorants, dietary fiber, protein, and fat, which might be used in different practical food definitions and would be beneficial for the environment is welcomed as a result of growing interest in novel quality food varieties and a focus on the medical benefits of waste. It has been possible to produce a product that is more protein and dietary fiber-rich by characterizing and using specific spice wastes, such as red chili as an innovative and functional nutrient rich, in baking products (biscuits, bread). Spice waste has been effectively used in composites to boost thermal stability as well as tensile strength, as weedicide, to make bioactive films, biofuel, and ceramics in addition to its potential application in food.

12.1 INTRODUCTION

In the majority of the world's nations, chili is a distinctive and well-liked spice that has a hot and pungent flavor. A bioactive substance known as capsaicin attaches to pain receptors to produce a strong burning sensation. The pigment capsanthin, a potent antioxidant, gives ripe fruits their red color. The origins of chili can be found in Central America and further south. High commercial importance and significant genetic variety exist for this crop. The crop is provided a comprehensive description that includes information on its botanical description, origin, distribution, management of biotic stresses, and usual crop husbandry practices. The enormous amounts of bio-waste produced by the agricultural industry pose serious ecological, economic, and societal problems (Arshadi et al., 2016). The generation of waste, which derives from the food supply chain (FSC), incorrect disposal, and alternative (non-food) uses of foods that are fit for human consumption, is significantly influenced by the food business (Gustavsson et al., 2013). Depending on the industry, a different amount of food is lost. The industrial production of fruits and vegetables generates over 50% of byproducts in the form of pomace, peels, damaged and unripe fruits, and vegetables, which are a good source of micro and macronutrients (Padayachee et al., 2017). About 5–30% of the total fruit production is wasted by the industry as byproducts or non-marketable products like stalks, peels, seeds, and unused flesh created at various processing stages (Sandoval-Castro et al., 2017). The agro-industrial sector's waste and byproducts could be used to recover and extract high-value chemicals (Delgado et al., 2021). Large volumes of waste are produced by the processing of spices, which causes the sector to suffer a significant economic loss (Kothari et al., 2010). Chilies are frequently used as a spice or colorant in powder form (paprika) in the spice industry. Chipotle paste, sauce, dried flakes, oil, and pickles are a few further items derived from chilies.

The waste produced is typically deposited on the ground, used as manure, or turned into animal feed. Proteins, phenolic compounds, organic acids, polysaccharides, vitamins, minerals, dietary fibers, and bioactive compounds are just a few of the useful substances that can be found in this chili waste. These substances are potential sources of secondary metabolites and nutrients which can be used for a variety of industrial applications, in place of synthetic additives, helping to reduce the amount of waste produced in the industries and benefiting producers financially (Viacava et al., 2013). A well-known domesticated member of the *Capsicum* genus belonging to the "Solanaceae" family is the chili pepper (*Capsicum annuum*). This spice has significant commercial value and is a staple in cuisines all over the world. Insipid cuisine can benefit from the pungency, tang, and flavor that chili contributes (Kothari et al., 2010). Considering its color and degree of pungency, chilies are regarded as the most significant spice and condiment. Due to the pigment capsanthin, certain types are well-known for their red color, while others are renowned for their capsaicin-related pungency. Genus *Capsicum* (C.) has about 31 species, of which *C. baccatum, C. annuum, C. chinense, C. pubescens,* and *C. frutescens* are five most commonly cultivated types (Batiha et al., 2020). Chilies have been around since the Stone Age. They are thought to have started off as a wild crop in tropical America as well as in Latin American nations of Guatemala and New Mexico. According to legend, the first crop ever domesticated in America was chili. Capsicum species naturally grow in the humid and tropical regions of South and Central America (Batiha et al., 2020).

In 1493, Columbus brought chili seeds to Spain. He was the first European to encounter them (in the Caribbean) who referred to them as "pepper." From Spain to Europe, the cultivation of capsicum and chilies spread quickly. In the year 1584, the Portuguese shipped capsicum from Brazil to India. Rapidly gaining popularity throughout all of Asia, chilies were also being grown by local Asians. Since its introduction in the 16th century, this crop has been extensively cultivated in South Asia due to the region's favorable environment. In India, chilies are the most widely consumed and least expensive spice. Andhra Pradesh, Maharashtra, Karnataka, and Tamil Nadu, which together account for about 75% of the total land, are the states in India with the highest chili production. Andhra Pradesh is the top producing state of dry chilies, which is followed by Tamil Nadu, Maharashtra, Orissa, and Karnataka. The chili plant is an annual sub-herb that can reach heights of 0.75–1.8 m. Depending on the commercial cultivar, the fruits might differ in size, shape, color, and pungency (Batiha et al., 2020). Annual, perennial, or herbaceous capsicum plants are available. They are up to 15 cm long, oval with sharply pointed leaves which are dark green on the upper surface and pale green on the under surface. The tiny, white blooms are borne singly or in groups of two or three in the axils of the leaves. The shapes and sizes of the fruits vary depending on the cultivar. Fruit is typically 10–20 mm long, 4–7 mm wide, thick, fleshy, blocky to thin, and long, conical. The pericarp is slender, withered, and orange-red to dark reddish in color. The membrane division of the pericarp results in the formation of two cells. About 5 to 10 small, flattened, whitish-yellow seeds make up each cell (Pawar et al., 2011).

In the world, chilies are grown over 1832 thousand hectares, yielding 2959 thousand tons. India, Indonesia, China, Pakistan, Korea, Sri Lanka, and Turkey are major chili producing nations. Other chili producing countries are, Ghana, Tunisia, Nigeria, Egypt, Mexico, United States of America, Spain, Yugoslavia, Italy, Romania, Hungary, Bulgaria, Argentina, and Peru. The most widely grown spice in India is the chili. India is the world's largest producer of chilies as well as its biggest consumer. It is a universal spice of India. Chilies are widely grown throughout India. India has 36% contribution to the total world chili production. Andhra Pradesh being largest producer contributes about 26%, followed by Maharashtra (15%), Karnataka (11%), Orissa (11%), Madhya Pradesh (7%), and other states contributing nearly 22% to the total area under chili cultivation in India. In Jammu and Kashmir, the area planted with green chilies had a 2018–19 output of 15.76 thousand metric tons, while the area planted with dried chilies had a production of 0.35 thousand metric tons (Hussain et al., 2021).

12.2 CHEMICAL COMPOSITION AND BIO-ACTIVE COMPOUNDS OF CHILIES

Chilies contain a large number of bioactive substances. The amount of bioactive chemicals in fruit varies depending on variety, ripening stage, fruit part (placenta, pericarp, and seeds), processing techniques, climatic and storage conditions (Jayaprakash et al., 2012). The majority of the phenolic chemicals in chilies are flavonoids and phenolic acid derivatives (Singh et al., 2009). The primary phytochemicals included in most types of chilies are carotenoids. According to Arimboor et al. (2015), mature chili fruits total carotenoid concentrations ranges from 0.69 to 30 mg g^{-1} dry weight, to 15 to 320 mg g^{-1} fresh weight. Red pigments capsanthin and capsorubin are specific to the *Capsicum* genus and account for 40–60% of the total carotenoids in various cultivars, with capsanthin being the main cause of the red color. The xanthophyll family member capsanthin is thought to have the highest antioxidant capacity. The pigments, capsanthin, cryptoxanthin, capsorubin, carotene, antheraxanthin, zeaxanthin, as well as lutein in trace amounts are accumulated by chilies (Villa-Rivera and Ochoa-Alejo, 2020). They are also an excellent source of vitamins, particularly high levels of vitamins C, E, and provitamin A (Martinez et al., 2005). Phosphorus (P), copper (Cu) boron (B), magnesium (Mg), potassium (K), sodium (Na), calcium (Ca), iron (Fe), manganese (Mn), zinc (Zn), and selenium (Se) and are the important minerals most frequently found in chili. Higher concentrations of K, Mg, P, Fe, Cu, Zn, Mn, and B are found in red chilies (Rubio et al., 2002). Chili is a potential source of dietary fiber (44.4% total dietary fiber (TDF), 34.9% insoluble dietary fiber, 9.5% soluble dietary fiber) and 19.7% protein (Sowbhagya et al., 2015).

12.3 BIO-WASTE PRODUCED FROM CHILI, EXTRACTION METHOD, AND THEIR APPLICATIONS

During the processing of spices, a large amount of waste, including seeds, peels, stems, and pulp, is produced. The recovery and extraction of high-value chemicals may benefit from these agro-industrial waste and byproducts (Delgado-Antequera et al., 2021). The byproducts from chili processing are rich in healthy substances like fiber, proteins, phenolic compounds, and organic acids. Additionally, they contain polysaccharides, vitamins, minerals, dietary fibers, and bioactive substances like flavonoids and carotenoids, all of which are beneficial food ingredients and are very nutrient-dense (Pugliese et al., 2014). However, between 5 and 30% of the total fruit production is rejected by industries as non-marketable products or byproducts, including stalks, seeds, peels, and unused flesh that are produced at various stages of industrial processing (Sandoval-Castro et al., 2017). Chilies are typically consumed as spice in powdered form or as colorant (paprika). The chili byproducts can be used as sources of nutrients and secondary metabolites for a variety of industrial applications, replacing synthetic and chemical additives and lowering the amount of waste generated in the industry. This benefits producers financially (Viacava et al., 2013). Different extraction techniques are used to isolate, characterize, and ultimately apply for industrial use the compounds found in peppers. The two categories of these techniques are classical and contemporary. Traditional techniques, like maceration, Soxhlet extraction, and magnetic stirring, offer high compound recovery but necessitate large amounts of solvents and prolonged extraction times, raising extraction cost and restricting its use due to environmental and safety concerns. Therefore, modern techniques such as pressurized liquid extraction (PLE), enzyme-assisted extraction (EAE), supercritical fluid extraction (SFE), and ultrasound-assisted extraction (UAE) have lessened the extraction time and used less organic solvents to achieve a smaller environmental impact. In comparison to more conventional techniques, these methods did produce extracts with a higher level of purity and a higher overall yield of the desired compounds (Saini and Keum, 2018). Chili seed and processed chili skin make up the majority of the bio-waste produced by the processing of chili. The fruit includes chili seeds, which are a strong source of capsaicin (89% of it) and are attached to the placenta at the top of the fruit. Minerals, vitamins, amino acids, carotenoids, phenolic compounds, vitamins, and polyunsaturated fatty acids (PUFA) are abundant in chili seeds. Waste from chili seeds may be a source of bioactive substances (Ananthan et al., 2014). The exocarp, or shiny and waxy layer, is the pericarp's outermost layer, which is called the chili skin. This layer is removed as waste during the preparation of the chili. This processed chili skin can be used as a source of carotenoids, which has the potential to replace synthetic food colors, as well as fiber, which can be used in a variety of food products to improve their texture, nutritive composition, and water-holding capacity of doughs (Arimboor et al., 2015). Additionally, there are a number of non-food uses for chili waste, including the creation of biofuel, ceramics, and composting.

12.4 BIOLOGICAL AND FUNCTIONAL PROPERTIES OF BIOACTIVE COMPOUNDS EXTRACTED FROM CHILI WASTE

Foods contain phytochemicals known as bioactive substances, which have the ability to modulate metabolic functions and promote improved health. Bioactive chemicals are extra-nutritional elements that are found in trace amounts in food and offer health advantages above and beyond the product's fundamental nutritional value. The physiological, behavioral, and immunological effects of bioactive substances on health are currently the subject of extensive research. Bioactive substances have become important food constituents for maintaining health and preventing disease. Carotenoids, capsaicinoids, phenolic compounds, PUFA, vitamins C and A are among the bioactive substances identified in chilies (Carbonell-Capella et al., 2014). Since these compounds are a great source of vitamins and have anti-inflammatory, antibacterial, antioxidant, anti-hypertensive, anti-obesity, anti-mutagenic, and anti-cancer properties, and are good for cardiovascular health and preventing stomach ulcers, they can be used to make a variety of food products with increased nutritional value. The primary phytochemicals in chili cultivars are carotenoids, which give them a distinctive flavor, vibrant color, and powerful antioxidant qualities. The primary carotenoid in *Capsicum* species, capsanthin, is a member of the xanthophyll class and is thought to have the highest antioxidant capacity. With high quantities of vitamin C, E, and provitamin A, chilies are also excellent providers of vitamins. The dietary fibers included in chili are also a good source of health benefits, including intestinal health maintenance, blood sugar control, lowering cholesterol levels, preventing colon cancer, and assisting in maintaining a healthy weight (Anderson et al., 2009).

12.5 NANOCELLULOSE ISOLATED FROM WASTE CHILI STEMS PROCESSING AS A DIETARY FIBER

The copious amounts of straw produced during the preparation of chilies have turned into a significant vegetable waste source. Currently, the majority of waste chili stems (CS) are thrown away, while a small portion is burned and put back on the fields. Crop growth and development may be hampered by the allelochemicals produced during the degradation of CS. As a result, returning them to fields is inappropriate (Lin and Alain, 2014). The traditional method of disposal entails burning CS before putting them back in fields. Although this method can completely negate the impact allelochemicals have on crops, it pollutes the air. The majority of CS in China is stored haphazardly in fields and has not been used. CS does naturally degrade or has been thrown out as waste. The eco-friendly handling of CS and plant straw has garnered a lot of interest. Additionally, a lot of academics have concentrated on creating novel approaches for applying CS to a variety of applications (Biswas et al., 2017; Ma et al., 2021). A natural polymer and biomass resource with remarkable qualities, such as high degradability and outstanding renewability, straw contains a significant amount of cellulose (Motte et al., 2014; Vassilev et al., 2012). A common study issue across the globe is the transformation of cellulose derived from basic materials into high value-added compounds. Additionally, the creation of nanocellulose from cellulose offers incredible application potential (Motte et al., 2014; Xu et al., 2013). CS's cell wall is primarily made up of cellulose, hemicellulose, and lignin. However, it is difficult to separate cellulose from hemicellulose and lignin because these elements are bound together by various forces. Hydrogen bonds are used to join the molecules of cellulose to create bundles. It is therefore challenging to depolymerize cellulose because 70% of it exists in the form of crystals (Ghasemi et al., 2016; Spagnol et al., 2012). The biomedical, pharmaceutical, material science, cosmetics, and papermaking industries can all benefit from nanocellulose's many exceptional qualities, including its natural regeneration ability, biocompatibility, biodegradability, high specific surface area, high aspect ratio, high strength, and good thermal stability (Favier et al., 1995). Nanocellulose has recently been employed in the food sector as a type of thickener, stabilizer, and fat replacement. Studies have shown that nanocellulose can reduce ice crystal development and enhance ice cream flavor (Lin and Alain, 2014).

As a result, several researchers have concentrated on extending the uses for nanocellulose. Nanocellulose can be prepared using a variety of techniques, primarily chemical techniques (like solid acid hydrolysis, TEMPO oxidation, and liquid acid hydrolysis), physical techniques (like grinding, freeze grinding, electrostatic silk prevention, and high-pressure homogenization), and biological techniques (like enzymatic hydrolysis and biosynthesis) (Cherian et al., 2010). Nanocellulose was prepared by sulfuric acid hydrolysis process. The sulfuric acid hydrolysis procedure produced nanocellulose that was suspended in the aqueous solution and collected. Due to its straightforward functioning, it is frequently employed in industry. However, more study is required to determine how byproduct recovery can be applied sustainably to environmental, economic, or product performance. Currently, rice straw, corn straw, cotton, and wheat straw were the most common raw materials used to make nanocellulose. The availability of nanocellulose raw materials was constrained due to the impact of culture area and planting conditions. Although cellulose made up to 37% of the composition of CS, it had not yet been fully utilized. As a result, the method used in this study to manufacture nanocellulose from CS had greater application. Whole wheat biscuits are becoming more well-known and in demand as public awareness of whole wheat foods and healthy eating grows. The amount of oil and sugar that are currently added to whole wheat cookies on the market frequently has a significant detrimental impact on blood sugar (Ajila et al., 2008). The benefits of using whole wheat raw materials have been countered by this processing approach. In addition, many whole wheat biscuits include significantly more wheat flour than whole wheat flour, which means that they fall short of meeting consumer expectations for high fiber.

12.6 TECHNOLOGICAL ADVANCES FOR THE EXTRACTION OF CAPSAICINOIDS

The extraction process is the most important and crucial step among the various upstream procedures for obtaining bioactive chemicals from agricultural leftovers. The proper release of phytochemicals from the residue matrix, which provides the necessary parameters to control the kind and quantity of bioactive compounds to extract, depends on the sample preparation. There are two primary categories of industrial procedures (pretreatment and extraction) to generate capsaicinoids: traditional approaches and cutting-edge methods (Sagar et al., 2018). By employing various procedures such as solvent, hydro-distillation, Soxhlet, and maceration extraction, the conventional approaches concentrate on using organic solvents as the extraction vehicle of capsaicinoids. The quantity, purity, and bioactivity of the extracted chemicals are all influenced by the extraction conditions. The most popular technique for recovering capsaicinoids relies on extracting them with hexane, which is extremely toxic and leaves residues that may degrade the compounds' purity. As an alternative to hexane, methanol, ethanol, acetonitrile, and water, new extraction techniques for capsaicinoids have been devised (Lu et al., 2020). The time required for extraction is inversely related to the solvent's polarity, and the standard procedures, which use organic solvents, yield between 70 and 80% of the total capsaicinoids (Abdurahman and Olalere, 2016). The hydro-distillation process is another extraction technique; while it is the simplest way to extract capsaicinoids from leftover materials, it has a low yield. Another widely used conventional approach with superior extraction yields is the *Soxhlet* process. Santos et al. (2015) used Soxhlet extraction with dichloromethane as a solvent in order to extract the capsaicinoids from Malagueta pepper (*Capsicum frutescens* L.) 92.1% of the capsaicinoids could be recovered thanks to this solvent alteration. Although these methods offer higher extract yields and total capsaicinoids contents, they also require large amounts of solvents and prolonged extraction times, which raises the cost and restricts their use due to environmental and health regulations.

12.7 CHEMISTRY AND BIOSYNTHESIS OF CAPSAICINOIDS

Alkaloid substances called capsaicinoids are what give Capsicum fruit its pungent flavor. With a melting point of 60–65°C and a molecular weight of 305.4 Da, capsaicinoids are a highly volatile, pungent, hydrophobic, colorless, and odorless white crystalline powder. The basic chemical structure of these molecules is an acid amide of vanillyl amine coupled with branched-chain fatty acids that range in length from 7 to 13 carbons (Aza-González et al., 2011). Their pungency is primarily influenced by the benzene ring but also by the length of the acyl chain. Although there are more than 10 different structures, capsaicin and dihydrocapsaicin make up the majority of capsaicinoids, making up nearly 90% of them (Meghvansi et al., 2010).

12.8 CONCLUSION

Industrial wastes are extremely desirable for various food applications. Byproducts from chili are a good source of bioactive substances, functional components, dietary fiber, minerals, and protein. The inclusion of chemicals derived from chilies as a possible source of functional ingredients can enhance the nutritional value of foods. The bread business could be one of the main uses for chili residue as a possible source of functional ingredients. By increasing the protein, mineral, and fiber content of bakery products, several physiological health benefits could be achieved. There is a lot of potential for the use of spices to be used properly, hence minimizing environmental pollution, as India is one of the biggest producers of spices and spice products.

KEYWORDS

- **bioactive compounds**
- **capsaicinoids extraction**
- **cellulose separation**
- **chili stems**
- **nanocellulose**
- **value-added applications**

REFERENCES

Abdurahman, N. H., & Olalere, O. A. (2016). Taguchi-based optimization technique: Piperine from black pepper. *Australian Journal of Basic and Applied Sciences, 10*(9), 293–299.

Ajila, C. M., Leelavathi, K., & Prasada Rao, U. J. S. (2008). Improvement of dietary fiber content and antioxidant properties in soft dough biscuits with the incorporation of mango peel powder. *Journal of Cereal Science, 48*, 319–326.

Ananthan, R., Subash, K., & Longvah, T. (2014). Assessment of nutrient composition and capsaicinoid content of some red chilies. *International Food and Nutrition Technology, 72*, 1–4.

Anderson, J. W., Baird, P., Davis, R. H., Ferreri, S., Knudtson, M., Koraym, A., & Williams, C. L. (2009). Health benefits of dietary fiber. *Nutrition Reviews, 67*(4), 188–205.

Arimboor, R., Natarajan, R. B., Menon, K. R., Chandrasekhar, L. P., & Moorkoth, V. (2015). Red pepper (*Capsicum annuum*) carotenoids as a source of natural food colors: Analysis and stability—A review. *Journal of Food Science and Technology, 52*(3), 1258–1271.

Arshadi, M., Attard, T. M., Lukasik, R. M., Brncic, M., da Costa Lopes, A. M., Finell, M., & Yuste-Córdoba, F. J. (2016). Pre-treatment and extraction techniques for recovery of added value compounds from wastes throughout the agri-food chain. *Green Chemistry, 18*(23), 6160–6204.

Aza-González, C., Núñez-Palenius, H. G., & Ochoa-Alejo, N. (2011). Molecular biology of capsaicinoid biosynthesis in chili pepper (*Capsicum* spp.). *Plant Cell Reports, 30*, 695–706.

Baenas, N., Belović, M., Ilić, N., Moreno, D. A., & García-Viguera, C. (2019). Industrial use of pepper (*Capsicum annuum* L.) derived products: Technological benefits and biological advantages. *Food Chemistry, 274*, 872–885.

Batiha, G. E. S., Alqahtani, A., Ojo, O. A., Shaheen, H. M., Wasef, L., Elzeiny, M., & Hetta, H. F. (2020). Biological properties, bioactive constituents, and pharmacokinetics of some *Capsicum* spp. and capsaicinoids. *International Journal of Molecular Sciences, 21*(15), 5179.

Biswas, B., Pandey, N., Bisht, Y., Singh, R., Kumar, J., & Bhaskar, T. (2017). Pyrolysis of agricultural biomass residues: Comparative study of corn cob, wheat straw, rice straw and rice husk. *Bioresource Technology, 237*, 57–63.

Carbonell-Capella, J. M., Barba, F. J., Esteve, M. J., & Frigola, A. (2014). Quality parameters, bioactive compounds and their correlation with antioxidant capacity of commercial fruit-based baby foods. *Food Science and Technology International, 20*(7), 479–487.

Chen, D., Liu, D., Zhang, H., Chen, Y., & Li, Q. (2015). Bamboo pyrolysis using TG-FTIR and a lab-scale reactor: Analysis of pyrolysis behavior, product properties, and carbon and energy yields. *Fuel, 148*, 79–86.

Cherian, B. M., Leão, A. L., Souza, S. F. D., Thomas, S., Pothan, L. A., & Kottaisamy, M. (2010). Isolation of nanocellulose from pineapple leaf fibers by steam explosion. *Carbohydrate Polymers, 81*, 720–725.

Coelho, M., Pereira, R., Rodrigues, A. S., Teixeira, J. A., & Pintado, M. E. (2019). Extraction of tomato byproducts bioactive compounds using ohmic technology. *Food and Bioproducts Processing, 117*, 329–339.

Cvetković, T., Ranilović, J., & Jokić, S. (2022). Quality of pepper seed byproducts: A review. *Foods, 11*(5), 748.

Delgado-Antequera, L., Gémar, G., Molinos-Senante, M., Gómez, T., Caballero, R., & Sala-Garrido, R. (2021). Eco-efficiency assessment of municipal solid waste services: Influence of exogenous variables. *Waste Management, 130*, 136–146.

Esayas, K., Shimelis, A., Ashebir, F., Negussie, R., Tilahun, B., & Gulelat, D. (2011). Proximate composition, mineral content and antinutritional factors of some *Capsicum annuum* varieties grown in Ethiopia. *Bulletin of the Chemical Society of Ethiopia, 25*(3).

Favier, V., Chanzy, H., & Cavaille, J. Y. (1995). Polymer nanocomposites reinforced by cellulose whiskers. *Macromolecules, 28*, 6365–6367.

Ghasemi, S., Behrooz, R., & Ghasemi, I. (2016). Extraction and characterization of nanocellulose structures from linter dissolving pulp using ultrafine grinder. *Journal of Nanoscience and Nanotechnology, 16*(6), 5791–5797.

Gustafsson, J., Cederberg, C., Sonesson, U., & Emanuelsson, A. (2013). *The Methodology of the FAO Study: Global Food Losses and Food Waste—Extent, Causes and Prevention*. FAO.

Hussain, A., Kamboj, N., Podgurski, V., Antonov, M., & Goliandin, D. (2021). Circular economy approach to recycling technologies of postconsumer textile waste in Estonia: A review. *Proceedings of the Estonian Academy of Sciences, 70*(1), 80–90.

Jayaprakash, A., Choi, J. K., Chahine, G. L., Martin, F., Donnelly, M., Franc, J. P., & Karimi, A. (2012). Scaling study of cavitation pitting from cavitating jets and ultrasonic horns. *Wear, 296*(1, 2), 619–629.

Klemm, D., Cranston, E. D., Fischer, D., Gama, M., Kedzior, S. A., & Kralisch, D. (2018). Nanocellulose as a natural source for groundbreaking applications in materials science: Today's state. *Materials Today, 21*, 720–748.

Kothari, S. L., Joshi, A., Kachhwaha, S., & Ochoa-Alejo, N. (2021). Chili peppers—A review on tissue culture and transgenesis. *Biotechnology Advances, 28*(1), 35–48.

Lin, N., & Dufresne, A. (2014). Nanocellulose in biomedicine: Current status and future prospect. *European Polymer Journal, 59*, 302–325.

Lu, M., Chen, C., Lan, Y., Xiao, J., Li, R., Huang, J., & Ho, C. T. (2020). Capsaicin—The major bioactive ingredient of chili peppers: Bio-efficacy and delivery systems. *Food & Function, 11*(4), 2848–2860.

Ma, Y. J., Bao, H. L., Hu, X. J., Wang, R. J., & Dong, W. B. (2021). Productions of phenolic rich bio-oil using waste chili stem biomass by catalytic pyrolysis: Evaluation of reaction parameters on products distributions. *Journal of Energy Institute, 97*, 233–239.

Mastilović, J., Kevrešan, Ž., Vukić, M., Ivanović, M., Radovanović, J., & Džinić, D. (2018). Possibilities for utilization of dietary fiber-rich supplement from pepper (*Capsicum annuum*) processing waste in bakery products. *Journal of Engineering Process Management, 10*, 28–33.

Meghvansi, M. K., Siddiqui, S., Khan, M. H., Gupta, V. K., Vairale, M. G., Gogoi, H. K., & Singh, L. (2010). Naga chili: A potential source of capsaicinoids with broad-spectrum ethnopharmacological applications. *Journal of Ethnopharmacology, 132*(1), 1–14.

Motte, J. C., Escudié, R., Beaufils, N., Steyer, J. P., Bernet, N., & Delgenès, J. (2014). Morphological structures of wheat straw strongly impact its anaerobic digestion. *Industrial Crops and Products, 52*, 695–701.

Padayachee, A., Day, L., Howell, K., & Gidley, M. J. (2017). Complexity and health functionality of plant cell wall fibers from fruits and vegetables. *Critical Reviews in Food Science and Nutrition, 57*(1), 59–81.

Pawar, S. S., Bharude, N. V., Sonone, S. S., Deshmukh, R. S., Raut, A. K., & Umarkar, A. R. (2011). Chilies as food, spice and medicine: A perspective. *International Journal of Pharmaceutical and Biological Sciences, 1*(3), 311–318.

Pugliese, A., O'Callaghan, Y., Tundis, R., Galvin, K., Menichini, F., O'Brien, N., & Loizzo, M. R. (2014). *In vitro* investigation of the bio accessibility of carotenoids from raw, frozen and boiled red chili peppers (*Capsicum annuum*). *European Journal of Nutrition, 53*, 501–510.

Rubio, C., Hardisson, A., Martín, R., Báez, A., Martín, M., & Álvarez, R. (2002). Mineral composition of the red and green pepper (*Capsicum annuum*) from Tenerife Island. *European Food Research and Technology, 214*, 501–504.

Sagar, N. A., Pareek, S., Sharma, S., Yahia, E. M., & Lobo, M. G. (2018). Fruit and vegetable waste: Bioactive compounds, their extraction, and possible utilization. *Comprehensive Reviews in Food Science and Food Safety, 17*(3), 512–531.

Saini, R. K., & Keum, Y. S. (2018). Carotenoid extraction methods: A review of recent developments. *Food Chemistry, 240*, 90–103.

Sandoval-Castro, C. J., Valdez-Morales, M., Oomah, B. D., Gutiérrez-Dorado, R., & Shakir, A. Y. (2021). Identification and characterization of bioactive compounds in chile pepper waste for potential use in functional foods (Doctoral dissertation). New Mexico State University.

Santos, P., Aguiar, A. C., Barbero, G. F., Rezende, C. A., & Martínez, J. (2015). Supercritical carbon dioxide extraction of capsaicinoids from malagueta pepper (*Capsicum frutescens* L.) assisted by ultrasound. *Ultrasonics Sonochemistry, 22*, 78–88.

Singh, R. K., Pandey, K. B., & Rizvi, S. I. (2012). Medicinal properties of some Indian spices. *Annals of Phytomedicine, 1*(1), 29–33.

Sowbhagya, H. B., Soumya, C., Indrani, D., & Srinivas, P. (2015). Physico-chemical characteristics of chili spent residue and its effect on the rheological, microstructural and nutritional qualities of bread. *Journal of Food Science and Technology, 52*(11), 7218–7226.

Spagnol, C., Rodrigues, F., Pereira, A. G. B., Fajardo, A. R., Rubira, A. F., & Muniz, E. C. (2012). Superabsorbent hydrogel composite made of cellulose nanofibrils and chitosan-graft-poly(acrylic acid). *Carbohydrate Polymers, 87*, 2038–2045.

Vassilev, S. V., Baxter, D., Andersen, L. K., Vassileva, C. G., & Morgan, T. J. (2012). An overview of the organic and inorganic phase composition of biomass. *Fuel, 94*, 1–33.

Viacava, G. E., Ansorena, M. R., Roura, S. I., González-Aguilar, G. A., & Ayala-Zavala, J. F. (2013). Fruit processing byproducts as a source of natural antifungal compounds. In *Antifungal Metabolites from Plants* (pp. 447–461).

Villa-Rivera, M. G., & Ochoa-Alejo, N. (2020). Chili pepper carotenoids: Nutraceutical properties and mechanisms of action. *Molecules, 25*(23), 5573.

Xu, F., Jian, S., Lv, W., Yu, Z., & Li, Y. (2013). Comparison of different liquid anaerobic digestion effluents as inocula and nitrogen sources for solid-state batch anaerobic digestion of corn stover. *Waste Management, 33*, 26–32.

CHAPTER 13

Aquaculture-Based Wastes, Byproducts, and Their Utilization

HARSHAVARDHAN DHULIPALLA, OLIVIA FRANCIS, ANEETA BABU, MRINMOY ROY, RADHESHYAM B. BAJAD, and IRSHAAN SYED*

Department of Food Technology, School of Agriculture and Food Technology, Vignan's Foundation for Science, Technology and Research, Vadlamudi, Guntur, Andhra Pradesh, India

*Corresponding author

ABSTRACT

Globally, seafood production has increased steadily over the past few decades. Rising population and health consciousness have increased demand for seafood. To meet demand, intensification of fish and aquaculture farming has been observed in recent years. Consequently, this has led to a significant incline in fish processing waste. Furthermore, disposal of fish processing waste has led to environmental impact, including water pollution, pathogen growth, and foul odors. One viable solution is to focus on valorization and the waste-to-value chain for utilizing fish waste. Processing of fish biomass into biomethane, biodiesel, and biofertilizers has become possible through advanced processing methods. Additionally, physical, biochemical, and thermochemical processes enable recovery of value-added compounds such as fish protein hydrolysates, pigments, chitosan, and collagen. Similarly, increase in nutrient loads in aquaculture effluent and related environmental impacts have led to a shift to effective wastewater treatment techniques. Microalgae represent a promising option for phytoremediation of aquaculture effluents owing to high nutrient removal efficiency. However, the ability to utilize or convert resulting

Utilization of Food Waste and Byproducts. Mudasir Ahmad Malik and Kshirod Kumar Dash (Eds.)

DOI: 10.1201/9781779641151-13

microalgal biomass into aquaculture feed plays a crucial role in enhancing sustainability. This book chapter focuses on sources, elements, management strategies, and processing approaches of aquaculture-based wastes.

13.1 INTRODUCTION

Aquaculture is the "husbandry of submarine creatures and plants" such as fish, seaweed and algae, mollusks, shrimps, and crabs. Aquaculture has the potential to bring a revolutionary change that could secure global food consumption. Over the past 50 years, the production of fish and seafood has increased rapidly. Global production of aquaculture and fisheries products was 214 million tons in 2020, which has risen sharply compared to the 1990s (FAO, 2020) (Figure 13.1). It includes 178 and 36 million tons of aquatic animals and algae (wet basis). A healthy, balanced diet should consist of fish and other seafood since they are essential sources of food and nourishment 2015 saw a new high for global fish consumption, 20.2 kg per person (FAO, 2018).

Fish waste (20–80%, depending on kind and processing circumstances) includes edible body meat and leftover body parts such as skin, heads, tails, frames, viscera, lungs, and bones. Massive amounts of particle organic and inorganic excretory waste are produced by large fish farms (Ward and Løes, 2011). Fish processing leaves behind a residue of almost 50% of the weight used to create byproducts. Fish waste, including viscera, fins, bones, skin, tails, frames, heads, and lungs, comprise the remaining garbage. However, the consumption demand of other aquatic animals and fish and the production of wastes from such aquatic sources trigger the need for waste processing, which can cause severe environmental impacts. Decaying organic waste generates large amounts of greenhouse gases, such as carbon dioxide and methane, contributing to climate change.

Massive amounts of organic and inorganic excretory waste are produced by large fish farms (Ward and Løes, 2011). After processing and treatments, the remaining residual parts represent fish waste near an average of 50% of the total weight. Consequently, the growing demand for edible seaweed has risen recently due to its enormous functional properties such as anti-microbial, anti-inflammatory, anti-diabetic, anti-hypertensive, and antioxidant properties and rich in bioactive compounds.

Government agencies are compelled to apply for waste operation programs due to the rising waste output and requirements for using renewable energy sources. In addition, less landfilling for residual disposal must be done in the waste operation sector. The good news is that people are becoming more interested in energy recovery from fish processing waste. Anaerobic digestion (AD) is a common method for treating fish waste in developed nations. One of the most important aspects of waste reduction initiatives is the recovery of marketable byproducts from aquaculture wastes. The fishing industry's industrial waste could be utilized as a sustainable source (FAO, 2016). Fish biomass has been converted into biomethane, biodiesel, and biofertilizer using specialized methods and procedures. It is possible to produce useful substances using a variety of physical, biochemical, and thermochemical processing methods, including natural pigments, chitosan, fish protein hydrolysate, and collagen.

The use of and energy recovery from fish waste have grown to be important economic sectors. Utilizing fish processing byproducts and underutilized marine species are given considerable attention in this area. This chapter discusses the essential compounds that could be derived from byproducts of fish waste processing, edible algae, and shellfish and emphasizes their potential uses.

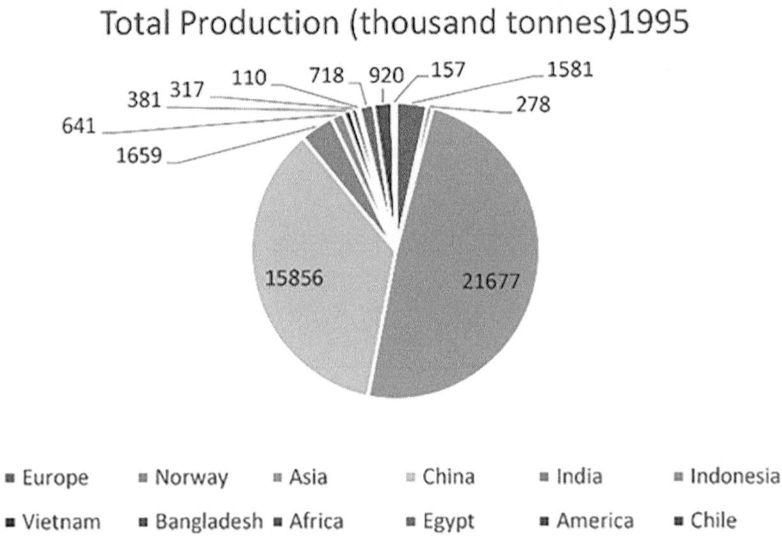

(a)

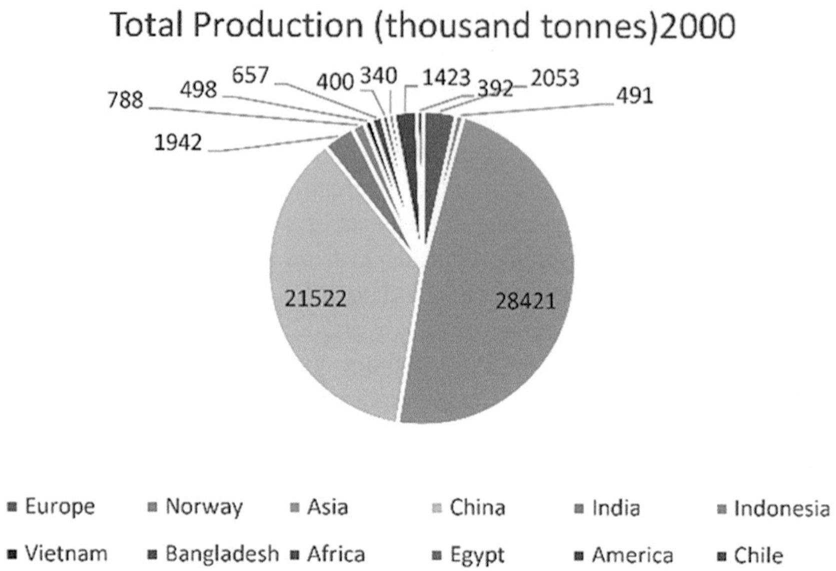
Total Production (thousand tonnes)2000
788
498
657
400
340
1423
392
2053
491
1942
21522
28421
Europe
Norway
Asia
China
India
Indonesia
Vietnam
Bangladesh
Africa
Egypt
America
Chile

(b)

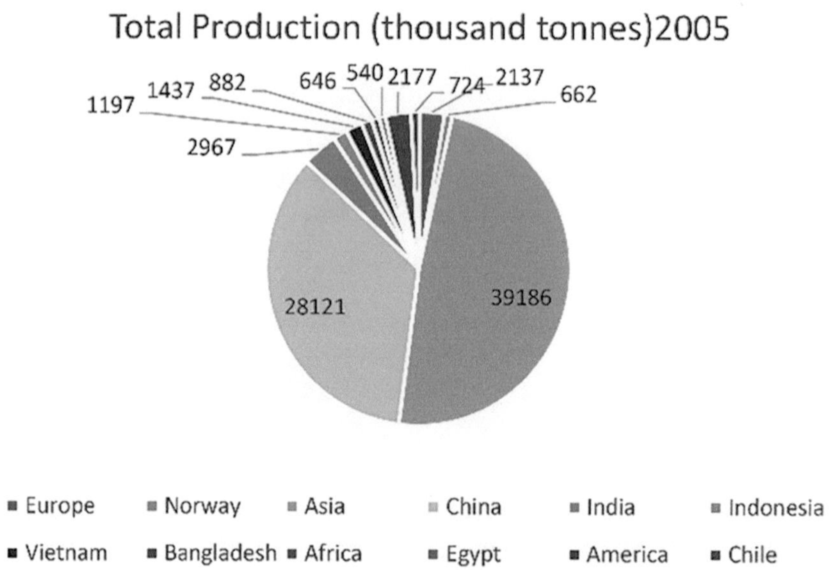
Total Production (thousand tonnes)2005
1197
1437
882
646
540
2177
724
2137
662
2967
28121
39186
Europe
Norway
Asia
China
India
Indonesia
Vietnam
Bangladesh
Africa
Egypt
America
Chile

(c)

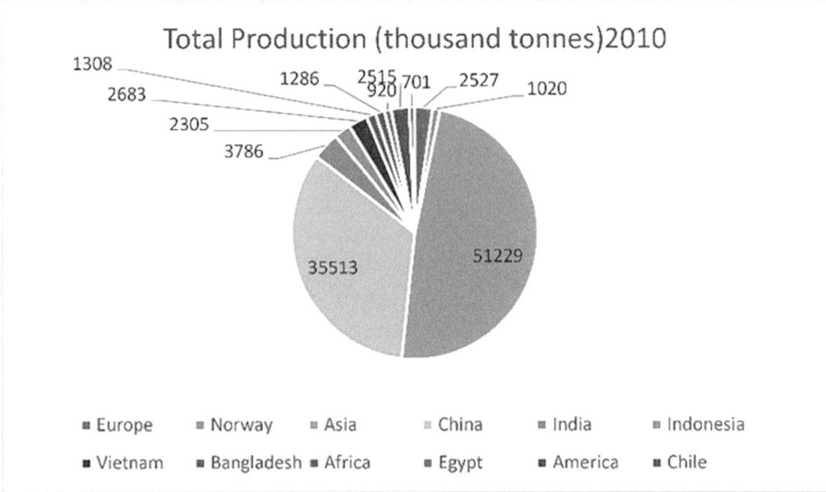

Total Production (thousand tonnes)2010
1308
2683
2305
3786
1286
2515
920
701
2527
1020
35513
51229
Europe
Norway
Asia
China
India
Indonesia
Vietnam
Bangladesh
Africa
Egypt
America
Chile

(d)

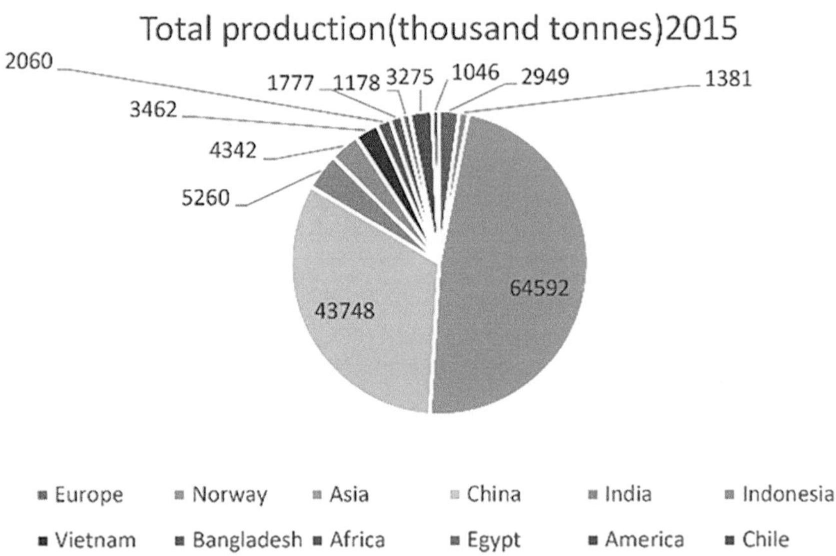

Total production(thousand tonnes)2015
2060
3462
4342
5260
1777
1178
3275
1046
2949
1381
43748
64592
Europe
Norway
Asia
China
India
Indonesia
Vietnam
Bangladesh
Africa
Egypt
America
Chile

(e)

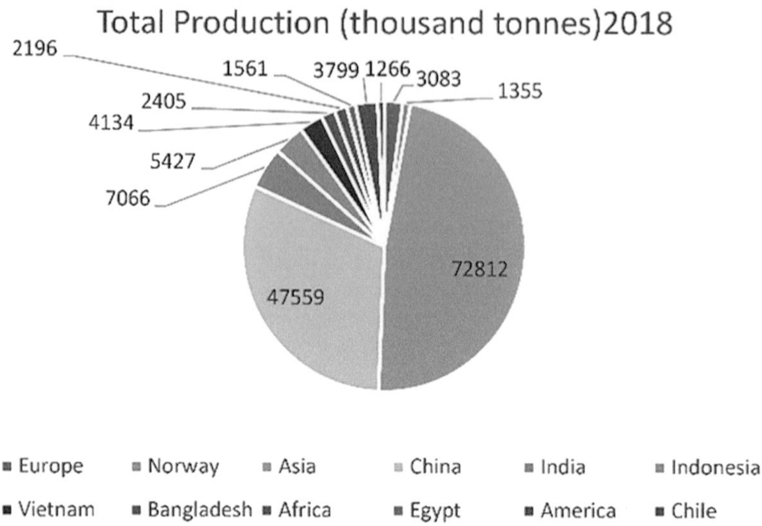

(f)

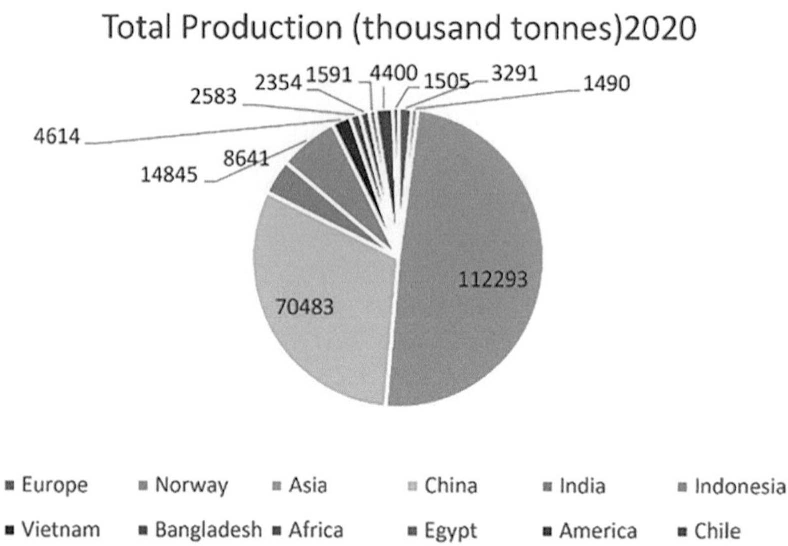

(g)

FIGURE 13.1 Annual growth of aquaculture production in various regions (in parentheses, represented as worldwide percentages and millions of tons).

Source: Adapted from FAO (2018, 2020).

13.2 SOURCES OF AQUACULTURE WASTE

Aquaculture is similar to other production industries, with different inputs to produce different natural and artificial products. After production or during production and cultivation, there is always waste in such a system. These wastes are generated in numerous ways: unutilized inputs or byproducts. Aquaculture wastes are categorized into three main categories, which are as follows as in sub section

13.2.1 FEED

It is the most important aspect of aquaculture production, and its relative significance depends upon the method of culture employed (Biswas et al., 2006, Dauda et al., 2017). In a whole system, the kind of environment has a significant impact on the feed given to fish. The fish are distributed sparsely over a vast aquaculture zone, and they mostly consume organisms that are in motion naturally. Pond fertilization helps to increase natural productivity (Ajani et al., 2011). The present demand for aquaculture products cannot be met by such an exterior-focused growing method. Fish in semi-intensive aquaculture are populated at densities ranging from moderate to considerably high levels, necessitating the provision of supplementary feed from the farmers in addition to their natural food sources (Dauda et al., 2017). Natural feed production is typically not suitable in intensive systems; instead, high-quality synthetic aqua feeds are employed to target the fish's rapid growth (Ajani et al., 2011). The main source of waste in aquaculture is feed, which is affected greatly by several factors such as nutritional content, manufacturing method (extruded or pelleted), the feeding method, the amount of fee consumed per minute, proportion of feed to fish size, and the length of storage (Miller and Semmens, 2002).

13.2.2 CHEMICALS

Chemicals are rigorously limited in use in a fish farm of current aquaculture practices. Nevertheless, certain substances are still employed as particular medications, cleaning agents, and anti-fouling agents (Read and Fernandes, 2003). They are employed for chemotherapy intentions, encompassing the utilization of antibiotics for both prophylaxis and therapeutic objectives, anesthetics, ectoparasitic ides, endo parasiticides, and usage of vaccines for the control and treatment of spongers (internal and external), also for infections caused by the microbes (Ajadi et al., 2016; Costello et al., 2001). In aquaculture, a variety of chemicals are utilized, including lime, which is used to neutralize the acidity of pond bottoms during pond preparation; salts, which primarily help to relieve stress in fish; and other substances thought to be safe for fish. Despite the fact that some chemicals are necessary for fish culture, they may potentially have adverse environmental effects (Boyd and McNevin, 2015).

13.2.3 PATHOGENS

The pathogen category of waste in the aquaculture system is seldom taken into account, often when it is below the position that impacts the farmed fish, despite the fact that releasing diseases with wastewater may harm the underwater species in natural water bodies (Goldburg et al., 1997). Pathogenic burdens in natural water bodies already exist, and adding supplementary pathogenic loads from fish culture may cause strain or even kill subaquatic creatures. The widespread release of effluents from ponds in semi-intensive pond aquaculture—from which organic fertilizers are used in aquaculture—results in a substantial rise in the amount of pathogens. This occurs more frequently in Africa. fecal streptococci are present in large concentrations in organic fertilizers made from blood, cow faces, cow dung, pig manure, and chicken manure (Ampofo and Clerk, 2003).

13.3 TYPES OF AQUACULTURE WASTE

Aquaculture wastes are categorized into two main categories as in sub sections.

13.3.1 SOLID WASTES

It is generated from the unconsumed feed of aquatic organisms, dead fishes, shellfish, some other dead aquatic organisms, and their fecal matter. Solid wastes can be again categorized into two groups, settled solids and suspended solids. The latter is difficult to remove from aquaculture systems after employed sedimentation due to their fine particle sizes and due to it being dissolved within water (Cripps and Bergheim, 2000). The settled solids can easily be removed from the aquaculture system due to their bigger particle size settling led down to the culture system in less time (Ebeling and Timmons, 2012).

Solid waste is considered hazardous waste as compared to any other waste, specifically in the case of sizable and stationary particles, as they have the potential to obstruct fish gills and result in mortality (Olusegun et al., 2016). To avoid such incidents within fish culture, they should be effectively removed as soon as possible from the culture system (Timmons and Summerfelt, 1997). The total suspended solid and total dissolved solid increase when it is left to decay over an extended period, which may cause an increase in nitrogen constituents and stress in the cultured fish (Olusegun et al., 2016). Furthermore, chemical and biochemical oxygen demand (BOD) increases with the increase in aerobic bacterial activity, reducing oxygen inside the culture column (Timmons, 1994). With proper farm management, which includes correct feeding, efficient use of feed, appropriate storage, around 30% of the feed will become solid waste (Miller and Semmens, 2002).

13.3.2 LIQUID OR DISSOLVED WASTE

It includes metabolic food products in fish or decomposed, unconsumed feed. Phosphorus and nitrogen products are the two significant factors of concern in dissolved wastes (Boyd and Massaut, 1999). Protein is the main element of fish feed, and it is constituted of two elements, that is, nitrogen and phosphorus, which are essential protein factors. Fish of all types, regardless of their species, bear significantly high salutary CP, which ranges between 25 to 50%. Fish feeds contain high levels of phosphorus and nitrogen with a high protein content, but the proportion of phosphorus and nitrogen that remains in the body of a fish is less than 50% (Piedrahita, 2003). Therefore, when eventually it has been released, a significant proportion is transferred to the water in which the culture is maintained, where it becomes irritating and has a lot of environmental impacts.

Ammonia forms in aquaculture water due to the release of nitrogen. It may be further decomposed to nitrate and nitrite in the culture column based on biological activity (Dauda et al., 2014; Piedrahita, 2003). If ammonia is not handled before being discharged into the environment, it is mostly toxic to fish culture in the system and those accessing water bodies (Romano and Zeng, 2013). Ammonia oxidation usually produces a hazardous intermediate product, nitrite, which is ideally kept below 0.5 mg/L in fish farming (Ajani et al., 2011). But, the unstable nature of nitrite leads to oxidation to form nitrate. Another critical metabolite or decomposed aquaculture feed product that is insufficiently used is phosphorus. Phosphorus does not harm farmed fish the way ammonia does. However, varying in its attention, degree of discharge and the size of the incoming water body, its discharge into the environment enhances the nutrient content water bodies, leading to eutrophication (Wong, 2001).

13.4 FISH WASTE MANAGEMENT AND UTILISATION

Production, treatment, planning, prepayment, processing, recycling, and final residual disposal of fish waste are the main areas of fish waste management. In the last three decades, a large number of directives have been launched under the aegis of the European Union. These recommendations deal with the environmental effects of fishing products and the management of aquaculture waste (Arvanitoyannis and Tserkezou, 2014). Guidelines for fishing programs must be implemented due to the growing application of byproducts. The ideal waste management strategy, however, is dependent on projected costs and benefits. Waste is processed by aquaculture and its connected sectors, which is seen to be a serious hazard to the ecology. Utilizing the right technology will help reduce pollution. The primary goal of fisheries management is to modify these wastes while concurrently revitalizing key materials before disposal (Uddin et al., 2010).

13.4.1 BIOGAS PRODUCT

Methane (CH_4), carbon dioxide (CO_2) and a fractional quantity of hydrogen sulfide (H_2S) and hydrogen (H_2) are the different compounds that consist in biogas that is generated through the breakdown using AD. The organic carbon content is high in fish waste; hence it is a potentially relevant source of methane products. On the other hand, fishery biomass limits biogas products due to elevated ammonia and nitrogen. Anaerobic treatment of fish disposal can be applied by using co-digestion. The co-digestion method's crucial problem is the co-substrate mixture's accurate composition. The parameters like carbon-nitrogen ratio, pH, macro- and micronutrients, biodegradable organic matter, dry matter, and poisonous composites are essential to balance (Tomczak-Wandzel et al., 2013). A general method of biogas product is shown in Figure 13.2. Organic residues, animal manure, energy crops, external solid waste and sewage sludge are the different types of biodegradable waste that treat biogas plants. The co-substrate mixture needed recycling, like chopping, liquefying, mixing, and inoculation. The relative amounts of CO_2, H_2S, CH_4 and NH_3 present in biogas are indicative of its quality. Biochemical Methane Potential (BMP) test is the primary method utilized to determine the ability of a feedstock to undergo anaerobic degradation. BMP test utilizes a batch digestion method where anaerobic microorganisms are incubated with substrates. A way to determine the speed of feedstock breakdown is to measure the exact volume of biogas produced (Veluchamy and Kalamdhadm, 2017). The main benefits of this system are its affordability, repeatability, simplicity, and quickness in determining if a substrate is suitable for AD. Standard temperature and pressure (STP) ml CH_4 per 1 g of VS injected is used to represent the methane eventuality (Hansen et al., 2004). Many lab-scale analyses of BMPs for fish waste have recently been reported. The majority of topics concern the concurrent digestion of fish byproducts and animal manure, organic waste, sewage sludge, and different wastes from the food sector. These studies sought to choose the best co-digestion substrates to boost the generation of biogas from aquaculture waste. The investigation's findings show the ideal proportion of several co-substrates for enhancing methane eventuality (Kafle et al., 2013). Numerous solid fish wastes' potential biochemical methane production was also investigated. The feedstock with the greatest methane output contained 1% of the total waste solids. Due to the enhanced fat content, it resembled excrement from sardines, tuna, and needlefish (Eiroa et al., 2012). There are just a few artificial operations, despite the promising results of using fish waste as an energy source for the production of biogas. The probable causes could be the significant expenses associated with constructing biogas facilities and the restricted biogas yield due to the elevated levels of ammonia nitrogen found in fish waste.

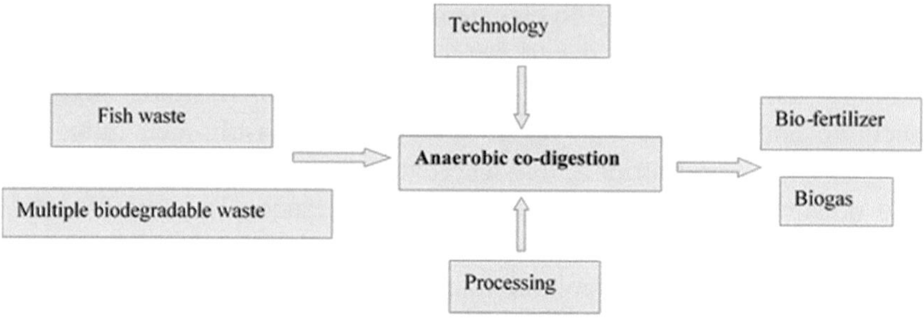

FIGURE 13.2 Anaerobic co-digestion of fish waste.

13.4.2 BIO-FERTILIZER

As the popularity of biogas increases, the byproducts obtained after AD are also utilized as a crucial source of nutrient-rich material. Digestate can be used as a bio-fertilizer after undergoing several unit procedures. The characteristics of the digestate are determined by the nature and makeup of the substrates that were digested (Drosg et al., 2015). Digestate is made up of inedible waste, dead microbes, process interposers, and structural plant materials. More than 90% of the feedstock is included in the entire volume. The digestate still contains substances that resemble potassium, phosphorus, and nitrogen. High-quality digestate found in the nutrient-rich excrement of fish may be converted into fertilizer for farming. The initial stage of metamorphosis involves combining organic waste and the byproduct of digestion with the equivalent minerals to achieve the necessary levels of NPK (nitrogen, phosphorus, and potassium). The average nutritional levels of digestate are comparable, although the composition varies on the source ingredients and the digestive method. The energy from the biogas plants is also used to dry a combination of thick liquid thickness. Small granulation of the bio-fertilizer, which is necessary for practical use, transportation, and storage, is the process's final phase. In spite of the fact that a raw slurry makes nutrients more accessible to stores, bio-fertilizer distributes nutrients slowly and lessens their influence on the ground.

13.4.3 BIO-OIL/ BIODIESEL PRODUCTION

The fish processing industry produces a considerable amount of fish oil. This waste product may be converted into sustainable energy. Due to its low carbon and high hydrogen content, fish oil has been the subject of several research studies about its potential as an energy source. It has an advanced flash point and a substantially reduced kinematic density (Yahyaee et al., 2013). As a fuel for diesel engines, bio-oil offers useful qualities. It is superior in quality and heating value to methyl-esterified vegetable oil waste and regular diesel fuel. At low temperatures, diesel engines might make significant use of biodiesel made from fish waste. The likelihood that fish waste will be converted into biofuel depends on the site and size of the processing facility, the properties of the fish waste and the type of energy conditions.

Numerous chemical, biological, thermochemical, and physical techniques are used to convert biomass into biofuel (Jayasinghe and Hawboldt, 2012). The pyrolysis of biomass is the initial procedure. In anaerobic circumstances, the thermal decomposition occurs at high temperatures (between 350 and 700°C). Bio-oil (alcohols, esters, organic acids, phenols, aldehydes, nitrogen compounds, ketones, and alkenes), biochar (hydrogen and carbon) and biogas (CO_2, CO, and CH_4) are the byproducts of pyrolysis. The most promising thermal conversion technique for biomass right now is rapid pyrolysis. Heat and mass transmission characteristics are crucial since fast pyrolysis lasts for many seconds. The choice of clothing is crucial as well. Figure 13.3 depicts the procedure, the required pharmaceutical route, and the sorts of products obtained.

The fermentation and hydrolysis processes can then utilize the oil that was recovered from the waste biomass. Applying sugars or organic acids to bacteria causes silage, from which oil can be extracted. Oil is extracted and refined in many ways to produce pure bio-oil. Numerous operation units, including pressing, filtration, degumming, neutralizing, bleaching, hydrogenation, and deodorizing, relate to them. Processing biomass for biofuel is becoming increasingly popular. It is anticipated that fish processing oil will be a crucial source of biodiesel. Notwithstanding the vast capacity of biofuels, there are obstacles in terms of both technology and finance.

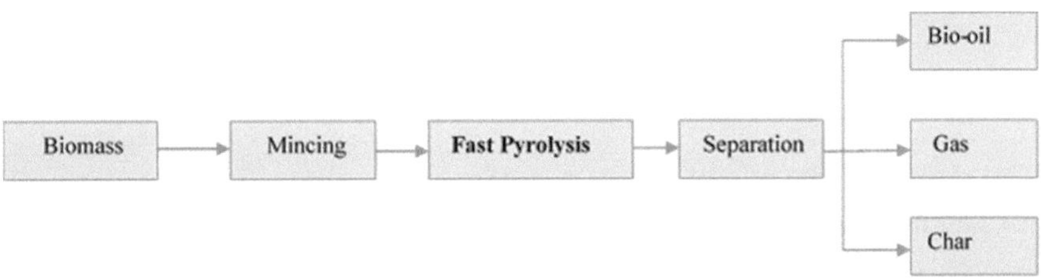

FIGURE 13.3 Fast pyrolysis process.

Source: Adapted from Meier et al. (2013).

13.5 FISH BYPRODUCTS AND THEIR APPLICATION

13.5.1 BIOACTIVE COMPOUND

Fish processing results in the production of a sizable number of byproducts. Approximately 27% of the entire fish consists of fish byproducts, such as tail and head. In contrast, 25% of the whole fish is made up of skin, bones, blood, and intestines, as seen in Figure 13.4. The majority of these wastes are presently utilized to manufacture fish meal, which is an excellent source of fish oil and protein (70% protein, 10% water, 10% fat, and 10% mineral) (Sanchez et al., 2018). The leftover seafood can be processed into a highly valuable commodity, such as fish food, that can either be sold or utilized as fertilizer. Fish oil can be used for both food and inedible purposes, such as in hydraulic oil, glycerol, or soap (Ferraro et al., 2010).

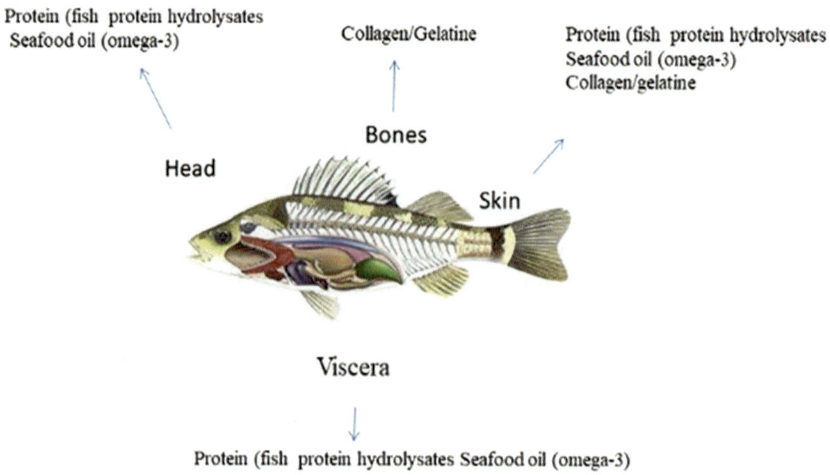

FIGURE 13.4 Main byproducts from fish processing.

12.5.2 FISH PROTEIN HYDROLYSATE (FPH)

The development of innovative techniques is being driven by the need for value-added products, including bioactive composites, gelatin, hydrolysates, and collagen, in the aquaculture sector. The intricate process of processing fish involves a series of steps, including stunning, grading, slime removal, beheading, disemboweling, fin slicing, and separation of bones. Depending on the kind of fish and the stage of processing, there is a significant amount of waste produced throughout the process of processing fish (20–80%) (Ghaly et al., 2013). Fish waste may be utilized for creating proteins, amino acids, collagen, oil, gelatin, and peptides, which are the industry's standard goods and a top priority globally. In an effort to develop novel uses for the fish waste generated by the aquaculture industry, several researches have been carried out during the past few years. The skull, skin, and viscera of fish that are not utilized for human food might be an intriguing source of proteins. Heat, chemicals, and enzymatic treatment easily achieve the denaturation of these numerous proteins, creating a gel to generate surimi and related compounds (Le Gouic et al., 2018). Bones and skin from fish may also be used to make collagen and gelatin, which can be used in place of human collagen in culinary items (Karim and Bhat, 2009). Fish viscera contain high protein and fat content, which is an excellent source for producing FPH (Villamil et al., 2017). Producing peptides with improved functional and bioactive characteristics has made FPH more attractive.

13.6 CRUSTACEANS BYPRODUCTS AND THEIR APPLICATION

13.6.1 BIOACTIVE EMULSION

In aquaculture, crustaceans are widely raised, particularly prawns, which are sold in markets all over the world. The head and exoskeleton (shell and tail) of prawns are the byproducts of prawn processing. With respect to the species, the waste generated for the head accounts for 28–38.9%, the tail accounts for 8–11.5%, and the shell accounts for 2.3–4% (Heu et al., 2003, Sánchez-Camargo et al., 2011). Machinery is employed to sort the byproducts into head and tail sections, which yield byproducts with notable variation in their chemical composition. Protein, fat, ash, carotenoid (especially astaxanthin), and chitin make up both types of discards. Overall, however, a shrimp head is more abundant than a shrimp shell in terms of carotenoid content (20.6 vs. 5.3 mg/100g), ash (21.2 vs. 18.77%), protein (54.4 vs. 43.98%), and fat (11.9 vs. 1.53%). According to Trung and Phuong (2012) and Sánchez-Camargo et al. (2011), in terms of chitin content, the shell surpasses the head as a superior source (34.92 vs. 9.3%).

13.6.2 CHITIN AND CHITOSAN

Chitosan, a vital product of chitin, holds immense importance in many sectors, including cosmetics, food, biomedical, and pharmaceutical industries, owing to its essential uses. It is a cost-effective, eco-friendly, non-hazardous material that finds its utility in the pharmaceutical industry to produce biomaterials and excipients (Kim and Thomas, 2007). Chitosan is a straight-chain polysaccharide obtained by partially deacetylating chitin under alkaline circumstances. It consists of N-acetylglucosamine, an amino sugar, linked together by beta (1–4) glycosidic bonds. Chitin, specifically found in the exterior shells of crustaceans like crabs, prawns, lobsters, and mollusks like squid pens, is the second-most significant polymer after cellulose. Contrary to other commercially available polysaccharides (including carrageenan, alginate, agar, cellulose, and pectin, among others), chitosan exhibits an alkaline or neutral behavior (Sing and Ray, 2000). This unique quality facilitates it to carry out a wide range of specialty tasks in the food sector as texture, emulsifier, and flavor additions (Kim and Thomas, 2007); as an operative emulsion, it operates as an antioxidant and an anti-microbial agent in edible films (Elsabee et al., 2008; No et al., 2007) or as an excipient in the pharmaceutic business (Rinaudo, 2006) or a medication supply method (Wong, 2009). Chitosan has been touted as a powerful lipid-immersion agent that prevents gastritis while also lowering blood cholesterol (Liao et al., 2007; Muzzarelli et al., 2007).

13.6.3 CAROTENOIDS

A large range of plants, algae, bacteria, and fungi, as well as other photosynthetic creatures, have pigments called carotenoids as the basis of their color. Carotenoid denovo synthesis is not possible in creatures. As a result, the carotenoid has to be consumed through food. In terms of its biological function in humans as a precursor to pro-vitamin A, vitamin A is the most significant carotenoid. Marine creatures (such as fish and crustaceans) may modify carotenoids via metabolic activities, producing -carotene metabolites such as alloxanthin, astaxanthin fucoxanthin, and peridinin, among others. Due to its beneficial effects on health, astaxanthin is a very pleasant carotenoid among these compounds. In general, carotenoids possess more antioxidant activity compared to that vitamin C and tocopherol (vitamin E). Astaxanthin demonstrates antioxidant and anti-inflammatory action (Stepnowski et al., 2004).

Astaxanthin is mostly obtained from algae and crustaceans, accounting for 74–98% of all pigments; the number of carotenoids in crab shells ranges from 2.3–33.1 g per 100 g of carotenoids (Shahidi and Brown, 1998). Therefore, the byproducts from shrimps, prawns, and other crustaceans have gained significant attention and are utilized not just for chitin separation (Ferraro et al., 2010). Astaxanthins, along with its esters including both monoester and diester, were found in prawn byproduct oil and were found to be more effective antioxidants than -tocopherol (Gómez-Estaca et al., 2017; Núñez-Gastélum et al., 2016).

13.6.4 LIPIDS OMEGA-3 POLYUNSATURATED ADIPOSE ACIDS

The fish byproducts section provided a preliminary explanation of the physiological impacts of eicosatetraenoic acid (EPA) and docosahexaenoic acid (DHA) on blood pressure, lipoproteins, and overall cardiovascular health. An intriguing source of n-3 polyunsaturated fatty acids (PUFA) is *Antarctic krill* (Heu et al., 2003, Ju and Harvey, 2004, Sánchez-Camargo et al., 2011). Spotted shrimps are increased to 10.7% and 10.9%, respectively, by shrimp byproducts' roughly 8.9% and 10.7% EPA and DHA contents, as shown in Table 13.1 (Heu et al., 2003).

TABLE 13.1 The Proportions of Eicosatetraenoic Acid (EPA) and Docosahexaenoic Acid (DHA) in Various Prawn Parts

		Oil Composition			
Species	**Shrimp Part**	**Oil Extraction Method**	**EPA (C20:5n-3) % FA in total FA**	**DHA (C22:6n-3) % FA in total FA**	**References**
Black tiger shrimp (*Penaeus monodon*)	Edible muscle	Soxhlet extraction	8.58	14.9	Sriket et al. (2007)
White shrimp (*Penaeus vannamei*)	Edible muscle	Soxhlet extraction	9.46	9.99	Sriket et al. (2007)
Red spotted shrimp (*Farfantepenaeus paulensis*)	Heads, shells, and tails	Soxhlet extraction	11.69	12.24	Sánchez-Camargo et al. (2011)
Northem pink shrimp (*Pandalus borealis*)	Heads, shells, and tails	Soxhlet extraction	8.9	10.7	(Heu et al. (2003)
Spotted shrimp (*Trachypena curvirostris*)	Heads, shells, and tails	Soxhlet extraction	10.7	10.9	Heu et al. (2003)

13.7 SEAWEED BYPRODUCTS AND THEIR UTILIZATIONS

13.7.1 BIOACTIVE COMPOSITES

Around 28% of all aquaculture products produced worldwide are marine microalgae or seaweed (Paul et al., 2013). Seaweeds can be categorized into three: green seaweeds (Monostroma/Ulva, Caulerpa, and Codium), red seaweeds (Kappaphycus/Eucheuma, Pyropia/Porphyra, and Gracilaria), and brown seaweeds (Sargassum, Saccharina, and Undaria).

Additionally, seaweeds act as a great reservoir of bioactive compounds such as polyphenols, PUFAs, peptides, proteins, and carbs. There is a sizable supply of bioactive materials with a variety of natural properties (anti-inflammatory, anti-diabetic, anti-cancer, anti-microbial, antioxidant, etc.), as depicted in Table 13.2 (Charoensiddhi et al., 2017). Compared to advanced plants, seaweeds have more bioactive metabolites, particularly sulfate-containing polysaccharides (Cunha and Grenha, 2016).

TABLE 13.2 Marine Macroalgae's Bioactive Compounds

	Species	Bioactive Compounds	References
Brown algae	–	Fucoidan	Negishi et al. (2013)
	–	Fucoxanthin and phloroglucinol	Lopes-Costa et al. (2017)
	–	Fucoidan and fucoxanthin	Hwang et al. (2016)
	–	Laminarin	Lee et al. (2012)
	–	Phloroglucinol, fucoxanthin and fucoidan	Pádua et al. (2015)
	–	Phlorotannin's	Sanjeewa et al. (2016)
	Ascophyllum nodosum	Total phenolics, uranic acid, and fucose	Kadam et al. (2015)
	Ascophyllum nodosum	Phlorotannin	Corona et al. (2016)
	Ascophyllum nodosum	Polyphenolic component	Zhang et al. (2007)
	Fucaceae	Phlorotannin's	Catarino et al. (2017)
	Fucus vesiculosus	Phlorotannin's	Wang et al. (2012)
	Gracilaria corticata	Sulfated polysaccharides	Seedevi et al. (2017)
	Hizikia fusiformis	Fucoxanthin	Yan et al. (1999)
	Laminaria digitata	Laminarin and fucoidan	Moroney et al. (2015)
	Laminaria ochroleuca	Fucoidan, alginate and phlorotannin's	Flórez-Fernández et al. (2019)
	Lobophora variegate	Fucans	Park et al. (2011)
	Sargassum binderi	Alginates and carrageenan's	Youssouf et al. (2017)

TABLE 13.2 *(Continued)*

	Species	Bioactive Compounds	References
	Sargassum thunbergii	Phlorotannin's	Bae (2011)
	Turbinaria ornata	Alginates and carrageenan's	Youssouf et al. (2017)
	Undaria pinnatifida	Fucoidan	Cho et al. (2009)
	Undaria pinnatifida	Fucoxanthin	Zaharudin et al. (2019)
Green algae	*Caulerpa*	Antioxidants and flavonoids	Tanna et al. (2018)
	Ulva armoricana	Ulvan	Berri et al. (2017)
	Ulva lactuca	Ulvan	Thanh et al. (2016)
	Ulva reticulata	Ulvan	Tran et al. (2018)
Red algae	*Chondrus ocellatus*	Carrageenan	Zhou et al. (2005)
	Eucheuma	Carrageenan	Hornshøj et al. (2015)
	Eucheuma cottonii	Carrageenan	Nogueira et al. (2016)
	Euchema denticulatum	Alginates and carrageenan	Youssouf et al. (2017)
	Gelidiumn sesquipedale	Nanocellulose and agar	de Oliveira et al. (2019)
	Gelidium sesquipedale	Sulfated polysaccharides	de Oliveira et al. (2019)
	Gracilaria corticata	Sulfated polysaccharides	Seedevi et al. (2017)
	Laminaria ochroleuca	Phycocyanin, phenolic compounds, phycoerythrin and carotenoid	Uribe et al. (2018)
	Osmundaria obtusiloba	β-carotene, phenolic content	de Alencar et al. (2016)
	Porphyra sp.	Phycoerythrobilin	Yabuta et al. (2010)
	–	Phenolic compounds, phycocyanin, phycoerythrin, and carotenoid substances	Uribe et al. (2018)
	Pterocladiella capillacea	phenolic content and β-carotene	de Alencar et al. (2016)
	Kappaphycus alvarezii	Alginates and carrageenan's	Youssouf et al. (2017)
	Undaria pinnatifida	Phycocyanin, phenolic compounds, phycoerythrin and carotenoid	Uribe et al. (2018)

Polysaccharides, which make up a large portion of seaweeds, are the most consumable compounds from seaweeds. This material may account for 75% (on a dry basis) of the overall mass (Jiménez-Escrig and Sánchez-Muniz, 2000). In green, red, and brown algae, the amount of polysaccharide makes up about 29–67%, 10–59%, and 10–75% of the dry weight, respectively. Glucans, laminarins, alginate, and fucoidans are the primary polysaccharides found in brown algae, despite the fact that their composition varies widely depending on the taxonomy.

Agar, carrageenan, mannans, porphyrin, and glucans are all present in red algae. The composition of glucans, xyloglucans, mannans, glucuronan's, and ulvans vary among green algae species. The majority of them are now readily available in markets and are utilized as prebiotics (Sardari and Karlsson, 2018). Additionally, green seaweed lectins show some potential as anticancer and antiviral agents, although little research has been done on them (Watanabe et al., 2013).

13.7.2 ALGINATE

Many different types of brown seaweeds have alginate built within them. Alginate is commonly used to stabilize dispersions, increase the viscosity of mixtures, and make beverages for those with low blood sugar. Californian Giant kelp (*Macrocystis pyrifera*) and the North Atlantic *Ascophyllum nodosum* are the primary industrial producers of alginates. *Sargassum, Ecklonia, Laminaria, Durvillaea*, and other types of seaweed are also utilized. Among these species, Laminaria, a fat derived from *Saccharina japonica* (kombu) eating, is extensively farmed; the others are obtained from the ocean. Thus, in 2015, the annual output of alginates was solely 644 tons, priced at USD 345 million and was retailed for USD 14 for each kilogram at an average rate (Porse and Rudolph, 2017).

13.7.3 CARRAGEENAN

Carrageenan composites are linear, highly molten chains of partly sulfated galactan. In particular, they are repeating units of α-D-glucopyranose attached to alternate 1, 3, and β-D-glucopyranose residues linked through positions 1, 4 to form hydrophilic sulfated linear galectins. This form of structure may be influenced by the type of seaweed, the environment in which it grows, and the type of industrial processing. The three primary isoforms of carrageenan are lambda, iota, and kappa, with only minor structural differences. About 50 carrageenan's in dry bases are generated mostly from Gigartina, Chondrus, Hypnea, and Eucheuma in the red seaweed species (McHugh, 1987). In terms of biological function, carrageenan is commonly employed as a gelling agent and hydrocolloid whose characteristics have crucial functions in the food business. Sulfated galactan insulated from carrageenan has reportedly been reported to have comparable natural properties to anticoagulant, fibrinolytic, and antiaggregant action as other oligo polysaccharides from seaweed (Güven et al., 1991). According to reports, these polysaccharides cause cancer cell lines to go into apoptosis (Kwon and Nam, 2007). In particular, carrageenan's immunomodulating effect can reduce the formation of rodent tumors (Yuan et al., 2006). In regard to this, several carrageenan's from *Chondrus ocellatus* were linked to anti-cancer and immunomodulation activities (Zhou et al., 2005). In reality, osteoinductive tools were created utilizing carrageenan in a $CaCO_3$ environment to create bias in medicine (Nogueira et al., 2016).

13.7.4 AGAR

Another crucial gelatinizing substance, agar, similar to carrageen, is mostly derived from Gracilaria and Gonidium (red algae). Agars are galectins that are hydrophilic and conform to (1–4) and (9-1-3)-D-galactose and include sulfates in a small quantity, just like carrageenan's (Pal, Kamthania et al., 2014). Methoxy group, sulfate ester, and pyruvate ketal substitutes are present in the main backbone to varying degrees in agar and agarose. As a result, agar has stronger gelatinizing characteristics than agarose because of its chemical makeup. Through a three-dimensional helix structure, enzymatic desulfation might affect the gel conformation (Lee, Lim et al., 2017). Agars do, in fact, become thick at low temperatures and soluble at high temperatures. As a result, agar's qualities for preserving density are undeniably comparable to those of carrageens. The *Gracilaria chilensis* species, which is forming in Chile, is used in the commercial development of Gracilaria. In Asia (China, Indonesia, and Vietnam), *Gracilaria* sp. is being grown in significant quantities in ponds and estuaries. Annually, 14,500 tons of agar were produced globally, with an approximate value of USD 246 million and a price of 17 USD per kg at an average rate in 2015 (Porse and Rudolph, 2017).

In addition to its usual use as gelling or thickening agent in the food industry and as an excipient in pharmaceutical capsules, it can also be used for food packaging because of its antioxidant and anti-microbial characteristics (de Oliveira et al., 2019; Seedevi et al., 2017). It also lowers blood sugar levels, stops red blood cells from clumping together, inhibits the generation of cytokines that promote inflammation, and possesses anti-cancer effects (Tanna and Mishra, 2019). It has been demonstrated that agar polysaccharides aid – glucosidase in avoiding antioxidant and anti-cancer actions (Sun et al., 2018).

13.7.5 FUCOIDAN, ULVAN, LAMINARIN

For their biological properties, seaweed polysaccharides are used extensively in tissue engineering, medicine, and damage repair. Pertaining to this, ulvan, fucoidan, and laminarin sulfate have also demonstrated significant use in biomedical research and the field of remedial medicine. These aquatic items are plentiful and inexpensive, and they have been devoured for show. A polysaccharide with sulfur found in brown seaweeds called fucoidan is composed mostly of sulfated L-fucose and a minute quantity of other sugars including glucose, galactose, xylose, uranic acid, and mannose that are joined by glycosidic linkages. According to reports, the pharmaceutical, cosmeceutical, and nutraceutical businesses have several biomedical activities (Senthilkumar et al., 2017).

Ulvan from green seaweed is largely sulfated because it includes xylose, xylose 2-sulfate, rhamnose 3-sulfate, iduronic acid, and glucuronic acid units. Some of these are uncommon sugars related to sulfate, iduronic acids, and rhamnose. As a result, it has a complicated chemical makeup and a changeable nature (Tziveleka et al., 2019). Several bioactivities, including antibacterial, immune-boosting, anti-cancer, antioxidant, antihyperlipidemic, antiviral, and anticoagulant conditioning, have been observed (Berri et al., 2017; Thanh et al., 2016; Tran et al., 2018). As a result, it provides significant prospective on goods related to biomaterials, human health, and agriculture (Kidgell et al., 2019).

"Laminarin" is a typical polysaccharide found in brown seaweed composed of glucose units linked together. It does not form gels and is not a thick emulsion like the other polysaccharides (Priyan et al., 2019). They possess antitumoral, anti-inflammatory, anticoagulant, and antioxidant activities, according to studies (Lee et al., 2012; Moroney et al., 2015). It is, therefore, widely used in medicine, cosmetic, and food products (Wijesekara and Karunarathna, 2017).

13.7.6 CAROTENOIDS

Additionally, bioactive carotenoids are also included in the seaweed. The most useful carotenoids are fucoxanthin in brown algae and lutein, violaxanthin, and beta-carotene in red and green algae. They are photosynthetic pigments that produce the colors red, orange, or yellow. Carotenoids function as antioxidants by being able to neutralize reactive oxygen species. A carotenoid present in coastal brown seaweed called fucoxanthin (photosynthetic pigments) features a unique allenoic link, a 6-monoepoxide, and nine conjugated double bonds. This structure is unstable due to the allenoic bond, providing a higher antioxidant activity (Zhang et al., 2015). Fucoxanthin demonstrated various properties which include anti-tumorigenic, anti-diabetic, anti-obesity, anti-inflammatory, hepatic defensive, cerebrovascular, antioxidant, and cardiovascular defensive properties (Hwang et al., 2016; Lopes-Costa et al., 2017, Zaharudin et al., 2019). Fucoxanthin derived from Undaria has been proven in *in vitro* experiments to drastically increase the mortality of colon cancer cells and the human prostate (Hosokawa et al., 2004, Kotake-Nara et al., 2005). Consuming seaweed is connected biologically to diseases of obesity, as well. There is an accord regarding the capacity of macroalgae fiber to increase bulkiness, resulting in a longer sensation of fullness after eating, thus reducing the intestinal metabolite release (Brownlee et al., 2005; Warrand, 2006). In an effort to produce healthier meals, some researchers have tried to incorporate seaweed fiber or protein into meat products (Marti-Quijal et al., 2018; Parniakov et al., 2018; Thirumdas et al., 2018; Žugčić et al., 2018).

13.7.7 POLYPHENOLS

Numerous polyphenols with seaweed origins have been shown to be advantageous for a variety of human ailments. The majority of polyphenols are reactive particles made up of an aromatic hydrocarbon group and a hydroxyl group (^{-}OH). Seaweed's anti-carcinogenic properties have long been recognized. In fact, bone cancer was treated with brown seaweed by the Chinese and the Egyptians. Numerous studies have proven *in vitro* antioxidant capability in a variety of seaweeds due to their high polyphenol content, which makes up their endogenous protective antioxidative defense mechanisms (Smit, 2004).

In terms of structure and polymerization, phlorotannin's constitute a distinctly diverse set of particles. These are the potent polyphenols that have been assembled and are now being researched as bioactive emulsions for the pharmaceutical and food industries. Anti-inflammatory, anti-diabetic, anti-cancer, and antioxidant properties make up their main conditioning (Catarino et al., 2017; Kim and Himaya, 2011; Sanjeewa et al., 2016). The primary physiological impact is the antioxidant properties, similar to other phenolic substances. For these reasons, gastrointestinal changes and bioavailability in the form of a meal matrix, capsule or extract were examined (Corona et al., 2016).

13.8 OTHER USES OF MACROALGAE

Wastes from aquaculture and underutilized byproducts can provide a crucial substrate for the production of single-cell protein (SCP). SCP is a protein derived from yeast, bacterial cultures, pure or mixed algae, or fungal. There are several different substrates that may be used to cultivate these tiny, single-celled organisms. They clearly have a high concentration of carbon, with the addition of minerals and nitrogen. Aquaculture wastes, byproducts, and effluents may therefore have an impact on really lucrative substrates for SCP goods. Given that fishmeal's quantity of animal protein cannot be fully replaced by this source of proteins, this SCP product is sustainable, and it can be given back to the aquaculture company that produced the wastes for the fishmeal component (Parisi et al., 2020).

Shells may be used as building materials, which is another potential use for aquaculture byproducts. The high calcium carbonate concentration in shells has made it possible to use them as a printing material or to partially substitute cement for conveying concrete (Razali et al., 2017).

13.9 CONCLUSIONS AND FUTURE TRENDS

In the future, the aquaculture industry will need to create more products with less waste. This entails an efficient balance between inputs, pollution control, and waste reduction on the one hand, and reserves, increasing profits, and sustainability practices on the other. Many different techniques might be used to remediate various wastes from aquaculture. The best approach for processing ill or dead fish and mixed wastes is the creation of biogas, biodiesel, and bio-fertilizer from non-recyclable fish waste. The harmful biomass is also recovered and transformed into useful heat, electricity, or energy. The waste that most closely resembles fish viscera in terms of consistency has the greatest chance of absorbing protein hydrolysate. Collagen, chitosan, and pigment isolation for the biomedical, pharmaceutical, culinary, and cosmetic industries are among the most well-known modern applications of aquaculture waste. The multiple waste processing procedures require additional inputs and outputs in order to recover energy and transport the necessary composites from undersea trash. Lab-scale testing should be done prior to developing and implementing the best alternative treatment method for a given kind of waste. However, it is crucial to calculate the advantages of operations in terms of lucrative income and take the environment into account. Macroalgae are essential for the digestion and recycling of fish byproducts, which helps to protect marine resources.

KEYWORDS

- **antioxidant activity**
- **caffeine**
- **coffee byproducts**
- **health benefits**
- **nutraceutical potential**
- **phenolic compounds**

REFERENCES

Ajadi, A., Sabri, M., Dauda, A. B., Ina-Salwany, M., & Hasliza, A. (2016). Immunoprophylaxis: A better alternative protective measure against shrimp vibriosis—A review. *Pertanika Journal of Science & Technology, 2*, 58–69.

Ajani, E., Akinwole, A., & Ayodele, I. (2011). *Fundamentals of Fish Farming in Nigeria* (1st ed.). Welecrown Ventures.

Ampofo, J., & Clerk, G. (2003). Bacterial flora of fish feeds and organic fertilizers for fish culture ponds in Ghana. *Aquaculture Research, 34*(8), 677–680.

Arvanitoyannis, I. S., & Tserkezou, P. (2014). Fish waste management. *Journal of Aquatic Food Product Technology*, 263–309.

Bae, J. S. (2011). Antithrombotic and profibrinolytic activities of phloroglucinol. *Food and Chemical Toxicology, 49*(7), 1572–1577.

Berri, M., Olivier, M., Holbert, S., Dupont, J., Demais, H., Le Goff, M., & Collen, P. N. (2017). Ulvan from *Ulva armoricana* (Chlorophyta) activates the PI3K/Akt signaling pathway via TLR4 to induce intestinal cytokine production. *Algal Research, 28*, 39–47.

Biswas, G., Jena, J. K., Singh, S. K., & Muduli, H. K. (2006). Effect of feeding frequency on growth, survival and feed utilization in fingerlings of *Catla catla* (Hamilton), *Labeo rohita* (Hamilton) and *Cirrhinus mrigala* (Hamilton) in outdoor rearing systems. *Aquaculture Research, 37*(5), 510–514.

Boyd, C. E., & Massaut, L. (1999). Risks associated with the use of chemicals in pond aquaculture. *Aquacultural Engineering, 20*(2), 113–132.

Boyd, C., & McNevin, A. (2015). *Aquaculture, Resource Use, and the Environment* (pp. 322–345).

Brownlee, I., Allen, A., Pearson, J., Dettmar, P., Havler, M., Atherton, M., & Onsøyen, E. (2005). Alginate as a source of dietary fiber. *Critical Reviews in Food Science and Nutrition, 45*(6), 497–510.

Catarino, A. I., Thompson, R., Sanderson, W., & Henry, T. B. (2017). Development and optimization of a standard method for extraction of microplastics in mussels by enzyme digestion of soft tissues. *Environmental Toxicology and Chemistry, 36*(4), 947–951.

Catarino, M. D., Silva, A. M., & Cardoso, S. M. (2017). Fucaceae: A source of bioactive phlorotannins. *International Journal of Molecular Sciences, 18*(6), 1327.

Charoensiddhi, S., Conlon, M. A., Franco, C. M., & Zhang, W. (2017). The development of seaweed-derived bioactive compounds for use as prebiotics and nutraceuticals using enzyme technologies. *Trends in Food Science & Technology, 70*, 20–33.

Cho, Y. S., Jung, W. K., Kim, J. A., Choi, I. W., & Kim, S. K. (2009). Beneficial effects of fucoidan on osteoblastic MG-63 cell differentiation. *Food Chemistry, 116*(4), 990–994.

Corona, G., Ji, Y., Anegboonlap, P., Hotchkiss, S., Gill, C., Yaqoob, P., Spencer, J. P., & Rowland, I. (2016). Gastrointestinal modifications and bioavailability of brown seaweed phlorotannins and effects on inflammatory markers. *British Journal of Nutrition, 115*(7), 1240–1253.

Costello, M., Grant, A., Davies, I., Cecchini, S., Papoutsoglou, S., Quigley, D., & Saroglia, M. (2001). The control of chemicals used in aquaculture in Europe. *Journal of Applied Ichthyology, 17*(4), 173–180.

Cripps, S. J., & Bergheim, A. (2000). Solids management and removal for intensive land-based aquaculture production systems. *Aquacultural Engineering, 22*(1, 2), 33–56.

Cunha, L., & Grenha, A. (2016). Sulfated seaweed polysaccharides as multifunctional materials in drug delivery applications. *Marine Drugs, 14*(3), 42.

Dauda, A. B., Akinwole, A. O., & Olatinwo, L. K. (2014). Biodenitrification of aquaculture wastewater at different drying times in water reuse system. *Jurnal Kelautan dan Perikanan Terapan, 4*(2), 6–12.

Dauda, A. B., Ibrahim, H. I., Bichi, A. H., & Tola-Fabunmi, A. S. (2017). Assessment of fish farming practices, operations, water resource management and profitability in Katsina state, Nigeria. *Journal of Northeast Agricultural University (English Edition), 24*(4), 89–96.

de Alencar, D. B., de Carvalho, F. C. T., Rebouças, R. H., Dos Santos, D. R., dos Santos Pires-Cavalcante, K. M., de Lima, R. L., Baracho, B. M., Bezerra, R. M., Viana, F. A., & dos Fernandes Vieira, R. H. S. (2016). Bioactive extracts of red seaweeds *Pterocladiella capillacea* and *Osmundaria obtusiloba* (Floridophyceae: Rhodophyta) with antioxidant and bacterial agglutination potential. *Asian Pacific Journal of Tropical Medicine, 9*(4), 372–379.

de Oliveira, J. P., Bruni, G. P., Fabra, M. J., da Rosa Zavareze, E., López-Rubio, A., & Martínez-Sanz, M. (2019). Development of food packaging bioactive aerogels through the valorization of *Gelidium sesquipedale* seaweed. *Food Hydrocolloids, 89*, 337–350.

Drosg, B., Fuchs, W., Al Seadi, T., Madsen, M., & Linke, B. (2015). Nutrient recovery by biogas digestate processing. (2015), 1.

Ebeling, J. M., & Timmons, M. B. (2012). Recirculating aquaculture systems. *Aquaculture Production Systems*, 245–277.

Eiroa, M., Costa, J., Alves, M., Kennes, C., & Veiga, M. C. (2012). Evaluation of the biomethane potential of solid fish waste. *Waste Management, 32*(7), 1347–1352.

Elsabee, M. Z., Abdou, E. S., Nagy, K. S., & Eweis, M. (2008). Surface modification of polypropylene films by chitosan and chitosan/pectin multilayer. *Carbohydrate Polymers, 71*(2), 187–195.

FAO (2018). *The State of World Fisheries and Aquaculture—2019 (SOFIA)*.

FAO (2020). *World Fisheries and Aquaculture* (pp. 1–244).

Fernando, P. S., Kim, K. N., Kim, D., & Jeon, Y. J. (2019). Algal polysaccharides: Potential bioactive substances for cosmeceutical applications. *Critical Reviews in Biotechnology, 39*(1), 99–113.

Ferraro, V., Cruz, I. B., Jorge, R. F., Malcata, F. X., Pintado, M. E., & Castro, P. M. (2010). Valorization of natural extracts from marine source focused on marine byproducts: A review. *Food Research International, 43*(9), 2221–2233.

Flórez-Fernández, N., Torres, M. D., González-Muñoz, M. J., & Domínguez, H. (2019). Recovery of bioactive and gelling extracts from edible brown seaweed *Laminaria ochroleuca* by non-isothermal autohydrolysis. *Food Chemistry, 277*, 353–361.

Ghaly, A., Ramakrishnan, V., Brooks, M., Budge, S., & Dave, D. (2013). Fish processing wastes as a potential source of proteins, amino acids and oils: A critical review. *Journal of Microbial & Biochemical Technology, 5*(4), 107–129.

Goldburg, R., Environmental Defense Fund, & Triplett, T. (1997). *Murky Waters: Environmental Effects of Aquaculture in the United States*.

Gómez-Estaca, J., Calvo, M., Álvarez-Acero, I., Montero, P., & Gómez-Guillén, M. (2017). Characterization and storage stability of astaxanthin esters, fatty acid profile and α-tocopherol of lipid extract from shrimp (*L. vannamei*) waste with potential applications as food ingredient. *Food Chemistry, 216*, 37–44.

Güven, K., Özsoy, Y., & Ulutin, O. (1991). Anticoagulant, fibrinolytic and antiaggregant activity of carrageenans and alginic acid.

Hansen, T. L., Schmidt, J. E., Angelidaki, I., Marca, E., la Cour Jansen, J., Mosbæk, H., & Christensen, T. H. (2004). Method for determination of methane potentials of solid organic waste. *Waste Management, 24*(4), 393–400.

Heu, M. S., Kim, J. S., & Shahidi, F. (2003). Components and nutritional quality of shrimp processing byproducts. *Food Chemistry, 82*(2), 235–242.

Hornshøj, B. H., Kobbelgaard, S., Blakemore, W. R., Stapelfeldt, H., Bixler, H. J., & Klinger, M. (2015). Quantification of free formaldehyde in carrageenan and processed *Eucheuma* seaweed using high-performance liquid chromatography. *Food Additives & Contaminants: Part A, 32*(2), 152–160.

Hosokawa, M., Kudo, M., Maeda, H., Kohno, H., Tanaka, T., & Miyashita, K. (2004). Fucoxanthin induces apoptosis and enhances the antiproliferative effect of the PPARγ ligand, troglitazone, on colon cancer cells. *Biochimica et Biophysica Acta (BBA)–General Subjects, 1675*(1–3), 113–119.

Hwang, P. A., Phan, N. N., Lu, W. J., Hieu, B. T. N., & Lin, Y. C. (2016). Low-molecular-weight fucoidan and high-stability fucoxanthin from brown seaweed exert prebiotics and anti-inflammatory activities in Caco-2 cells. *Food & Nutrition Research, 60*(1), 32033.

Jayasinghe, P., & Hawboldt, K. (2012). A review of bio-oils from waste biomass: Focus on fish processing waste. *Renewable and Sustainable Energy Reviews, 16*(1), 798–821.

Jiménez-Escrig, A., & Sánchez-Muniz, F. (2000). Dietary fiber from edible seaweeds: Chemical structure, physicochemical properties and effects on cholesterol metabolism. *Nutrition Research, 20*(4), 585–598.

Ju, S. J., & Harvey, H. R. (2004). Lipids as markers of nutritional condition and diet in the Antarctic krill *Euphausia superba* and *Euphausia crystallorophias* during austral winter. *Deep Sea Research Part II: Topical Studies in Oceanography, 51*(17–19), 2199–2214.

Kadam, S. U., Tiwari, B. K., Smyth, T. J., & O'Donnell, C. P. (2015). Optimization of ultrasound assisted extraction of bioactive components from brown seaweed *Ascophyllum nodosum* using response surface methodology. *Ultrasonics Sonochemistry, 23*, 308–316.

Kafle, G. K., Kim, S. H., & Sung, K. I. (2013). Ensiling of fish industry waste for biogas production: A lab scale evaluation of biochemical methane potential (BMP) and kinetics. *Bioresource Technology, 127*, 326–336.

Karim, A., & Bhat, R. (2009). Fish gelatin: Properties, challenges, and prospects as an alternative to mammalian gelatins. *Food Hydrocolloids, 23*(3), 563–576.

Kidgell, J. T., Magnusson, M., de Nys, R., & Glasson, C. R. (2019). Ulvan: A systematic review of extraction, composition and function. *Algal Research, 39*, 101422.

Kim, K. W., & Thomas, R. (2007). Antioxidative activity of chitosans with varying molecular weights. *Food Chemistry, 101*(1), 308–313.

Kim, S. K., & Himaya, S. (2011). Medicinal effects of phlorotannins from marine brown algae. In *Advances in Food and Nutrition Research* (Vol. 64, pp. 97–109).

Kotake-Nara, E., Sugawara, T., & Nagao, A. (2005). Antiproliferative effect of neoxanthin and fucoxanthin on cultured cells. *Fisheries Science, 71*, 459–461.

Kwon, M. J., & Nam, T. J. (2007). A polysaccharide of the marine alga *Capsosiphon fulvescens* induces apoptosis in AGS gastric cancer cells via an IGF-IR-mediated PI3K/Akt pathway. *Cell Biology International, 31*(8), 768–775.

Le Gouic, A., Harnedy, P., & FitzGerald, R. (2018). Bioactive peptides from fish protein byproducts. In J. M. Mérillon & K. G. Ramawat (Eds.), *Bioactive Molecules in Food* (pp. 1–35).

Lee, J. Y., Kim, Y. J., Kim, H. J., Kim, Y. S., & Park, W. (2012). Immunostimulatory effect of laminarin on RAW 264.7 mouse macrophages. *Molecules, 17*(5), 5404–5411.

Lee, W. K., Lim, Y. Y., Leow, A. T. C., Namasivayam, P., Abdullah, J. O., & Ho, C. L. (2017). Factors affecting yield and gelling properties of agar. *Journal of Applied Phycology, 29*, 1527–1540.

Liao, F. H., Shieh, M. J., Chang, N. C., & Chien, Y. W. (2007). Chitosan supplementation lowers serum lipids and maintains normal calcium, magnesium, and iron status in hyperlipidemic patients. *Nutrition Research, 27*(3), 146–151.

Lopes-Costa, E., Abreu, M., Gargiulo, D., Rocha, E., & Ramos, A. A. (2017). Anticancer effects of seaweed compounds fucoxanthin and phloroglucinol, alone and in combination with 5-fluorouracil in colon cells. *Journal of Toxicology and Environmental Health, Part A, 80*(13–15), 776–787.

Marti-Quijal, F. J., Zamuz, S., Galvez, F., Roohinejad, S., Tiwari, B. K., Gómez, B., Barba, F. J., & Lorenzo, J. M. (2018). Replacement of soy protein with other legumes or algae in turkey breast formulation: Changes in physicochemical and technological properties. *Journal of Food Processing and Preservation, 42*(12), e13845.

McHugh, D. J. (1987). Production, properties and uses of alginates. In *Production and Utilization of Products from Commercial Seaweeds* (*FAO Fisheries Technical Paper*, 288, pp. 58–115).

Meier, D., Van De Beld, B., Bridgwater, A. V., Elliott, D. C., Oasmaa, A., & Preto, F. (2013). State-of-the-art of fast pyrolysis in IEA bioenergy member countries. *Renewable and Sustainable Energy Reviews, 20*, 619–641.

Miller, D., & Semmens, K. (2002). Waste management in aquaculture, *8*(1).

Moroney, N. C., O'Grady, M. N., Lordan, S., Stanton, C., & Kerry, J. P. (2015). Seaweed polysaccharides (laminarin and fucoidan) as functional ingredients in pork meat: An evaluation of anti-oxidative potential, thermal stability and bio accessibility. *Marine Drugs, 13*(4), 2447–2464.

Muzzarelli, R. A., Morganti, P., Morganti, G., Palombo, P., Palombo, M., Biagini, G., Belmonte, M. M., Giantomassi, F., Orlandi, F., & Muzzarelli, C. (2007). Chitin nanofibrils/chitosan glycolate composites as wound medicaments. *Carbohydrate Polymers, 70*(3), 274–284.

Negishi, H., Mori, M., Mori, H., & Yamori, Y. (2013). Supplementation of elderly Japanese men and women with fucoidan from seaweed increases immune responses to seasonal influenza vaccination. *The Journal of Nutrition, 143*(11), 1794–1798.

No, H., Meyers, S. P., Prinyawiwatkul, W., & Xu, Z. (2007). Applications of chitosan for improvement of quality and shelf life of foods: A review. *Journal of Food Science, 72*(5), R87–R100.

Nogueira, L. F., Maniglia, B. C., Pereira, L. S., Tapia-Blácido, D. R., & Ramos, A. P. (2016). Formation of carrageenan-$CaCO_3$ bioactive membranes. *Materials Science and Engineering: C, 58*, 1–6.

Núñez-Gastélum, J. A., Sánchez-Machado, D. I., López-Cervantes, J., Rodríguez-Núñez, J. R., Correa-Murrieta, M. A., Sánchez-Duarte, R. G., & Campas-Baypoli, O. N. (2016). Astaxanthin and its esters in pigmented oil from fermented shrimp byproducts. *Journal of Aquatic Food Product Technology, 25*(3), 334–343.

Olusegun, A. A., Babatunde, D. A., & Abiodun, O. O. (2016). Hematological response of *Clarias gariepinus* juveniles reared in treated wastewater after waste solids removal using alum or *Moringa oleifera* seed powder. *International Journal of Fisheries and Aquaculture, 6*, 1–8.

Pádua, D., Rocha, E., Gargiulo, D., & Ramos, A. (2015). Bioactive compounds from brown seaweeds: Phloroglucinol, fucoxanthin and fucoidan as promising therapeutic agents against breast cancer. *Phytochemistry Letters, 14*, 91–98.

Pal, A., Kamthania, M. C., & Kumar, A. (2014). Bioactive compounds and properties of seaweeds—A review. *Open Access Library Journal, 1*(4), 1–17.

Parisi, G., Tulli, F., Fortina, R., Marino, R., Bani, P., Dalle Zotte, A., De Angelis, A., Piccolo, G., Pinotti, L., & Schiavone, A. (2020). Protein hunger of the feed sector: The alternatives offered by the plant world. *Italian Journal of Animal Science, 19*(1), 1204–1225.

Park, J., Craggs, R., & Shilton, A. (2011). Recycling algae to improve species control and harvest efficiency from a high-rate algal pond. *Water Research, 45*(20), 6637–6649.

Parniakov, O., Toepfl, S., Barba, F. J., Granato, D., Zamuz, S., Galvez, F., & Lorenzo, J. M. (2018). Impact of the soy protein replacement by legumes and algae-based proteins on the quality of chicken rotti. *Journal of Food Science and Technology, 55*, 2552–2559.

Paul, N. A., Tseng, C., & Borowitzka, M. (2013). Seaweed and microalgae. In *Aquaculture: Farming Aquatic Animals and Plants* (2nd ed., pp. 268–293).

Piedrahita, R. H. (2003). Reducing the potential environmental impact of tank aquaculture effluents through intensification and recirculation. *Aquaculture, 226*(1–4), 35–44.

Porse, H., & Rudolph, B. (2017). The seaweed hydrocolloid industry: 2016 updates, requirements, and outlook. *Journal of Applied Phycology, 29*(5), 2187–2200.

Razali, N., Aris, R., Razali, N., & Pa'ee, K. F. (2017). Revalorization of aquaculture waste: The performance of calcined mussel shells as partial cement replacement. *International Conference on Environmental Research and Technology*, 375–381.

Read, P., & Fernandes, T. (2003). Management of environmental impacts of marine aquaculture in Europe. *Aquaculture, 226*(1–4), 139–163.

Rinaudo, M. (2006). Chitin and chitosan: Properties and applications. *Progress in Polymer Science, 31*(7), 603–632.

Romano, N., & Zeng, C. (2013). Toxic effects of ammonia, nitrite, and nitrate to decapod crustaceans: A review on factors influencing their toxicity, physiological consequences, and coping mechanisms. *Reviews in Fisheries Science, 21*(1), 1–21.

Sanchez, A., Blanco, M., Correa, B., Perez-Martin, R. I., & Sotelo, C. G. (2018). Effect of fish collagen hydrolysates on type I collagen mRNA levels of human dermal fibroblast culture. *Marine Drugs, 16*(5), 144.

Sánchez-Camargo, A. P., Meireles, M. Â. A., Lopes, B. L. F., & Cabral, F. A. (2011). Proximate composition and extraction of carotenoids and lipids from Brazilian red spotted shrimp waste (*Farfantepenaeus paulensis*). *Journal of Food Engineering, 102*(1), 87–93.

Sanjeewa, K. K. A., Fernando, I. P. S., Samarakoon, K. W., Lakmal, H. H. C., Kim, E. A., Kwon, O. N., Dilshara, M. G., Lee, J. B., & Jeon, Y. J. (2016). Anti-inflammatory and anti-cancer activities of sterol rich fraction of cultured marine microalga *Nannochloropsis oculata. Algae, 31*(3), 277–287.

Sanjeewa, K. K. A., Kim, E. A., Son, K. T., & Jeon, Y. J. (2016). Bioactive properties and potential cosmeceutical applications of phlorotannins isolated from brown seaweeds: A review. *Journal of Photochemistry and Photobiology B: Biology, 162*, 100–105.

Sardari, R. R., & Nordberg Karlsson, E. (2018). Marine poly- and oligosaccharides as prebiotics. *Journal of Agricultural and Food Chemistry, 66*(44), 11544–11549.

Seedevi, P., Moovendhan, M., Viramani, S., & Shanmugam, A. (2017). Bioactive potential and structural characterization of sulfated polysaccharide from seaweed (*Gracilaria corticata*). *Carbohydrate Polymers, 155*, 516–524.

Senthilkumar, K., Ramajayam, G., Venkatesan, J., Kim, S. K., & Ahn, B. C. (2017). Biomedical applications of fucoidan, seaweed polysaccharides. In *Seaweed polysaccharides* (pp. 269–281).

Shahidi, F., & Brown, J. A. (1998). Carotenoid pigments in seafoods and aquaculture. *Critical Reviews in Food Science, 38*(1), 1–67.

Smit, A. J. (2004). Medicinal and pharmaceutical uses of seaweed natural products: A review. *Journal of Applied Phycology, 16*(4), 245–262.

Sriket, P., Benjakul, S., Visessanguan, W., & Kijroongrojana, K. (2007). Comparative studies on chemical composition and thermal properties of black tiger shrimp (*Penaeus monodon*) and white shrimp (*Penaeus vannamei*) meats. *Food Chemistry, 103*(4), 1199–1207.

Stepnowski, P., Olafsson, G., Helgason, H., & Jastorff, B. (2004). Recovery of astaxanthin from seafood wastewater utilizing fish scales waste. *Chemosphere, 54*(3), 413–417.

Sun, J., Ren, F., Chang, Y., Wang, P., Li, Y., Zhang, H., & Luo, J. (2018). Formation and structural properties of acid-induced casein–agar double networks: Role of gelation sequence. *Food Hydrocolloids, 85*, 291–298.

Tanna, B., & Mishra, A. (2019). Nutraceutical potential of seaweed polysaccharides: Structure, bioactivity, safety, and toxicity. *Comprehensive Reviews in Food Science and Food Safety, 18*(3), 817–831.

Tanna, B., Choudhary, B., & Mishra, A. (2018). Metabolite profiling, antioxidant, scavenging and anti-proliferative activities of selected tropical green seaweeds reveal the nutraceutical potential of *Caulerpa* spp. *Algal Research, 36*, 96–105.

Thanh, T. T. T., Quach, T. M. T., Nguyen, T. N., Luong, D. V., Bui, M. L., & Van Tran, T. T. (2016). Structure and cytotoxic activity of ulvan extracted from green seaweed *Ulva lactuca. International Journal of Biological Macromolecules, 93*, 695–702.

Thirumdas, R., Brnčić, M., Brnčić, S. R., Barba, F. J., Gálvez, F., Zamuz, S., Lacomba, R., & Lorenzo, J. M. (2018). Evaluating the impact of vegetal and microalgae protein sources on proximate composition, amino acid profile, and physicochemical properties of fermented Spanish "chorizo" sausages. *Journal of Food Processing and Preservation, 42*(11), e13817.

Timmons, M. B., & Michael, T. M. L. (1994). *Aquaculture Water Reuse System: Engineering Design and Management* (*Developments in Aquaculture and Fisheries Science*, Vol. 346).

Timmons, M., & Summerfelt, S. (1997). Advances in circular culture tanks engineering hydraulics, solids removal and fish management. *Recent Advances in Aquacultural Engineering*, 1–10.

Tomczak-Wandzel, R., Levlin, E., & Ekengren, Ö. (2013). Biogas production from fish wastes in co-digestion with sewage sludge. *IWA Specialist Conference Holistic Sludge Management*, 6–8.

Tran, T. T. V., Truong, H. B., Tran, N. H. V., Quach, T. M. T., Nguyen, T. N., Bui, M. L., Yuguchi, Y., & Thanh, T. T. T. (2018). Structure, conformation in aqueous solution and antimicrobial activity of ulvan extracted from green seaweed *Ulva reticulata. Natural Product Research, 32*(19), 2291–2296.

Trung, T. S., & Phuong, P. T. D. (2012). Bioactive compounds from by-products of shrimp processing industry in Vietnam. *Journal of Food and Drug Analysis, 20*(1), 64.

Tziveleka, L. A., Ioannou, E., & Roussis, V. (2019). Ulvan, a bioactive marine sulphated polysaccharide as a key constituent of hybrid biomaterials: A review. *Carbohydrate Polymers, 218*, 355–370.

Uddin, M. S., Ahn, H. M., Kishimura, H., & Chun, B. S. (2010). Production of valued materials from squid viscera by subcritical water hydrolysis. *Journal of Environmental Biology, 31*(5), 675–679.

Uribe, E., Vega-Gálvez, A., Heredia, V., Pastén, A., & Di Scala, K. (2018). An edible red seaweed (*Pyropia orbicularis*): Influence of vacuum drying on physicochemical composition, bioactive compounds, antioxidant capacity, and pigments. *Journal of Applied Phycology, 30*, 673–683.

Veluchamy, C., & Kalamdhad, A. S. (2017). Biochemical methane potential test for pulp and paper mill sludge with different food/microorganisms' ratios and its kinetics. *International Biodeterioration & Biodegradation, 117*, 197–204.

Villamil, O., Váquiro, H., & Solanilla, J. F. (2017). Fish viscera protein hydrolysates: Production, potential applications and functional and bioactive properties. *Food Chemistry, 224*, 160–171.

Wang, T., Jónsdóttir, R., Liu, H., Gu, L., Kristinsson, H. G., Raghavan, S., & Ólafsdóttir, G. (2012). Antioxidant capacities of phlorotannins extracted from the brown algae *Fucus vesiculosus*. *Journal of Agricultural and Food Chemistry, 60*(23), 5874–5883.

Ward, A. J., & Løes, A. K. (2011). The potential of fish and fish oil waste for bioenergy generation: Norway and beyond. *Biofuel, 2*(4), 375–387.

Warrand, J. (2006). Healthy polysaccharides. *Food Technology & Biotechnology, 44*(3).

Watanabe, Y., Naganuma, T., Ogawa, T., & Muramoto, K. (2013). Lectins of marine origin and their clinical applications. In *Antitumor Potential and Other Emerging Medicinal Properties of Natural Compounds* (pp. 33–54).

Wijesekara, I., & Karunarathna, W. (2017). Usage of seaweed polysaccharides as nutraceuticals. In *Seaweed Polysaccharides* (pp. 341–348).

Wong, K. B. (2001). Enhanced solids removal for aquacultural raceways (pp. 1–24).

Wong, T. W. (2009). Chitosan and its use in design of insulin delivery system. *Recent Patents on Drug Delivery & Formulation, 3*(1), 8–25.

Yabuta, Y., Fujimura, H., Kwak, C. S., Enomoto, T., & Watanabe, F. (2010). Antioxidant activity of the phycoerythrobilin compound formed from a dried Korean purple laver (*Porphyra* sp.) during *in vitro* digestion. *Food Science and Technology Research, 16*(4), 347–352.

Yahyaee, R., Ghobadian, B., & Najafi, G. (2013). Waste fish oil biodiesel as a source of renewable fuel in Iran. *Renewable and Sustainable Energy Reviews, 17*, 312–319.

Yan, X., Chuda, Y., Suzuki, M., & Nagata, T. (1999). Fucoxanthin as the major antioxidant in *Hijikia fusiformis*, a common edible seaweed. *Bioscience, Biotechnology, and Biochemistry, 63*(3), 605–607.

Youssouf, L., Lallemand, L., Giraud, P., Soulé, F., Bhaw-Luximon, A., Meilhac, O., D'Hellencourt, C. L., Jhurry, D., & Couprie, J. (2017). Ultrasound-assisted extraction and structural characterization by NMR of alginates and carrageenans from seaweeds. *Carbohydrate Polymers, 166*, 55–63.

Yuan, H., Song, J., Li, X., Li, N., & Dai, J. (2006). Immunomodulation and antitumor activity of κ-carrageenan oligosaccharides. *Cancer Letters, 243*(2), 228–234.

Zaharudin, N., Staerk, D., & Dragsted, L. O. (2019). Inhibition of α-glucosidase activity by selected edible seaweeds and fucoxanthin. *Food Chemistry, 270*, 481–486.

Zhang, H., Tang, Y., Zhang, Y., Zhang, S., Qu, J., Wang, X., Kong, R., Han, C., & Liu, Z. (2015). Fucoxanthin: A promising medicinal and nutritional ingredient. *Evidence-Based Complementary and Alternative Medicine, 2015*.

Zhang, J., Tiller, C., Shen, J., Wang, C., Girouard, G. S., Dennis, D., Barrow, C. J., Miao, M., & Ewart, H. S. (2007). Antidiabetic properties of polysaccharide- and polyphenolic-enriched fractions from the brown seaweed *Ascophyllum nodosum*. *Canadian Journal of Physiology and Pharmacology, 85*(11), 1116–1123.

Zhou, G., Xin, H., Sheng, W., Sun, Y., Li, Z., & Xu, Z. (2005). *In vivo* growth-inhibition of S180 tumor by mixture of 5-Fu and low molecular λ-carrageenan from *Chondrus ocellatus*. *Pharmacological Research, 51*(2), 153–157.

Žugčić, T., Abdelkebir, R., Barba, F. J., Rezek-Jambrak, A., Gálvez, F., Zamuz, S., Granato, D., & Lorenzo, J. M. (2018). Effects of pulses and microalgal proteins on quality traits of beef patties. *Journal of Food Science and Technology, 55*, 4544–4553.

CHAPTER 14

Enzymatic Treatment of Food Processing Wastes

ANWESA SARKAR,[1,*] VIVEK KUMAR,[1] and MOIRANGTHEM KALPANA DEVI[2]

[1]*Department of Food Processing Technology, Ghani Khan Choudhury Institute of Engineering and Technology, Malda, West Bengal, India*

[2]*Department of Agricultural and Food Engineering, Indian Institute of Technology, Kharagpur, West Bengal, India*

*Corresponding author

ABSTRACT

Enzymes are biocatalysts that have immense potential to transform agro-industrial food waste into a sustainable zero-waste industry through the commercial production of value-added products. The use of enzymes in food preparations is an age-old process, but with the advancement of technology and the production of novel enzymes, the door has opened for enzymes to a wide range of applications. One of those is the enzymatic treatment for the valorization of industrial wastes. The present chapter discusses the recent advancements in enzymatic treatment in food industries for the utilization of food processing waste and converting it to useful products. A comprehensive list of enzymes used in food industries and their application in different types of food waste has been discussed.

14.1 INTRODUCTION

Out of all industries, the food industry generates the highest quantity of waste, posing a significant environmental concern. Around 40% of all

Utilization of Food Waste and Byproducts. Mudasir Ahmad Malik and Kshirod Kumar Dash (Eds.)

DOI: 10.1201/9781779641151-14

food produced goes to waste. Nevertheless, the waste produced by food processing industries contains substantial amounts of lipids, carbohydrates, and proteins. These substances offer considerable economic opportunities, particularly when converted into profitable sources like biofuels, enzymes, and other valuable ingredients. Numerous food waste streams contain high levels of protein, sugar, and fat, making them ideal raw materials for enzymatic valorization. Enzymes show higher specificity, consume less energy, and contribute to enhanced environmental sustainability compared to synthetic catalysts commonly used in chemical reactions. Nevertheless, their effectiveness in their native form is restricted by their limited stability and recovery abilities.

14.2 α-AMYLASES

α-Amylases (EC 3.2.1.1) have the ability to break down starch primarily, resulting in the production of short-chain dextrin's. They also hydrolyze α-1,4 glycosidic bonds of polysaccharides (Sindhu et al., 2017). These enzymes are widely distributed in various living organisms. The majority of α-amylases are metalloenzymes and depend on calcium ions for their activity, stability, and integrity (Sindhu et al., 2016). In the context of bioethanol production, α-amylases play a crucial role in converting starch into fermentable sugars, which are subsequently fermented into alcohol by *Saccharomyces cerevisiae*.

14.3 GLUCOAMYLASES

Glucoamylases (EC 3.2.1.3) belong to the category of exo-acting enzymes, primarily responsible for breaking down polysaccharide starch and releasing β-glucose from the non-reducing end. These saccharifying enzymes are naturally found in various living organisms, with *Aspergillus niger* and *Aspergillus awamori* being the main producers. However, for industrial applications, Glucoamylases derived from *Rhizopus oryzae* are commonly employed (Coutinho and Reilly, 1997). These enzymes demonstrate stability at low temperatures but experience reduced activity at higher temperatures due to conformational changes.

Glucoamylases find wide-ranging applications in the food industry, especially in the production of glucose from bio waste. Additionally, they play a crucial role in bioethanol production, where fermentation using *Saccharomyces cerevisiae* in combination with Glucoamylases results in the production of bioethanol.

14.4 PROTEASES

Protease enzymes facilitate the hydrolysis of peptide bonds present in proteins and polypeptides. These versatile enzymes find extensive use in detergent, pharmaceutical, and food industries. Currently, 60% of the industrial enzymes utilized in the market are proteases, with a compound annual growth rate (CAGR) of 5.3% (Singh et al., 2016). The demand for proteases is projected to further increase in the near future, driven by their newly discovered applications in leather processing and bioremediation processes.

Proteases can be categorized based on their origin, catalytic activity, and the nature of the reactive group at their catalytic site. Major sources of crude protease include animals, plants, and various microorganisms, including bacteria and fungi. These enzymes can be further classified into two groups based on their site of action on polypeptide chains: exo-peptidases, which act on the ends of polypeptide chains, and endo-peptidases, which act randomly within the inner regions of polypeptide chains.

Different plant proteases, such as bromelain, ficin, and papain, offer various applications in brewing, meat tenderization, milk coagulation, and as digestive aids in the food industry (Li et al., 2013). Furthermore, proteases can enhance the flavor, nutritional value, solubility, and digestibility of food proteins. As proteases continue to demonstrate their versatility and effectiveness, their usage is expected to expand even further, contributing to advancements and improvements in various industrial processes.

14.5 LACTASE (β-GALACTOSIDASE)

The hydrolysis of lactose is a crucial bioprocessing step in food production. This reaction is catalyzed by the enzyme β-galactosidase, which belongs to the hydrolase family and can be derived from various biological sources, including plants, animals, and microorganisms. Recently, the production of β-galactosidase from microorganisms like bacteria, fungi, and yeast has gained popularity due to its higher yield and cost-effectiveness (Santos, 1998).

Different microorganisms produce β-galactosidase with varying pH requirements. For instance, yeast-derived β-galactosidase works optimally at pH 6.5–7.0, while fungal β-galactosidase performs best at a pH of 3.0–5.0. Consequently, the selection of β-galactosidase depends on its intended application in food processing.

In food processing, β-galactosidase with a pH range of 6–7 is commonly used for lactose hydrolysis in whey, a byproduct of cheese production containing lactose, proteins, and minerals. Lactose contributes to high biological oxygen demand (BOD) and chemical oxygen demand (COD), leading to significant environmental concerns (Johansen, 2002; Khider, 2004). By utilizing lactase, the lactose in whey can be hydrolyzed, or galacto-oligosaccharides (GOS) can be formed, which can serve as prebiotic food ingredients (Gibson and Wang, 1994).

14.6 LIPASES

Lipase enzymes are capable of catalyzing the hydrolysis of long-chain triglycerides. In the stomach and pancreas of humans and other animal species, lipases occur naturally and play a role in fat and lipid digestion (Fink et al., 1984). The majority of the global lipase market, around 90%, is attributed to microbial lipases produced by bacteria, fungi, and yeast (Guerrand, 2021), with prominent usage in the food industry. Moreover, lipases find applications in various other sectors, including biofuel production, detergents, and animal feed. Additionally, lipases can be valuable in processing various waste materials generated by food industries (Ghosh et al., 1996).

14.7 PHOSPHOLIPASES

Phospholipases have a specific role in converting phospholipids into fatty acids and other lipophilic substances. These enzymes are categorized into four major classes (A, B, C, and D) based on their respective mechanisms of action (70). Phospholipase A1 (PLA1), phospholipase A2 (PLA2), and phospholipase B primarily target the carboxylic ester bonds of phospholipids, facilitating the displacement and replacement of the acyl group chain through various chemical reactions like hydrolysis, esterification, and transesterification. On the other hand, phospholipases C (PLC) and

D (PLD), which alter the polar head group, are also referred to as phosphodiesterase's, as they identify the phosphodiester linkage (Borrelli and Trono, 2015).

14.8 ESTERASES

In the process of fermenting and enzymatically treating waste from the food industry in an aqueous solution, esterases are responsible for breaking down esters into acids and alcohols. Notably, esterases exhibit a preference for hydrolyzing short-chain acylglycerols rather than long-chain ones, setting them apart from lipases. Esterases play a crucial role in both the food industry and alcoholic beverage industries, particularly in modifying the oil and fat content in various fruit juices and producing fragrances and flavors (Panda and Gowrishankar, 2005).

A specific type of esterase, known as *Feruloyl esterases*, is employed in waste treatment processes. These enzymes belong to the esterase family and are vital for breaking the ester bond between ferulic acid and different polysaccharides found in the plant cell walls, leading to the degradation of fruit and vegetable peels. Due to their ability to hydrolyze lignocellulosic biomass, feruloyl esterases are indispensable in second-generation biofuel production (Faulds, 2010).

14.9 LIPOXYGENASES

Lipoxygenases (LOX) are involved in the deoxygenation process of food waste containing polyunsaturated fatty acids (PUFA) in lipids. These enzymes consist of a single polypeptide chain organized into an N-terminal domain and a catalytic β-barrel domain. Lipoxygenases are categorized as iron-containing enzymes. Their catalyzed reactions yield various precursors responsible for producing diverse volatile and aroma-producing chemical substances in plants.

In the food industry and bakery sector, lipoxygenases play a significant role in aroma generation. Among the lipoxygenase enzymes, soybean lipoxygenases (LOX) have been extensively studied for agricultural residue treatment. Bacterial LOXs exhibit distinct specificities toward fatty acids. For instance, lipoxygenase from *Nostoc* sp., produced through substrate fermentation, oxygenates at specific sites in linoleic acid, while the lipoxygenase from *Anabaena* sp. demonstrates varying specificity during acid degradation (Gao et al., 2010; Koeduka et al., 2008).

14.10 CELLULASES

Cellulases are a group of enzymes that act on polymeric cellulose, breaking it down into free glucose units that microorganisms can use during fermentation. These cellulases are classified into three categories: endo-(1,4)-β-d-glucanase (EC 3.2.1.4), exo-(1,4)-β-d-glucanase (EC 3.2.1.91), and β-glucosidases (EC 3.3.1.21) (Schülein, 1988). Their catalytic modules belong to the glycoside hydrolase (GH) family, and they are further subdivided based on differences in amino acid sequences and three-dimensional structural features. Cellulase enzymes primarily employ an acid-base catalysis mechanism to cleave glycoside bonds in cellulose. This process involves two crucial residues, a proton donor, and a nucleophile, located in the enzyme's active site region (Davies and Henrissat, 1995). Depending on the spatial arrangement of these catalytic residues, cellulose hydrolysis occurs through retention or inversion mechanisms.

A diverse range of microorganisms is known to produce cellulases during their growth on cellulosic and lignocellulosic materials. Commercially, cellulases are mainly sourced from microbial origins, including bacteria and fungi, which can thrive in various environments. While aerobic bacteria and fungi share similar mechanisms of cellulose degradation, anaerobic bacteria utilize cellulosomes situated on the cell exterior, functioning through a distinct system. Cellulases derived from fungi (*Aspergillus* and *Trichoderma*) and bacteria (*Bacillus* and *Paenibacillus*) have significant potential for application in food production.

14.11 XYLANASES

Xylanases are enzymes produced by microorganisms to break down xylans, a key component of hemicellulose. The process requires the collaborative action of three major enzymes: endoxylanases, exoxylanases, and β-xylosidases, which work together to break down the xylan backbone in hemicellulose. Endoxylanases (EC 3.2.1.8) cleave the β-1,4 bonds of the xylan backbone. Exoxylanases (EC 3.2.1.37) hydrolyze the β-1,4 bonds of xylan from the non-reducing ends, releasing xylooligosaccharides. β-Xylosidases, on the other hand, cleave the xylobiose and xylooligosaccharides, releasing xylose (100).

Microbes, such as actinomycetes, bacteria, and fungi, are responsible for producing xylanases. Actinomycetes and bacteria, including *Streptomyces* sp., *Bacillus* sp., and *Pseudomonas* sp., are notable producers of xylanase (Bajaj and Singh, 2010; Sanghi et al., 2010; Sharma and Chand, 2012). Xylanase from these microbial sources exhibits effectiveness over a wide pH range (5.0–9.0), with the optimum temperature for enzyme activity ranging from 35 to 60°C. Among the microbial sources, fungi play a significant role as they have a higher content of xylanase, which is released extracellularly (Nair et al., 2008). *Aspergillus* sp., *Fusarium* sp., and *Penicillium* sp. are major fungal species producing xylanase. Fungal xylanases display higher activity compared to those produced by bacteria or yeast (Mandal, 2015).

14.12 PECTINASES

Pectinases are a type of enzyme that specifically target pectic substances, naturally present in fruits such as tomato, pineapple, orange, apple, lemon pulp, orange peel, and other citrus fruits, serving as their natural substrate. Consequently, when these fruits are used as raw materials in the food industry, it leads to the generation of considerable amounts of pectin-rich waste. The primary function of pectinases is to facilitate the hydrolysis of glycosidic bonds within the pectic polymers present in industrial waste. Functionally, pectinases can be categorized into various groups, including polygalacturonases that break down glycosidic α-(1-4) bonds, pectin esterases that remove acetyl and methoxyl groups from pectin, as well as pectin lyase and pectate lyase (Saadoun, 2013).

Pectinases can be produced by both natural and recombinant microbes, with ongoing research efforts focused on improving their stability at high temperatures and increasing their yield (Rebello et al., 2017). These enzymes also demonstrate the ability to act on either the smooth or hairy regions of pectin (Tapre et al., 2014). Depending on the pH conditions, pectinases can be further classified as either acidic or alkaline, and they can be grouped into endopectinases, which randomly cleave pectin, or exopectinases, which target the terminal ends of pectin.

14.13 GLUCOSE OXIDASE

Glucose oxidase (EC 1.1.3.4) is a flavoprotein classified within the oxido-reductases enzyme family. When in the presence of dissolved oxygen, glucose oxidase catalyzes a series of reactions that convert glucose into gluconic acid. This process involves the oxidation of glucose, forming gluconolactone, while molecular oxygen is simultaneously reduced to hydrogen peroxide. The gluconolactone then undergoes spontaneous hydrolysis, ultimately yielding gluconic acid (EL-Sherbeny, 2005).

Glucose oxidase can be obtained from various microorganisms, initially discovered in *Aspergillus niger* and *Penicillium glaucum*. Among these, *Aspergillus niger* is widely favored for glucose oxidase production due to its ability to yield higher quantities of the enzyme (Kona et al., 2001). Another common fungus used for extracellular glucose oxidase production is *Penicillium adametzii* (Mikhailova et al., 2000). Although bacterial species can also be employed to produce glucose oxidase, fungi are generally preferred for large-scale industrial production of the enzyme (Ramachandran et al., 2006).

Glucose oxidase is extensively applied across various industries, including pharmaceuticals, food, and biofuel cells (Zhu, 2006).

14.14 LACCASE

Laccases (EC 1.10.3.2) belong to a group of oxidases, constituting the largest category of multicopper enzymes. Commonly known as blue oxidases, they hold significant importance in various industrial sectors due to their capacity to oxidize phenolic compounds (Giardina et al., 2010; Madhavi et al., 2009; Morozova et al., 2007). These enzymes serve as potent biocatalysts and find diverse applications in chemical processes, bio bleaching of paper pulp, bioremediation, biosensing, wine stabilization, and cloth finishing. Laccases demonstrate distinct specificities towards various substrates, offering a wide range of oxidizable compounds, which depend on the microbial sources responsible for their production (Madhavi et al., 2009).

Laccases efficiently catalyze the oxidation of various compounds, including phenolics, aromatic amines, and ascorbate (Giardina et al., 2010; Madhavi et al., 2009). During this process, they bind dioxygen to substrates with four oxidized electrons while simultaneously providing four reduced electrons, which is facilitated by the presence of four copper atoms in laccases (Giardina et al., 2010).

The catalytic activity of laccases has been explored by Madhavi and Lele (Morozova et al., 2007). These enzymes are secreted extracellularly by different fungi as part of their secondary metabolism during fermentation, although their production is limited to numerous fungal species (Morozova et al., 2007). Notable laccase producers include Deuteromycetes, Ascomycetes, and Basidiomycetes (Gochev and Krastanov, 2007; Sadhasivam, 2008). *Funalia trogii*, a white spoilage fungus, demonstrates the capability to produce laccase through absorbent fermentation.

14.15 CATALASE

Catalase is a naturally occurring enzyme, existing as a tetrameric protein and commonly found in aerobic organisms. It is synthesized by various microbial sources such as *Aspergillus niger*, *Micrococcus luteus*, and is also extracted from bovine liver. Its primary role involves assisting in the breakdown of hydrogen peroxide. Microorganisms are often favored as the preferred sources for enzyme production due to their advantageous characteristics, including rapid growth, ease of handling, and the ability to genetically manipulate them to achieve specific desired products (Puri et al., 2012; Singh, 2017).

Aspergillus niger isolates have shown the highest catalase activity, reaching 400 mg/g (Bučková, 2005). Additionally, a catalase-positive entomopathogenic fungus, Metarhizium anisopliae strain Ma10 (CNRCB MaPL10), was discovered in *Geraeus senilis*, displaying promising catalase activity (Zeng et al., 2011).

14.16 PEROXIDASE

Peroxidases (EC 1.11.1.7) are a group of oxidoreductase proteins that contain iron and protoporphyrin as the prosthetic group. These enzymes play a crucial role in facilitating the reduction of peroxides and the oxidation of various inorganic and organic compounds found in industrial waste. This group encompasses distinct enzymes, including iodide peroxidase, NADH peroxidase, glutathione peroxidase, and other nonspecific enzymes

(Hamid and Khalil, 2009). Peroxidases are present in plants, microorganisms, and animals. Within plants, they participate in lignification processes (Wakamatsu and Takahama, 1993) and contribute to defense mechanisms against damaged or infected tissues (Biles and Martyn, 1993). Among all organisms, *Phanerochaete chrysosporium* is extensively studied as a peroxidase-secreting organism (Janusz, 2013).

One challenge in the industrial application of fungal peroxidases lies in the post-translational modification of proteins (Conesa and Punt, 2002). However, bacterial peroxidases offer advantages as they are easier to produce using waste from the food industry as a substrate, and they demonstrate improved stability and activity for industrial purposes. These enzymes are particularly employed in dye decolorization (Colpa et al., 2014). Peroxidases have been observed to be produced by Firmicutes, Proteobacteria, Actinobacteria, and Actinobacteria (Brown and Chang, 2014). Additionally, certain soil bacteria, actinomycetes, which resemble fungi in growth, also contribute to the production of peroxidases through lignin degradation (Colpa et al., 2014; Majumdar et al., 2014).

14.17 CONCLUSION

Using enzymes to valorize food processing waste presents a unique technological approach that improves the environmental and economic sustainability of food production. However, certain challenges persist, requiring further research to enhance their commercial viability. A significant hurdle lies in the diversity of enzyme application methods, each with specific operational requirements, making it impractical to identify a universal ideal method for all waste systems. Moreover, cost is a notable barrier to enzyme implementation in waste valorization, with approximately 47% of enzyme costs attributed to the support material. The use of purified enzymes further increases biocatalyst expenses. A comprehensive economic analysis of enzymatic treatment, considering innovative application technologies, would greatly benefit researchers and industry stakeholders.

The intricate nature of enzyme applications in food waste treatment calls for interdisciplinary collaborations to foster innovation. When applying enzymes in food applications, careful consideration of their source is essential, ensuring that the producing organism is recognized as generally recognized as safe (GRAS) or approved for use as a processing aid in food ingredient production, in accordance with 21 CFR Part 173. Although genetically modified enzymes can enhance performance, they may compromise the organic status of the resulting products. Therefore, experts in biochemistry, molecular biology, genetic engineering, food science, process engineering, agricultural economics, and food law/regulations must collaborate to drive enzymology research for waste treatment in the food industry. Exploring the use of multiple enzymes, instead of relying on a single enzyme, to treat complex food industry waste shows promise. These customized biocatalysts have the potential to enhance food industry sustainability by generating value-added products and significantly reducing food waste generated during processing.

KEYWORDS

- **enzymatic treatment**
- **food industry waste**
- **bioactive compounds**
- **antioxidants**
- **nutritional value**
- **functional foods**

REFERENCES

Andreou, A. Z., Vanko, M., Bezakova, L., & Feussner, I. (2008). Properties of a mini 9R-lipoxygenase from *Nostoc* sp. PCC 7120 and its mutant forms. *Phytochemistry, Vol.*, p. 7.

Bajaj, B. K., & Singh, N. P. (2010). Production of xylanase from an alkali tolerant *Streptomyces* sp. 7b under solid-state fermentation, its purification, and characterization. *Applied Biochemistry and Biotechnology, 162*, p. 18.

Biles, C. L., & Martyn, R. D. (1993). Peroxidase, polyphenol oxidase and shikimate dehydrogenase isozymes in relation to the tissue type, maturity and pathogen induction of watermelon seedlings. *Plant Physiology and Biochemistry, 31*(4), 499–506.

Borrelli, G. M., & Trono, D. (2015). Recombinant lipases and phospholipases and their use as biocatalysts for industrial applications. *International Journal of Molecular Sciences.*

Brown, M. E., & Chang, M. C. Y. (2014). Exploring bacterial lignin degradation. *Current Opinion in Chemical Biology*, *19*, 1–7.

Bučková, M., Godočíková, J., Šimonovičová, A., & Polek, B. (2005). Production of catalases by *Aspergillus niger* isolates as a response to pollutant stress by heavy metals. *Current Microbiology*, *50*(4), p. 9.

Colpa, D. I., Fraaije, M. W., & van Bloois, E. (2014). DyP-type peroxidases: A promising and versatile class of enzymes. *Journal of Industrial Microbiology and Biotechnology*, *4*, 1–7.

Conesa, A., Punt, P. J., & van den Hondel, C. A. M. J. J. (2002). Fungal peroxidases: Molecular aspects and applications. *Journal of Biotechnology*, *93*, 143–158.

Coutinho, P. M., & Reilly, P. J. (1997). Glucoamylase structural, functional and evolutionary relationships. *Proteins*, *29*(3), 334–347.

Davies, G., & Henrissat, B. (1995). Structures and mechanisms of glycosyl hydrolases. *Structure*, *3*, p. 9.

El-Sherbeny, G. A., Shindia, A. A., & Sheriff, Y. M. M. (2005). Optimization of various factors affecting glucose oxidase activity produced by *Aspergillus niger*. *International Journal of Agriculture and Biology*, *7*, 953–958.

Faulds, C. B. (2010). What can feruloyl esterases do for us? *Phytochemistry Reviews*, *9*, p. 32.

Fink, C. S., Hamosh, P., & Hamosh, M. (1984). Fat digestion in the stomach: Stability of lingual lipase in the gastric environment. *Pediatric Research*, *18*(3), p. 54.

Gao, B., Boeglin, W. E., & Brash, A. R. (2010). Omega-3 fatty acids are oxygenated at the n-7 carbon by the lipoxygenase domain of a fusion protein in the cyanobacterium *Acaryochloris marina*. *Biochimica et Biophysica Acta–Molecular and Cell Biology of Lipids*, *58*, p. 63.

Gao, B., Boeglin, W. E., Zheng, Y., Schneider, C., & Brash, A. R. (2009). Evidence for an ionic intermediate in the transformation of fatty acid hydroperoxide by a catalase-related allene oxide synthase from the cyanobacterium *Acaryochloris marina*. *Journal of Biological Chemistry*, *284*, p. 98.

Ghosh, P. K., Saxena, R. K., Gupta, R., Yadav, R. P., & Davidson, S. (1996). Microbial lipases: Production and applications. *Science Progress*, *79*, p. 57.

Giardina, P., Faraco, V., Pezzella, C., Piscitelli, A., Vanhulle, S., & Sannia, G. (2010). Laccases: A never-ending story. *Cellular and Molecular Life Sciences*, *67*, p. 85.

Gibson, G. R., & Wang, X. (1994). Regulatory effects of bifidobacteria on the growth of other colonic bacteria. *Journal of Applied Microbiology*, *77*, p. 20.

Gochev, V. K., & Krastanov, A. I. (2007). Isolation of laccase producing *Trichoderma* sp. *Bulgarian Journal of Agricultural Science*, *13*, 171–176.

Guerrand, D. (2021). Lipases industrial applications: Focus on food and agroindustries. *OCL*, *Vol.*, p. 24.

Hamid, M., & Khalil-ur-Rehman. (2009). Potential applications of peroxidases. *Food Chemistry*, *115*, p. 86.

Janusz, G., Kucharzyk, K. H., Pawlik, A., Staszczak, M., & Paszczynski, A. J. (2013). Fungal laccase, manganese peroxidase and lignin peroxidase: Gene expression and regulation. *Enzyme and Microbial Technology*, *52*(1), 1–12.

Johansen, A. G., Vegarud, G. E., & Skeie, S. (2002). Seasonal and regional variation in the composition of whey from Norwegian Cheddar-type and Dutch-type cheeses. *International Dairy Journal*, *12*, p. 621.

Khider, K., Akretche, D. E., & Larbot, A. (2004). Purification of water effluent from a milk factory by ultrafiltration using Algerian clay support. *Desalination*, *167*, p. 147.

Koeduka, T., Kajiwara, T., & Matsui, K. (2007). Cloning of lipoxygenase genes from a cyanobacterium, *Nostoc punctiforme*, and its expression in *Escherichia coli*. *Current Microbiology*, *54*, p. 9.

Kona, R. P., Qureshi, N., & Pai, J. S. (2001). Production of glucose oxidase using *Aspergillus niger* and corn steep liquor. *Bioresource Technology*, *78*, 123–126.

Li, Q., Yi, L., Marek, P., & Iverson, B. L. (2013). Commercial proteases: Present and future. *FEBS Letters*, p. 63.

Madhavi, V., & Lele, S. S. (2009). Laccase: Properties and applications. *Bioresources*, *4*, p. 717.

Majumdar, S., Lukk, T., Solbiati, J. O., Bauer, S., Nair, S. K., Cronan, J. E., & Gertl, J. A. (2014). Roles of small laccases from *Streptomyces* in lignin degradation. *Biochemistry*, *53*, 47–58.

Mandal, A. (2015). Review on microbial xylanases and their applications. *International Journal of Life Sciences*, *4*, 178–187.

Mikhailova, R. V., Shishko, Zh. F., Yasenko, M. I., & Lobanok, A. G. (2000). Effect of culture conditions on extracellular glucose oxidase production by *Penicillium adametzii* strain BIM-90. *Mikologiya i Fitopatologiya*, *34*, 48–53.

Morozova, O. V., Shumakovich, G. P., Shleev, S. V., & Yaropolov, Y. I. (2007). Laccase-mediator systems and their applications: A review. *Applied Biochemistry and Microbiology*, *43*(5), 523–535.

Nair, S. G., Sindhu, R., & Shashidhar, S. (2008). Fungal xylanase production under solid state and submerged fermentation conditions. *African Journal of Microbiology Research*, *2*, p. 8.

Panda, T., & Gowrishankar, B. S. (2005). Production and applications of esterases. *Applied Microbiology and Biotechnology*, *67*, p. 9.

Puri, M., Sharma, D., Barrow, J. C., & Tiwary, A. K. (2012). Optimization of novel method for the extraction of steviosides from *Stevia rebudiana* leaves. *Food Chemistry*, *132*, p. 20.

Ramachandran, S., Fontanille, P., Pandey, A., & Larroche, C. (2006). Gluconic acid: Properties, applications and microbial production. *Food Technology and Biotechnology*, *44*, 185–195.

Rebello, S., Anju, M., Aneesh, E. M., Sindhu, R., Binod, P., & Pandey, A. (2017). Recent advancements in the production and application of microbial pectinases: An overview. *Reviews in Environmental Science and Bio/Technology*, *16*, 381–394.

Saadoun, I., Dawagreh, A., Jaradat, Z., & Ababneh, Q. (2013). Influence of culture conditions on pectinase production by *Streptomyces* sp. (strain J9). *International Journal of Life Sciences and Medical Research*, *3*, 148.

Sadhasivam, S., Savitha, S., Swaminathan, K., & Lin, F. H. (2008). Production, purification and characterization of mid-redox potential laccase from a newly isolated *Trichoderma harzianum* WL1. *Process Biochemistry*, *43*, 736–742.

Sanghi, A., Garg, N., Gupta, V. K., Mittal, A., & Kuhad, R. C. (2010). One step purification and characterization of a cellulose free xylanase produced by alkalophilic *Bacillus subtilis* ASH. *Brazilian Journal of Microbiology*, *41*, p. 76.

Santos, A., Ladero, M., & García-Ochoa, F. (1998). Kinetic modeling of lactose hydrolysis by a β-galactosidase from *Kluyveromyces fragilis*. *Enzyme and Microbial Technology*, *22*, p. 558.

Schülein, M. (1988). Cellulases of *Trichoderma reesei*. *Methods in Enzymology*, *160*, p. 42.

Sharma, P. K., & Chand, D. (2012). Production of cellulase free thermostable xylanase from *Pseudomonas* sp. XPB-6. *International Research Journal of Biological Sciences*, *1*(5), 31–41.

Sindhu, R., Binod, P., & Pandey, A. (2016). α-Amylases. In A. Pandey, S. Negi, & C. R. Soccol (Eds.), *Current Developments in Biotechnology and Bioengineering–Production, Isolation and Purification of Industrial Products* (p. 3). Amsterdam, the Netherlands: Elsevier B.V.

Sindhu, R., Binod, P., Madhavan, A., Beevi, U. S., Mathew, A. K., Abraham, A., Pandey, A., & Kumar, V. (2017). Molecular improvements in microbial α-amylases for enhanced stability and catalytic efficiency. *Bioresource Technology*, *245*(Part B), p. 8.

Singh, R. S., Sooch, B. S., & Puri, M. (20017). Optimization of medium and process parameters for the production of inulinase from a newly isolated *Kluyveromyces marxianus* YS-1. *Bioresource Technology*, *98*, p. 25.

Singh, R., Mittal, A., Kumar, M., & Mehta, P. K. (2016). Microbial protease in commercial applications. *Journal of Pharmaceutical, Chemical and Biological Sciences*, *4*(3), 365–374.

Tapre, A. R., & Jain, R. K. (2014). Pectinases: Enzymes for fruit processing industry. *International Food Research Journal*, *21*, 447–453.

Wakamatsu, K., & Takahama, U. (1993). Changes in peroxidase activity and in peroxidase isozymes in carrot callus. *Physiologia Plantarum*, *68*, 167–171.

Zeng, H. W., Cai, Y. J., Liao, X. R., Zhang, F., Li, Y. L., & Zeng, X. K. (2011). *Serratia marcescens* SYBC08 catalase isolated from sludge containing hydrogen peroxide shows increased catalase production by regulation of carbon metabolism. *Engineering in Life Sciences*, *11*, 37–43.

Zhu, Z., Momeu, C., Zakhartsev, M., & Schwaneberg, U. (2006). Making glucose oxidase fit for biofuel cell applications by directed protein evolution. *Biosensors and Bioelectronics*, *21*, 46–51.

CHAPTER 15

Microbiological Treatment of Food Processing Wastes

PINKI SAINI* and ANCHAL SINGH

Center of Food Technology, University of Allahabad, Prayagraj, Uttar Pradesh, India

*Corresponding author

ABSTRACT

Rapid urbanization and changing lifestyles worldwide have resulted in a rise in the production of food waste from diverse industrial, agricultural, and household sources. The Food and Agriculture Organization (FAO) reports that roughly one-third of the global food production is wasted each year. The organic nature of food and agriculture waste not only poses a significant concern to the environment but also contributes to the generation of toxic gases and severe human health hazards when disposed of through conventional practices such as land-filling and incineration. This issue has become particularly critical in densely populated agricultural-based countries such as India, China, and the USA, and it requires urgent attention for sustainable development. Reprocessing food waste offers substantial advantages since it contains many components that can be converted into other useful products. Converting agricultural waste into beneficial products brings economic benefits, including improved food security, reduced production costs, the production of biodegradable products, and environmental sustainability. Bioconversion is one of the most favorable method to tackle the issue of organic waste. Microorganisms are extensively researched for their potential in valorizing such waste through bioconversion. This chapter aims to highlight and discuss the different

Utilization of Food Waste and Byproducts. Mudasir Ahmad Malik and Kshirod Kumar Dash (Eds.)

DOI: 10.1201/9781779641151-15

origins of food processing waste generation, the utilization of microorganisms in waste bioconversion, the functional ingredients derived during the process, and the potential challenges in implementing this technology on an industrial scale. Additionally, the review provides a brief outline of food waste management through the generation of biofuels and electrical energy.

15.1 INTRODUCTION

Food holds immense importance in our survival, as well as our social and psychological well-being. However, mishandling and improper utilization of food at any point in its life cycle can result in severe social, economic, and environmental implications. The issue of food waste and its disposal has gained considerable recognition as a significant environmental challenge, capturing the attention of scientists, activists, and consumers worldwide. In fact, food wastage stands as the second largest category of solid waste that finds its way into landfills. Various sources contribute to the generation of organic waste, including commercial and residential kitchens, food processing facilities, restaurants, and cafeterias, spanning across different stages of food production. The FAO has stated that approximately 1.3 billion tons (FAO, 2012) of vegetables, meat products, fruits, bakery items, dairy items, and other food products are spoiled annually in the supply chain (Melikoglu et al., 2013). The food waste in urban areas has been projected to increase to 138 million tons by the year 2025, in comparison to 2005 (Paritosh et al., 2017). This food waste comprises vegetables, fruits, dairy items, bakery products, and meat. It originates from various sources including households, the hospitality sector, commercial kitchens, food processing industries, and agricultural waste (Karthikeyan et al., 2018; Kiran et al., 2014). Food waste contains a variety of organic components, with CHO's (69%), proteins (21.9%), oils, fats as well as organic acids (Kiran et al., 2014; Lin et al., 2014). The high moisture content present in food waste poses a challenge when it comes to its disposal. Incineration of food waste can result in the generation of dioxins, which are harmful substances. On the other hand, dumping food waste in open areas leads to environmental pollution and can also have adverse health effects (Figure 15.1). Additionally, food waste is a significant contributor to greenhouse gas emissions. It is estimated that food waste generates approximately 3.3

billion tons of CO_2 emissions annually, further exacerbating the issue of climate change (Paritosh et al., 2017).

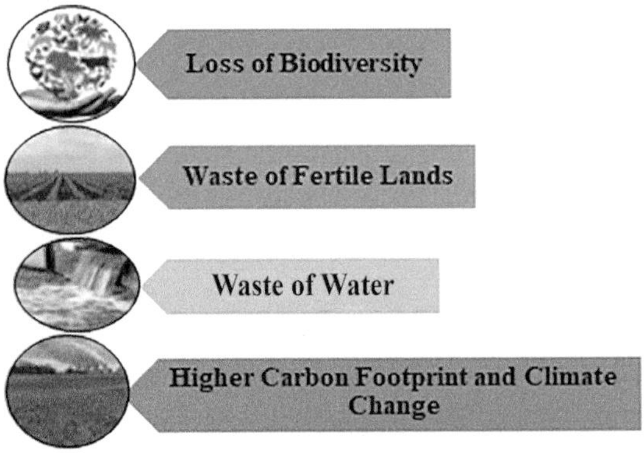

FIGURE 15.1 Impact of food waste on environment.

To tackle the significant global production of waste, it is crucial to implement effective and advanced waste management systems that can bridge the gap between waste production and disposal. The most suitable waste management system involves the beneficial use of microorganisms. Managing food waste through microorganisms includes composting, recycling, anaerobic treatment, bioremediation, biosorption, and the generation of bio-based products such as bio-chemicals, biofuels, bioenergy materials, and enzymes (He et al., 2012; Ohkouchi and Inoue, 2007; Zhang et al., 2013a, b). According to Tuck et al. (2012), the market demand for biofuels and biogas is greater than that of chemicals. Biofuels and bioenergy offer several benefits, including reducing dependence on crude oil, stabilizing food prices, and enhancing global energy security. Utilizing food waste to generate biogas is one way to achieve these benefits.

This chapter focuses on the valorization of food waste for the production of various biofuels such as biodiesel, ethanol, hydrogen, and methane. It explores the different sources and characteristics of food waste that can be efficiently converted into these value-added products. The chapter emphasizes the significance of biofuels, including biodiesel, ethanol, bio-hydrogen, methane, butanol, and the production of bio enzymes and electric power generation from food waste. By highlighting the potential of these biofuels, the chapter underscores the importance of utilizing food waste for sustainable energy production and waste management.

15.2 FOOD WASTES

Food loss and food waste occur at various stages of the food supply chain (FSC), starting from agricultural production to household usage, and the reason for these is depicted in Figure 15.2. Quantifying the exact amount of each wasted food item globally is challenging due to the vast variety. FAO categorizes food waste into eight commodities: cereals, starchy roots, oil crops and pulses, fruits, meat, fish and seafood, milk and eggs, and vegetables (Food and Agriculture Organization of the United Nations, 2014). Among these categories, vegetables, starchy roots, and fruits are the primary contributors to total food wastage across all regions.

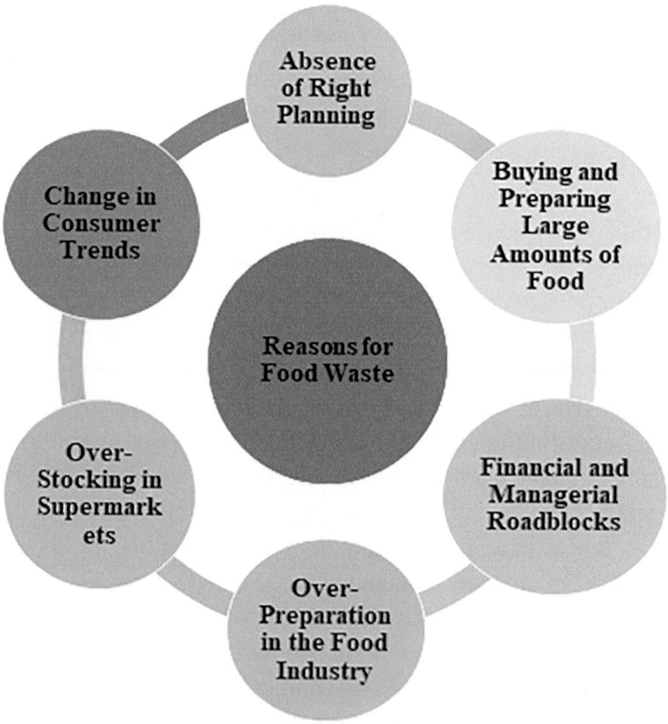

FIGURE 15.2 Reasons for food waste.

In fact, vegetable products alone account for approximately 85% of total food waste, while the remaining 15% comes from animal products. However, it is important to note that the wastage of meat is higher in medium- and high-income regions compared to low-income regions. This disparity may be attributed to various factors such as consumption patterns, production processes, and cultural preferences. Nonetheless, efforts to minimize food waste and promote sustainable consumption practices should be implemented across all regions to address this global issue. These regions not only contribute significantly to the volume of food waste but also to the loss of water and land, along with greenhouse gas emissions resulting from food waste (Food and Agriculture Organization of the United Nations, 2014b).

Waste refers to anything that does not give value to a product; however, a byproduct is defined as a secondary product that is obtained during the manufacturing process of the primary product, which has a specific market value. Several byproducts need further processing, such as membrane filtration, before they can be utilized effectively (Castro et al., 2017). This processing adds value to the byproduct. Agro-based byproducts are primarily derived during the stages of agricultural production, harvesting, and pre-processing in farming areas. This may include processes like oilseed extraction, brewery, malt production, grain milling, and fruit and vegetable processing units. These secondary products serve as valuable sources of proteins and nutrients for animal feed, and can also be converted into products like biofuels, bioenergy, and other economically valuable resources (Kathirvel, 2017). Figure 15.3 describes the hierarchy and principles of food and agri-waste management, providing guidance on effective waste management practices.

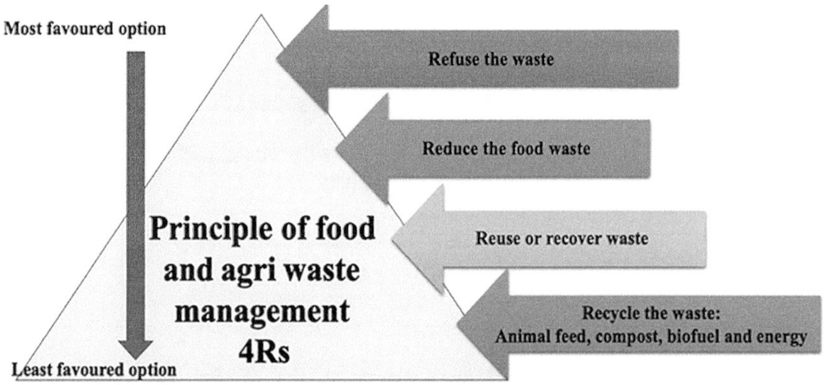

FIGURE 15.3 Food waste hierarchy and principles of waste management.

15.3 SOURCES OF FOOD WASTE

15.3.1 FRUITS AND VEGETABLES

Vegetables, starch-rich roots, and fruits indeed make huge contributions to food waste. Figure 15.4 provides an overview of the waste generated from various sectors such as the canneries, frozen industry, dehydration industry, fruit pulping, fruit and vegetable juice concentrate, and fruit concentrate. These sectors generate waste in their processes, including vegetable washing water that often contains soil and organic content. According to Valta et al. (2017), cannery wastewater shares similarities in composition with household kitchen waste. Waste from fruit and vegetable processing industries often contains significant amounts of solid suspensions and has a high biochemical oxygen demand (BOD). These wastes typically consist of hydrocarbons like sugars, nitrogen compounds, cellulose fibers, and relatively small quantities of proteins and fats. In addition to these components, the wastewater may also contain dissolved compounds such as pesticides, herbicides, and cleaning chemicals. It is essential to properly manage and treat such wastewater to minimize its environmental impact and ensure the protection of water resources.

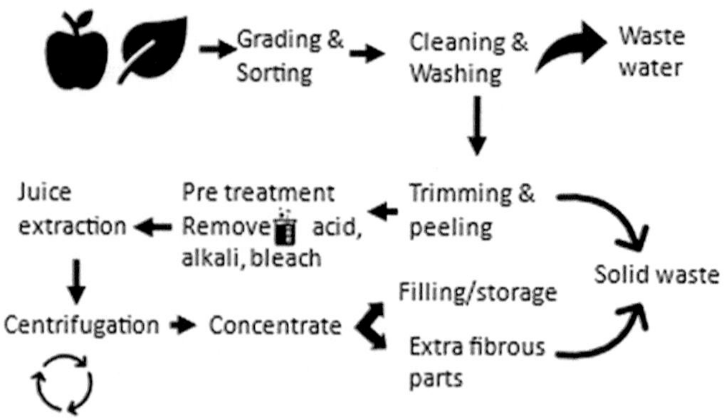

FIGURE 15.4 Waste produced from fruit and vegetable industries.

15.3.2 DAIRY INDUSTRY

Dairy industries have a significant impact on the quality of surface water and soil. One of the major sources of pollution from dairy industries is the chemically induced liquid waste generated by them. This waste carries a high organic load, including fatty substances and other compounds. It exhibits significant variations in supply, with fluctuations in pH ranging from 4.2 to 9.4, with a high amount of suspended solids, typically ranging from about 400 to 2000 mg/L. Dairy wastewater is characterized by high concentrations of organic components like whey, lactose, fats, and mineral content. The waste generated from milk processing is either neutral or slightly alkaline. However, due to fermentation of lactose, it can become highly acidic within a short period (Hansen and Cheong, 2019). The presence of contaminants in dairy wastewater highlights the significance of implementing appropriate management and treatment measures to minimize their environmental impact.

15.3.3 MEAT AND POULTRY INDUSTRY

The meat and poultry industry generates significant amounts of waste in the food sector. Such industries encompass slaughterhouses where meat is prepared, cut into pieces, and processing units for various treatments such as freezing, cooking, curing, smoking, or sausage production. Among these units, slaughterhouses have a greater impact on environmental pollution. The waste generated in these units consists of diverse components including blood, fats, intestinal residues, and manure (Cournoyer, 1996). It is advantageous to separate the waste into solid waste and wastewater. Solid waste includes intestines, meat scraps, bones, and can undergo further processing to be used as animal feed. The wastewater generated exhibits high moisture content (90–95%) and nitrogen levels, along with a high organic content and BOD. Various applications are considered for managing slaughterhouse wastewater, including its use in animal feed, oxidation processes, heat separation, biodegradable composites, and fabric production. In addition to slaughterhouse waste, the poultry feathers collected are also utilized in the development of biofuels following hydrolysis (Seidavi et al., 2019). Effective management and utilization of waste generated by the meat, poultry, and fish industries are crucial for minimizing their environmental impact and promoting sustainability.

15.3.4 FERMENTATION INDUSTRY

The fermentation industry can be categorized as three major divisions: brewing industry, distilling, and wine manufacturing. All of these sectors generate liquid waste with characteristics like high BOD and COD.

However, the quantity of organic compounds in the waste can vary, influencing the selection of appropriate biological treatment methods. In winery waste treatment, a significant challenge is posed by the presence of vinasse, also known as molasses. Vinasse requires biological treatment over a period of 4 to 8 days to achieve a 90% reduction in COD (Boudouropoulos and Arvanitoyannis, 2000). This treatment process is crucial for effectively mitigating the environmental impact of winery waste.

15.3.5 EDIBLE OIL

The edible oil industry generates waste throughout different stages of the oil refining process, including steps like degumming, neutralization, bleaching, deodorization, and the occurrence of oxidative as well as hydrolytic rancidity. For instance, hydrolytic rancidity results from the oxidation of lipids due to factors such as aging, moisture, oxygen presence, and is manifested through the discharge of effluents containing significant amounts of carbohydrates, proteins, and fatty acids (Delgado et al., 2017). The edible oil industry produces a substantial amount of byproducts, such as oil cake and oil meal, which amounts to approximately 350.9 million tons annually. These byproducts serve as concentrated sources of protein and can be further processed for the preparation of value-added products, animal feed, and fertilizer after undergoing appropriate pre-treatment (Chang et al., 2018). Historically, effluents generated from the processing of edible oil were directly discharged into soil and groundwater, leading to the formation of oil films on the surface of water. This practice created a significant problem to the existence of aquatic flora and fauna. Additionally, the methanization of oil in water caused a greenhouse effect, exacerbating environmental concerns (Okino-Delgado et al., 2017). To address these issues, microbial treatments, such as biotransformation and biodegradation, can be employed to effectively treat the effluents.

15.4 MICROBIOLOGICAL FOOD WASTE TREATMENT TECHNOLOGY

15.4.1 ANAEROBIC DIGESTION (AD)

It is a process that involves the decomposition of organic residue using bacteria within an oxygen-free environment. To facilitate this process, optimal conditions like alkaline pH, controlled temperature, and the availability of essential inorganic minerals like phosphorus, nitrogen, and potassium are required. Anaerobic digestion (AD) offers an effective solution for the management of food waste, allowing the recovery of energy through the generation of methane. It is an economically viable process that minimizes the production of residue waste and uses food waste in the form of a renewable source of energy (Nasir et al., 2012; Morita and Sasaki et al., 2012). The AD process can be broadly classified into three phases: enzymatic hydrolysis, acid formation, and gas production, as depicted in Figure 14.5. Bacterial genera such as *Streptococcus* and *Enterobacter* cause the hydrolysis of sugar peptides and fatty acids during the enzymatic hydrolysis phase (McInerney et al., 1979). Acetogenic bacteria, which include genera like *Syntrophomonas* and *Syntrophobacter*, convert the acids produced in the previous phase into acetates and hydrogen. This stage is crucial as it contributes to the production of around 70% of biomethane (Melikoglu et al., 2013). Methanogens, which are archaea, such as Methanosaeta, Methanolobus, Methanohalobium, and Halomethanococcus, are responsible for the final step of methane production. However, the methanogens only contribute to around 30% of methane production. Overall, AD is capable of recovering about 60% more energy as compared to the direct combustion process (Valorgas, 2014). Additionally, the process yields a nutrient-rich fertilizer that can replenish soils.

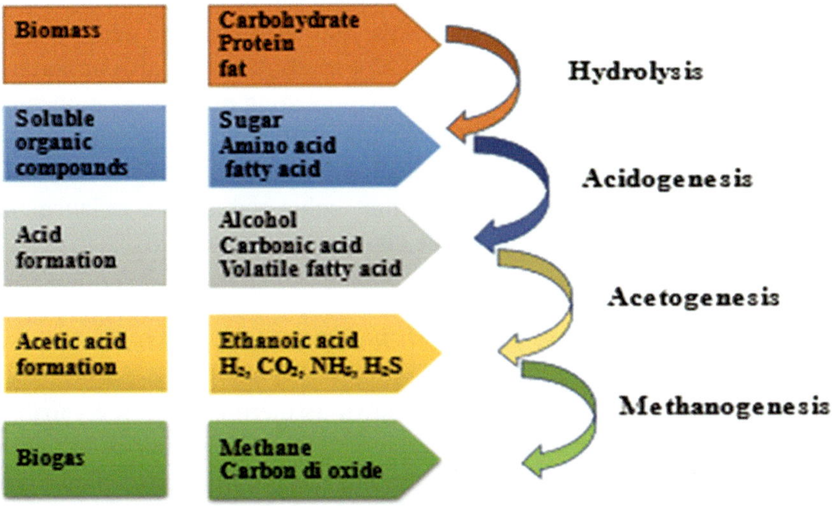

FIGURE 15.5 Major steps of anaerobic digestion.

15.4.2 COMPOSTING

Composting is a natural and relatively simple process that includes the decomposition of organic matter through the activity of microbes like bacteria, actinomycetes, mold, yeast, protozoa, and small invertebrates. Controlled conditions, that is, pH, moisture content, temperature, carbon-nitrogen ratio, and particle size of residue, are necessary for effective composting (Mondal and Palit, 2019). The process is mostly used for the treatment of solid waste. During composting, the biodegrading mass produces heat, increasing the temperature up to 70°C, which helps to enhance the biodegradation and natural fermentation process of the biomass. Various composting techniques are available, ranging from small-scale methods like backyard composting to larger-scale approaches such as aerated in-vessel composting. Each technique has its own advantages and considerations.

One of the main advantages of composting is the production of good quality organic compost, which may be used to improve soil fertility and enhance plant yield. Compost provides essential nutrients, improves soil structure, and promotes beneficial microbial activity. Additionally, composting helps divert organic waste from landfills, reducing the volume of waste and associated environmental impacts. However, there are some disadvantages to consider. Composting does not directly recover energy from the organic waste being processed. Furthermore, careful management of contaminants such as plastics, metals, and chemicals are necessary to avoid contamination of the compost. Odor management is also important to prevent unpleasant smells during the composting process. Composting is a valuable waste management technique that promotes organic waste recycling, soil restoration, and sustainable agriculture, along with being a feasible and eco-friendly method that helps reduce the reliance on synthetic fertilizers and promotes the circular economy.

15.4.3 MICROBIAL ENZYMES/THERMOPHILIC ENZYMES

Thermophilic microorganisms, also known as thermophiles, are capable of surviving and thriving in extreme temperatures. They have been found to be highly effective in treating environmental pollution generated from food and agricultural waste. Cellulose, a polysaccharide commonly found in food and agricultural waste, serves as a major nutritional source for growth of microbes and production of enzymes. Mesophiles and thermophiles utilize waste materials with cellulose content by converting them into glucose and soluble sugars through the action of cellulase enzymes (Camassola and Dillon, 2007). Several microorganisms have been identified as predominant cellulose-digesting bacteria, including *Bacillus* sp, *Cellulomonas*, *Thermoactinomycetes*, and *Pseudomonas*. Cellulase enzymes are complex multi-enzyme systems consisting of different soluble enzymes present in the extracellular space. These enzymes cover 1, 4-β-exoglucanase, 1, 4-β endoglucanase, and β-glucosidase (Bansal et al., 2012). Thermophiles are capable of utilizing the nutrient residues present in lignocellulosic biomass at higher temperatures, thereby facilitating the conversion process to bioenergy (Blumer et al., 2008).

The ability of thermophiles to utilize food and agricultural waste is attributed to their capacity to produce thermophilic primary cellulases, secondary cellulases such as endoglucanases, and enzyme complexes such as cellulosomes (Sedhai, 2022). These enzymes enable the conversion of cellulose into simpler sugars, which can then be further fermented or metabolized by the microorganisms for the production of bioenergy. The utilization of thermophilic microorganisms and their cellulose-degrading enzymes in the degradation of food and agricultural waste offers promising opportunities for the creation of renewable energy and the reduction of waste. By harnessing the capabilities of thermophiles, we can explore more sustainable and efficient ways to manage and utilize organic waste resources.

15.4.4 HIGH-TEMPERATURE DECONSTRUCTION

These processes involve the use of extreme heat and pressure to break down solid biomass into liquid or gaseous intermediates. There are three primary routes in this pathway: pyrolysis, gasification, and hydrothermal liquefaction. Pyrolysis is a thermal decomposition process that rapidly heats biomass at high temperatures (500°C–700°C) in the absence of oxygen. The heat causes the biomass to break down into three main products: pyrolysis vapor, gas, and char. The char is separated, while the vapors are cooled and condensed into a liquid known as "bio-crude" oil. Gasification involves exposing biomass to even higher temperatures (>700°C) in the presence of a controlled amount of oxygen. This process produces a gas called syngas, which is a mixture of carbon monoxide and hydrogen. Syngas can be used as a fuel or further processed into valuable chemicals and biofuels.

Hydrothermal liquefaction utilizes wet feedstocks, such as algae. In this process, biomass is subjected to moderate temperatures (200°C–350°C) and elevated pressures in the presence of water. Under these conditions, the biomass is converted into a liquid bio-crude oil. These high-temperature deconstruction techniques provide potential pathways for converting biomass into valuable energy carriers. They enable the production of biofuels, chemicals, and other useful intermediates from biomass resources. By utilizing these processes, we can explore more efficient and sustainable ways to utilize biomass and reduce our dependence on fossil fuels.

15.5 ROLE OF MICROBES IN TREATMENT AND MANAGEMENT OF FOOD WASTE

The field of environmental microbiology has received prominence worldwide due to the detrimental impact of environmentally transmitted diseases caused by microbes on health of humans. The remarkable efficiency of microorganisms present in the environment has inspired extensive utilization of these organisms for the remediation of harmful and contaminated waste, primarily derived from organic food waste. Microorganisms possess unique characteristics, an unpredictable nature, and biosynthetic capabilities, which make them promising candidates for addressing challenges in various life science and related fields.

Bacteria, in particular, has a significant role in the degradation of organic matter into substances which are stable and less polluting, contributing to natural cycling processes. The growing recognition of microorganisms in biotechnology has sparked renewed interest in their potential applications. Microbes not only demonstrate survival capabilities in harsh environments but also possess the ability to immobilize, degrade, or detoxify environmental contaminants. Microorganisms play a crucial role in waste management by utilizing waste materials as a food source and facilitating the complete elimination or mineralization of toxic compounds. They have the ability to convert these compounds into basic elements such as carbon dioxide and water through their metabolic processes.

However, it is important to note that in certain cases, microorganisms may only partially degrade or transform the waste or contaminants, resulting in a less complex form. Additionally, while microbial assistance can contribute to the immobilization of contaminants, it is rarely a permanent or final solution to waste management challenges.

Given the escalating fuel crisis and growing concerns about climate change, the development of alternative and sustainable energy sources has been increasingly imperative. Conventional procedures of food waste discard, like incineration, composting and landfilling, have proven to be economically and environmentally unsustainable, leading to emissions of toxic gas and water contamination (Paritosh et al., 2017). However, food waste holds significant potential as a valuable resource for the production of biopolymers and biofuels through various processes, due to its organic nature and nutrient-rich composition. This has stimulated interest in changing diverse forms of food waste into bioenergy or biofuels, including methane, hydrogen, ethanol, biodiesel, biogas, electrical energy, and polyhydroxyalkanoates. It is essential to explore these innovative approaches for harnessing the energy and resource potential of food waste, thereby promoting a sustainable and eco-friendly strategy to waste management and energy generation.

15.5.1 PHAs FROM FOOD WASTE USING PURE CULTURES

Bordes et al. (2009) and Chanprateep (2010) reported polyhydroxyalkanoates (PHAs) are extensively researched bioplastics because of their potential as eco-friendly alternatives to petrochemical plastics, given their biodegradability and biocompatibility. PHAs (Polyhydroxyalkanoates) are carbon-based polymers that microorganisms naturally produce to store surplus carbon sources and maintain their energy balance. These polymers serve as intracellular reserves of carbon and energy for microorganisms (Escapa et al., 2012). Food waste represents an excellent and cost-effective carbon source for PHA production, given its broad availability and the prospects to address crucial waste management challenges (Bhatia et al., 2020). Numerous systems have been advocated for conversion of food waste into PHAs, considering the diverse waste lines generated from food production to processing as well as consumption. Every food has its own complex system, requiring specific pre-processing, microbial strains, culture conditions, and methods of downstream processing. In many cases, the organic constituents present in food waste are composite and are not utilized by PHA-producing micro-organisms (Anderson and Dawes, 1990). Therefore, pre-processing methods are required for conversion of these composites into PHA precursors, which may be simple sugar molecules such as glucose or lactose as well as fatty acids like propionic and acetic acids.

Various food wastes, including whey solids, oils, starch, legumes, lignocellulosic materials, and molasses, have shown potential as sources for PHA production (Escapa et al., 2012). The versatility of PHA production is demonstrated by the existence of more than 155 unique monomer subunits of PHA, highlighting the ability of microorganisms to produce diverse PHA polymers (Agnew and Pfleger, 2013). AD using mixed microbial cultures (MMCs) is one method employed for conversion of complex food wastes into precursors of PHA. The typical process for PHA production from food waste involves three stages, that is, acidogenic fermentation, culture selection, followed by PHA accumulation (Pandey et al., 2021). PHAs have shown potential utilization in various fields, such as commercial packaging, agricultural purposes, and medical uses, with the aim of replacing petrochemical plastics (Akaraonye et al., 2010; Bhardwaj et al., 2014; Bugnicourt et al., 2014; Chen and Patel, 2011; Dinjaski and Prieto, 2015; Keshavarz and Roy, 2010; Mozejko-Ciesielska and Kiewisz, 2016; Wu et al., 2009; Zhu et al., 2016).

Different food waste generation sources may be utilized for PHA production. Starch derived from plants like potatoes, maize, rice, wheat, and cassava, as well as whey from the dairy industry, serve as possible cheap carbon sources for production of PHA (Girotto et al., 2015; Poomipuk et al., 2014). Spent oils produced from industrial applications and household use may be directly incorporated to the culture media as carbon substrates sans any pre-treatment (Obruca et al., 2014). Waste coffee grounds, mostly composed of lignin-cellulose complex substances, are another viable source for PHA production. They can be combusted for generation of heat or hydrolyzed and transformed into PHAs by a specific microorganism *Burkholderia cepacia* (Obruca et al., 2014). Lignocellulosic materials, which consist of pectin, cellulose, hemicellulose, and lignin, are plant-based materials commonly found in food industry waste streams, including rice straw, bran, husk, wheat straw, and bagasse (Cesario et al., 2014; Munoz and Riley, 2008; Sindhu et al., 2013;

Van-Thuoc et al., 2008; Yu and Stahl, 2008; Zhang et al., 2013). Bagasse, a waste derived sugar cane industry, is a particularly promising source as it contains xylose, which is suitable for production of PHA (Sindhu et al., 2016; Silva et al., 2014).

15.5.2 BIOFUELS

Biofuels have gained significant attention as an alternative to conventional fuels due to their potential to minimize greenhouse gas emissions as well as to mitigate the impact of climate change in the transportation sector. The two main generations of biofuels are conventional biofuels (first generation) and advanced biofuels (second generation). Conventional biofuels are produced from food or feed crops, including sugar, starch, and vegetable oils. The production of these biofuels involves established methods like fermentation, distillation, and transesterification (Harish et al., 2020).

To address concerns related to food security and competition for agricultural land, second-generation biofuels, also known as advanced biofuels, have been developed. These biofuels are produced from sustainable feedstocks and food wastes, which do not compete with the food supply. Examples of feedstocks used for advanced biofuels include dedicated energy crops like miscanthus and switchgrass, agricultural residues, forest residues, and other waste materials such as used cooking oil and municipal solid waste (MSW) (Harish et al., 2020). The development and utilization of second-generation biofuels contribute to the sustainable production of bioenergy while minimizing negative environmental and social impacts. These biofuels offer the potential to reduce reliance on fossil fuels, decrease GHG emissions, and promote the circular economy by utilizing waste materials as feedstocks for energy production.

Third-generation biofuels are often associated with microalgae-based biodiesel production. In this case, biofuels are generated using microalgae and typically undergo traditional processes like transesterification or hydro-treatment of algal oil. Biofuels derived from microalgae are considered third-generation biofuels and represent a promising avenue for future biofuel development. The transition of biofuels from first-generation to second- and third-generation aims to improve sustainability by utilizing non-food feedstocks and waste materials, reducing land-use conflicts, and enhancing overall environmental performance (Harish et al., 2020). These advanced biofuels have the prospects to develop a more sustainable and diversified energy system while reducing reliance on fossil fuels and mitigating GHG emissions.

15.5.3 BIODIESEL

The ever-increasing pollution, growing demands of fuel, and reduction of fossil fuels have highlighted the need for substitute fuels. The Biodiesel Association of India (BDAI) is a non-profit association that is actively promoting the use of biofuels, particularly biodiesel and compressed biogas. It aims to support viable and sustainable agricultural growth, energy security, rural development, and environmental protection. Biodiesel is a type of liquid fuel developed from renewable resources, primarily fats and oils. It falls under the category of first-generation biofuels or green fuels. In comparison to fossil-based diesel, biodiesel is found to have lower toxicity, lower environmental impact, and can contribute to reducing greenhouse gas emissions (Daud et al., 2015; Karmee et al., 2015). The production of biodiesel can be expensive due to the high feedstock cost. To address this issue, using food source waste as an alternative feed for biodiesel production is considered a positive alternative, as it can help reduce costs (Karmee et al., 2015). The synthesis of biodiesel at an industrial scale typically involves transesterification, a process where the fatty acids present in the feedstock are changed into fatty acid methyl esters (FAMEs) using oleaginous microorganisms or homogeneous base catalysts (Atadashi et al., 2013; Kiran et al., 2014). Biodiesel composition consists of these FAMEs, which are obtained by catalytic reaction of lipids with alcohol (Qi et al., 2016). Feedstocks with high lipid content, such as sludge, hold promise as substrates for biodiesel production (Qi et al., 2016).

Siddiquee and Rohani (2011) and Vyas et al. (2010) have reported four primary processes used for biodiesel production, that is, direct use, micro emulsions, pyrolysis, and transesterification (Figure 15.6). The direct blending process involves diluting vegetable oils directly with diesel, aiming to lower viscosity and enhance performance (Mahmudul et al., 2017). This process does not require chemical processing and typically uses a blending ratio of 1:10 or 2:10 (Abbaszaadeh et al., 2012). Micro emulsion formation was developed to address the high viscosity of vegetable oil. It involves combining vegetable oil with organic solvents like ethanol, methanol, butanol, hexanol, and so forth (Abbaszaadeh et al., 2012; Mahmudul et al., 2017). Pyrolysis, a heat treatment method, converts the feedstock into biodiesel under anaerobic conditions. While effective and pollution-free, this process generates low-value byproducts due to the absence of oxygen (Abbaszaadeh et al., 2012; Singh and Singh, 2010). The most commonly used process in biodiesel production is trans-esterification, which is favored for being simple, inexpensive, and for viscosity reduction of the oil, making it suitable for use in engines and equipment (Abbaszaadeh et al., 2012; Koutsouki et al., 2016). Biodiesel produced through this process typically exhibits lower viscosity in comparison to biodiesel obtained from other three processes (Capodaglio and Callegari, 2017). During transesterification, triglycerides, which are commonly found in vegetable oils or animal fats, are reacted with an alcohol, typically methanol or ethanol. This reaction leads to the conversion of triglycerides into fatty acid alkyl esters (biodiesel) and glycerol (Koutsouki et al., 2016).

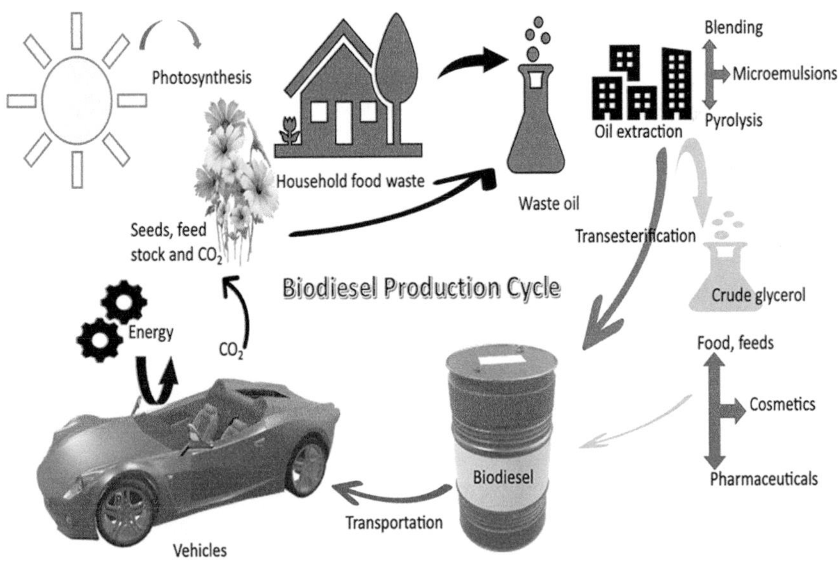

FIGURE 15.6 Production process of biodiesel.

15.5.4 BIOBUTANOL

Butanol, a four-carbon alcohol, is widely recognized as an advanced biofuel with distinct advantages over ethanol, as highlighted in studies by Tao et al. (2014). Its higher energy content makes it an appealing choice for gasoline blending, as it offers improved combustion efficiency. Moreover, butanol exhibits lower vapor pressure, enhanced ability to dissolve in vegetable oils, reduced viscosity, and increased hydrophobicity, allowing for efficient transportation through existing pipelines (Lan et al., 2013; Tao et al., 2014).

While the production of butanol from food-grade feedstocks such as glucose and corn starch can be financially burdensome, researchers have focused on developing a more cost-effective approach using cellulosic biomass sources. These sources include wheat straw, corn stover, barley straw, cassava bagasse, switchgrass, and miscanthus, as extensively discussed in studies by Gao et al. (2014), Lu et al. (2011), Qureshi et al. (2008, 2014), Wang and Chen (2011), and Zhang et al. (2014). By leveraging these cellulosic biomass sources, the production of butanol becomes not only economically viable but also environmentally sustainable. This approach addresses the challenges associated with reliance on food-grade feedstocks, providing a more efficient and cost-effective pathway towards the production of advanced biofuels like butanol.

Producing butanol from food waste presents significant advantages over utilizing expensive cellulosic biomass sources. Food waste is inherently rich in sugars, starch, and other functionalized molecules like proteins, fatty acids, and minerals, which serve as valuable nutrients for the growth of butanol-producing cultures, commonly involving the bacterium *Clostridium* (Girotto et al., 2015; Huang et al., 2015; Ujor et al., 2014). In contrast, the utilization of cellulosic biomass requires energy-intensive pre-treatment under controlled conditions (Humbird et al., 2011). Numerous studies have successfully demonstrated the potential of food waste as a feedstock for biobutanol production. For instance, lactose-rich waste whey yielded an impressive conversion of 0.3 grams of butanol per gram of carbohydrates, showcasing the remarkable value of this waste stream in biobutanol production (Girotto et al., 2015). Similarly, starch-rich industrial food waste, including bread liquid, batter liquid, and inedible dough, has shown promising results as a viable source for biobutanol production (Ujor et al., 2014). The valorization of food waste for biobutanol production not only presents an efficient waste management strategy but also exhibits economic viability by transforming waste into a valuable and sustainable fuel product.

15.5.5 ETHANOL

The demand for ethanol in daily life is increasing worldwide, as it is recognized as a cleaner biofuel with zero net emissions of carbon dioxide during combustion (Vohra et al., 2014). Production of ethanol from food waste typically involves three steps that is, hydrolysis or saccharification to convert the substrates into glucose, fermentation which converts the produced glucose into ethanol and CO_2, and separation and purification of ethanol through distillation process (Ma et al., 2008). Various microorganisms have been employed for bioethanol production. Yeast, particularly *Saccharomyces cerevisiae*, is considered highly efficient for conversion of glucose to ethanol (Chen, 2011). Acid-tolerant bacteria like *Zymomonas mobilis* is also used for generation of ethanol from food waste (He et al., 2014). Co-cultures, such as *Sacchromyces cerevisiae* and *Enterobacter aerogenes*, have shown enhanced yield of ethanol from stillage (residual material after distillation) (Choonut et al., 2015). Co-cultures have also been used for the fermentation of pentose and hexose sugars obtained from lignocellulosic material. The benefits obtained from a co-culture system depend on the growth of different microorganisms in hydrolyzed mediums, compatibility with fermentation conditions, and factors like pH and temperature (Cardona and Sánchez, 2007).

Enzymatic hydrolysis is a commonly employed method for pre-treating food waste in ethanol production, enabling the efficient breakdown of complex carbohydrates. Recent research has shown promising results using in situ-produced fungal mash, which contains hydrolytic enzymes, for rapid hydrolysis of food waste (Ma et al., 2017). Ethanol derived from food waste is frequently utilized as a feedstock chemical for the synthesis of ethylene and polyethylene (Kiran et al., 2014). Bioethanol production from inexpensive feedstocks has garnered interest, and food waste rich in starch and cellulose content, such as sugar cane, rice, and potatoes, are considered suitable sources for bioethanol production (Thomsen et al., 2003). In the bioconversion process, food waste with high cellulose content undergoes saccharification, where it is converted into glucose, which is subsequently fermented into ethanol in the presence of *Saccharomyces cerevisiae*. Food waste with low cellulose content is often preferred for bioethanol production compared to waste with high cellulose content. The hydrolysis step becomes more challenging when waste contains a significant amount of cellulose feedstocks. Hence, researchers have explored the use of abundant and cost-effective waste materials such as municipal waste, lignocellulosic waste, and food waste as viable alternatives for ethanol production. Fresh and wet food waste has shown to be an efficient source for bioethanol production, with studies demonstrating the production of 36 grams of ethanol and 60 grams of reducing sugar from 100 grams of food waste in a 48-hour fermentation period (Hong and Yoon, 2011).

15.5.6 BIOHYDROGEN

Hydrogen has been a promising replenishable source with a high energy yield. Commercially, hydrogen is primarily produced through techniques such as steam reforming of hydrocarbons, partial oxidation of petroleum-based fuels, and coal gasification. However, these methods require high temperatures and complex machinery, resulting in high energy consumption and operational costs. When biomass is used as a feed for production of hydrogen, it is referred to as biohydrogen, which is considered a green, eco-friendly, and sustainable approach (Habashy et al., 2021; Jarunglumlert et al., 2018). Production of biohydrogen can be done through processes like direct or indirect biophotolysis, dark fermentation, photofermentation, or any combined method (Łukajtis et al., 2018). Microbe-mediated biological processes offer advantages such as operating at moderate conditions of temperature and pressure, the versatility of microbes, and the ability to use a wide variety of substrates. Food waste rich in carbohydrates is particularly suitable for production of biohydrogen. The substrate composition has a significant role, as carbohydrate-based food waste can produce 20 times more hydrogen in comparison to fat-based and protein-based waste (Show et al., 2012). Pre-treatments, such as the use of mixed cultures of *Clostridium* and *Enterobacter*, are used for the generation of hydrogen from food waste (Li and Fang, 2007). Food waste can serve as a source of microflora capable of producing hydrogen, with lactic acid bacteria being the predominant species in untreated waste, while hydrogen-producing bacteria dominate in pre-treated waste. Various fermentation systems, such as batch, semi-continuous, continuous, and one or multiple stages, have been developed for hydrogen production from food waste (Kiran et al., 2014). Anaerobic sequencing batch reactors (ASBR) and upflow anaerobic sludge blanket (UASB) reactors have demonstrated high rates of hydrogen production due to their concentrated biomass in the reactor (Karthikeyan et al., 2018).

Another approach for converting organic acids into hydrogen is through photofermentation. A combined dark and photofermentation system has been proposed to enhance overall hydrogen yield. In this process, lactic acid derived from food waste is utilized by photofermentative bacteria, primarily purple non-sulfur bacteria, to generate hydrogen, while the remaining residue is converted into methane (Show et al., 2012). However, this dark and photofermentation process still has certain limitations. Lignocellulosic biomass, considered a carbon-neutral energy source, represents the most abundant raw material on Earth for biohydrogen production (Kumar et al., 2015). It holds significant potential for sustainable and efficient biohydrogen production.

15.5.7 METHANE

AD is an effective and cost-efficient method for producing biogas, particularly methane, which offers multiple benefits for waste management. This process utilizes renewable energy sources and generates minimal residual waste. Methane, the primary component of biogas, can be utilized as a fuel for various applications, including ovens, water heaters, automobiles, turbines, and even as a soil conditioner or fertilizer (Kiran et al., 2014). With an energy content of 55.5 MJ/kg, methane represents a valuable source of energy. During AD, organic waste undergoes biodegradation and reduction, resulting in the production of methane. Several factors influence the methane production in this process, such as alkalinity, pH levels, organic loading rate (OLR), nutrient availability, reactor type, volatile fatty acids, carbon-to-nitrogen ratio, operating temperatures, ammonium ions, and the characteristics of the substrate (Park et al., 2018). By carefully controlling these parameters, optimal methane production can be achieved, maximizing the energy potential of the organic waste.

There are three main processes involved in biogas production: single-stage AD, two-stage AD, and packed bed reactors (PBR) or fixed bed systems. Single-stage AD is a commonly employed method for MSW treatment. In this process, all the reactions, including hydrolysis, acidogenesis, acetogenesis, and methanogenesis, occur simultaneously within a single reactor. This system is advantageous as it experiences fewer technical failures and requires lower investment costs compared to other systems (Forster-Carneiro et al., 2008). AD can be conducted in either wet or dry conditions. However, dry AD generally results in lower methane production compared to wet waste due to limitations in volatile fatty acid (VFA) transport (Nagao et al., 2012).

Two-stage AD involves the production of hydrogen and methane in separate reactors. In the first stage, fast-growing amidogens and hydrogen-producing microorganisms are enriched to produce hydrogen and VFAs. In the second stage, slow-growing acetogens and methanogens convert VFAs into methane and carbon dioxide. Research has demonstrated that the two-stage process achieves higher methane generation compared to other systems (Chu et al., 2008; Park et al., 2008). PBR or fixed bed systems are alternative approaches used for methane generation. These systems offer benefits such as high loading capacity, immobilization of microbial consortia, and stabilization of methanogenesis (Kastner et al., 2012). By utilizing PBR or fixed bed systems, efficient methane production can be achieved while maintaining the stability of the process.

Studies have reported the highest methane yields from food waste when employing a two-stage hydrogen and methane processing approach. For instance, Koike et al. (2009) achieved a biogas yield of approximately 850 L/g from food waste. Combining microbial electrolysis cells (MECs) with AD reactors has also been found to enhance methane production rates by 1.7 times compared to AD alone (Park et al., 2018).

Pre-treatment of food waste can significantly improve its digestibility, acidification rates, and overall methane recovery. Various pre-treatment methods are employed, including physical grinding, thermal treatment, acid and alkali treatment, high-pressure treatment, pulse discharge of high voltage, microwave-mediated treatment, micro-aeration, and biological treatment (Karthikeyan et al., 2018). Among these, thermal treatment followed by alkali treatment has been identified as the most effective strategy for enhancing AD of food waste. Alkali pre-treatment alone can increase methane yield by 25%, and when combined with thermal treatment, the yield can be further increased to 32% (Naran et al., 2016). Different types of fruits and vegetable waste have been found to produce methane in the range of 180–732 mL per gram of volatile solids. In a two-stage anaerobic digester, utilizing 95.1% of volatile solids from fruits and vegetable waste resulted in a methane yield of 530 mL per gram of volatile solids (Kiran et al., 2014). These findings highlight the potential for efficient methane production from food waste and the importance of optimizing pre-treatment strategies for improved biogas yields.

15.5.8 ELECTRIC POWER GENERATION

Bioelectricity, a sustainable and renewable form of electricity, can be generated from various biomass sources, including food waste like sugarcane residues, wood waste, rice hulls, and other biomasses. The recovery of energy or electricity from food waste offers numerous benefits, such as environmental friendliness, cost-effectiveness, pollution reduction, and long-term sustainability. One effective method to achieve this is through the use of microbial fuel cells (MFCs), specialized devices that employ microorganisms as catalysts for electricity generation from diverse waste materials, including industrial and domestic wastewater, as well as excess sludge (Cercado-Quezada et al., 2010; Feng et al., 2015; Li et al., 2016).

In MFCs, microorganisms oxidize the organic matter present in the food waste, transferring electrons to the anode. At the cathode, these electrons are either reduced by microorganisms or through abiotic processes involving oxygen (Cercado-Quezada et al., 2010). The organic-rich composition of food waste makes it an excellent energy source for electricity generation in MFCs (Li et al., 2016). For instance, an MFC utilizing food waste leachate with a concentration of 5000 mg/L achieved a maximum power output of 15.14 W/m^3 and an open circuit voltage of 1.12 V. However, it should be noted that the power output decreased when the substrate concentration was increased to 20,000 mg/L, mainly due to microbial inhibition in the anode chamber (Rikame et al., 2012).

A research study investigated the use of MFCs to generate electricity from food industry wastes, including wine lees, yogurt waste, and fermented apple juice, in combination with different inoculum sources. The study found that only yogurt waste inoculated with compost leachate demonstrated stable power density generation, reaching 44 mW/m^2 (Cercado-Quezada et al., 2010). Another study focused on the utilization of complex food waste from a canteen, specifically the organic fractions, as a substrate for electricity production in MFCs. The study observed that the OLR, which indicates the rate of organic matter supplied to the MFC, significantly influenced the electricity generation from food waste (Chandrasekhar et al., 2015).

These findings highlight the potential of MFCs as an effective and sustainable means of producing bioelectricity from food waste. By harnessing the power of anaerobic microorganisms present in the anodic biocatalyst, MFCs offer an environmentally friendly solution for both energy generation and waste management. The utilization of food waste in MFCs not only contributes to the production of clean energy but also helps in the efficient utilization of organic resources that would otherwise be discarded as waste.

15.6 CONCLUSION

The global population is growing very fast, resulting in the depletion of natural deposits. Concurrently, the agricultural and food sectors are experiencing advancements, leading to a significant amount of waste generation and subsequent land degradation. This waste not only pollutes ecosystems and threatens biodiversity but also highlights the need for effective techniques that can reduce waste while generating valuable byproducts. Converting agricultural waste into useful products can bring about economic benefits, such as improving food security, reducing production costs, promoting the use of biodegradable materials, and ensuring environmental sustainability. Through innovative research, various techniques have emerged to utilize and minimize food waste, highlighting the significance of implementing food waste management policies and infrastructure. Among these technologies, microbial treatment stands out as a promising and widely adopted biological approach for managing and minimizing food and agricultural waste, as well as for producing bioenergy and biofuels.

KEYWORDS

- **bioconversion**
- **bioelectricity**
- **bioenergy food waste management**
- **biofuels**
- **food waste**
- **valorization**

REFERENCES

Abbaszaadeh, A., Ghobadian, B., Omidkhah, M. R., & Najafi, G. (2012). Current biodiesel production technologies: A comparative review. *Energy Conversion and Management*, *63*, 138–148.

Agnew, D. E., & Pfleger, B. F. (2013). Synthetic biology strategies for synthesizing polyhydroxyalkanoates from unrelated carbon sources. *Chemical Engineering Science*, *103*, 58–67.

Akaraonye, E., Keshavarz, T., & Roy, I. (2010). Production of polyhydroxyalkanoates: The future green materials of choice. *Journal of Chemical Technology and Biotechnology*, *85*(6), 732–743.

Atadashi, I. M., Aroua, M. K., Aziz, A. A., & Sulaiman, N. M. N. (2013). The effects of catalysts in biodiesel production: A review. *Journal of Industrial and Engineering Chemistry*, *19*(1), 14–26.

Balat, M., & Balat, H. (2010). Progress in biodiesel processing. *Applied Energy*, *87*(6), 1815–1835.

Bansal, N., Tewari, R., Soni, R., & Soni, S. K. (2012). Production of cellulases from *Aspergillus niger* NS-2 in solid state fermentation on agricultural and kitchen waste residues. *Waste Management*, *32*(7), 1341–1346.

Blumer-Schuette, S. E., Kataeva, I., Westpheling, J., Adams, M. W., & Kelly, R. M. (2008). Extremely thermophilic microorganisms for biomass conversion: Status and prospects. *Current Opinion in Biotechnology*, *19*(3), 210–217.

Boudouropoulos, I. D., & Arvanitoyannis, I. S. (2000). Potential and perspectives for application of environmental management system (EMS) and ISO 14000 to food industries. *Food Reviews International*, *16*(2), 177–237.

Bryant, M. P. (1979). Microbial methane production—Theoretical aspects. *Journal of Animal Science*, *48*(1), 193–201.

Bugnicourt, E., Cinelli, P., Lazzeri, A., & Alvarez, V. (2014). Polyhydroxyalkanoate (PHA): Review of synthesis, characteristics, processing and potential applications in packaging. *Express Polymer Letters*, *8*, 791–808.

Camassola, M., & Dillon, A. J. P. (2007). Production of cellulases and hemicellulases by *Penicillium echinulatum* grown on pretreated sugar cane bagasse and wheat bran in solid-state fermentation. *Journal of Applied Microbiology*, *103*(6), 2196–2204.

Capodaglio, A. G., & Callegari, A. (2018). Feedstock and process influence on biodiesel produced from waste sewage sludge. *Journal of Environmental Management*, *216*, 176–182.

Cardona, C. A., & Sánchez, Ó. J. (2007). Fuel ethanol production: Process design trends and integration opportunities. *Bioresource Technology*, *98*(12), 2415–2457.

Castro, J., Lu, J., Zhang, G., Dong, Y., & Martínez, L. (2017). Opinion dynamics-based group recommender systems. *IEEE Transactions on Systems, Man, and Cybernetics: Systems*, *48*(12), 2394–2406.

Cercado-Quezada, B., Delia, M. L., & Bergel, A. (2010). Testing various food-industry wastes for electricity production in microbial fuel cell. *Bioresource Technology*, *101*(8), 2748–2754.

Cesário, M. T., Raposo, R. S., de Almeida, M. C. M., van Keulen, F., Ferreira, B. S., & da Fonseca, M. M. R. (2014). Enhanced bioproduction of poly-3-hydroxybutyrate from wheat straw lignocellulosic hydrolysates. *New Biotechnology*, *31*(1), 104–113.

Chauhan, B. S., Kumar, N., Du Jun, Y., & Lee, K. B. (2010). Performance and emission study of preheated *Jatropha* oil on medium capacity diesel engine. *Energy*, *35*(6), 2484–2492.

Chen, G. Q., & Patel, M. K. (2012). Plastics derived from biological sources: Present and future: A technical and environmental review. *Chemical Reviews*, *112*(4), 2082–2099.

Chen, R. H., Chiang, L. B., Chen, C. N., & Lin, T. H. (2011). Cold-start emissions of an SI engine using ethanol–gasoline blended fuel. *Applied Thermal Engineering*, *31*(8–9), 1463–1467.

Choonut, A., Yunu, T., Pichid, N., & Sangkharak, K. (2015). Ethanol production from reused liquid stillage. *Energy Procedia*, *79*, 808–814.

Chu, C. F., Li, Y. Y., Xu, K. Q., Ebie, Y., Inamori, Y., & Kong, H. N. (2008). A pH- and temperature-phased two-stage process for hydrogen and methane production from food waste. *International Journal of Hydrogen Energy*, *33*(18), 4739–4746.

Cournoyer, M. S. (1996). Sanitation and stabilization of slaughterhouse sludges through composting. In *Proceedings of the Canadian Meat Research Institute Technology Symposium* (pp. 1–7). Toronto, Canada: Canadian Meat Research Institute.

Daud, N. M., Abdullah, S. R. S., Hasan, H. A., & Yaakob, Z. (2015). Production of biodiesel and its wastewater treatment technologies: A review. *Process Safety and Environmental Protection*, *94*, 487–508.

Dhar, P., Bhardwaj, U., Kumar, A., & Katiyar, V. (2014). Cellulose nanocrystals: A potential nanofiller for food packaging applications. In *Food Additives and Packaging* (pp. 197–239). Washington, DC: American Chemical Society.

Dinjaski, N., & Prieto, M. A. (2015). Smart polyhydroxyalkanoate nanobeads by protein-based functionalization. *Nanomedicine: Nanotechnology, Biology and Medicine*, *11*, 885–899.

Escapa, I. F., García, J. L., Bühler, B., Blank, L. M., & Prieto, M. A. (2012). The polyhydroxyalkanoate metabolism controls carbon and energy spillage in *Pseudomonas putida*. *Environmental Microbiology*, *14*(4), 1049–1063.

Food and Agriculture Organization of the United Nations. (2012). *Towards the Future We Want: End Hunger and Make the Transition to Sustainable Agricultural and Food Systems*. Rome, Italy: FAO.

Forster-Carneiro, T., Pérez, M., & Romero, L. I. (2008). Influence of total solid and inoculum contents on performance of anaerobic reactors treating food waste. *Bioresource Technology*, *99*(15), 6994–7002.

Gao, K., Boiano, S., Marzocchella, A., & Rehmann, L. (2014). Cellulosic butanol production from alkali-pretreated *switchgrass* (*Panicum virgatum*) and *phragmites* (*Phragmites australis*). *Bioresource Technology*, *174*, 176–181.

Girotto, F., Alibardi, L., & Cossu, R. (2015). Food waste generation and industrial uses: A review. *Waste Management*, *45*, 32–41.

Hansen, C. L., & Cheong, D. Y. (2019). Agricultural waste management in food processing. In *Handbook of Farm, Dairy and Food Machinery Engineering* (pp. 673–716). Cambridge, MA: Academic Press.

Harish, B. A., Ramana, N. V., & Gnaneswar, K. (2020). A study on properties of recycled coarse aggregate and its concrete. In *IOP Conference Series: Materials Science and Engineering* (Vol. 998, No. 1, p. 012025). Bristol, UK: IOP Publishing.

He, Y., Fang, Z., Zhang, J., Li, X., & Bao, J. (2014). De-ashing treatment of corn stover improves the efficiencies of enzymatic hydrolysis and consequent ethanol fermentation. *Bioresource Technology*, *169*, 552–558.

Hong, Y. S., & Yoon, H. H. (2011). Ethanol production from food residues. *Biomass and Bioenergy*, *35*(7), 3271–3275.

Humbird, D., Davis, R., Tao, L., Kinchin, C., Hsu, D., Aden, A., & Dudgeon, D. (2011). *Process Design and Economics for Biochemical Conversion of Lignocellulosic Biomass to Ethanol: Dilute-Acid Pretreatment and Enzymatic Hydrolysis of Corn Stover* (No. NREL/TP-5100-47764). Golden, CO: National Renewable Energy Laboratory (NREL).

Jarunglumlert, T., Prommuak, C., Putmai, N., & Pavasant, P. (2018). Scaling-up bio-hydrogen production from food waste: Feasibilities and challenges. *International Journal of Hydrogen Energy*, *43*(2), 634–648.

Karmee, S. K., Linardi, D., Lee, J., & Lin, C. S. K. (2015). Conversion of lipid from food waste to biodiesel. *Waste Management*, *41*, 169–173.

Karthikeyan, O. P., Trably, E., Mehariya, S., Bernet, N., Wong, J. W., & Carrere, H. (2018). Pretreatment of food waste for methane and hydrogen recovery: A review. *Bioresource Technology*, *249*, 1025–1039.

Karthikeyan, S., Kalaimurugan, K., Prathima, A., & Somasundaram, D. (2018). Novel microemulsion fuel additive Ce–Ru–O catalysts with algae biofuel on diesel engine testing. *Energy Sources, Part A: Recovery, Utilization, and Environmental Effects*, *40*(6), 630–637.

Kastner, V., Somitsch, W., & Schnitzhofer, W. (2012). The anaerobic fermentation of food waste: A comparison of two bioreactor systems. *Journal of Cleaner Production*, *34*, 82–90.

Kathirvel, S., Raju, R., Fasla, K. K., & Shahana, V. P. (2017). An overview of sample preparation techniques for food analysis. *Asian Journal of Pharmaceutical Analysis*, *7*(1), 48–51.

Keshavarz, T., & Roy, I. (2010). Polyhydroxyalkanoates: Bioplastics with a green agenda. *Current Opinion in Microbiology*, *13*, 321–326.

Kiran, E. U., & Liu, Y. (2015). Bioethanol production from mixed food waste by an effective enzymatic pretreatment. *Fuel*, *159*, 463–469.

Kiran, E. U., Trzcinski, A. P., Ng, W. J., & Liu, Y. (2014). Bioconversion of food waste to energy: A review. *Fuel*, *134*, 389–399.

Klinke, H. B., Olsson, L., Thomsen, A. B., & Ahring, B. K. (2003). Potential inhibitors from wet oxidation of wheat straw and their effect on ethanol production of *Saccharomyces cerevisiae*: Wet oxidation and fermentation by yeast. *Biotechnology and Bioengineering*, *81*(6), 738–747.

Koike, Y., et al. (2009). Production of fuel ethanol and methane from garbage by high-efficiency two-stage fermentation process. *Journal of Bioscience and Bioengineering*, *108*(6), 508–512.

Koutsouki, A. A., Tegou, E., Badeka, A., Kontakos, S., Pomonis, P. J., & Kontominas, M. G. (2016). *In situ* and conventional transesterification of rapeseeds for biodiesel production: The effect of direct sonication. *Industrial Crops and Products*, *84*, 399–407.

Kumar, P., Sharma, P. K., Sharma, P. K., & Sharma, D. (2015). Micro-algal lipids: A potential source of biodiesel. *Journal of Innovative Pharmaceutical and Biological Sciences*, *2*(2), 135–143.

Lan, E. I., Ro, S. Y., & Liao, J. C. (2013). Oxygen-tolerant coenzyme A-acylating aldehyde dehydrogenase facilitates efficient photosynthetic n-butanol biosynthesis in cyanobacteria. *Energy & Environmental Science*, *6*, 2672–2681.

Li, H., Zhao, W., & Fang, Z. (2017). Hydrophobic Pd nanocatalysts for one-pot and high-yield production of liquid furanic biofuels at low temperatures. *Applied Catalysis B: Environmental*, *215*, 18–27.

Lin, C. S. K., Koutinas, A. A., Stamatelatou, K., Mubofu, E. B., Matharu, A. S., Kopsahelis, N., & Luque, R. (2014). Current and future trends in food waste valorization for the production of chemicals, materials and fuels: A global perspective. *Biofuels, Bioproducts and Biorefining*, *8*(5), 686–715.

Lu, C., Zhao, J., Yang, S. T., & Wei, D. (2012). Fed-batch fermentation for n-butanol production from cassava bagasse hydrolysate in a fibrous bed bioreactor with continuous gas stripping. *Bioresource Technology*, *104*, 380–387.

Łukajtis, R., Hołowacz, I., Kucharska, K., Glinka, M., Rybarczyk, P., Przyjazny, A., & Kamiński, M. (2018). Hydrogen production from biomass using dark fermentation. *Renewable and Sustainable Energy Reviews*, *91*, 665–694.

Ma, L., Ma, S. Y., Kang, H., Shen, X. F., Wang, T. T., Jiang, X. H., & Chen, Q. (2017). Preparation of Ag-doped ZnO-SnO_2 hollow nanofibers with an enhanced ethanol sensing performance by electrospinning. *Materials Letters*, *209*, 188–192.

Mahmudul, H. M., Hagos, F. Y., Mamat, R., Adam, A. A., Ishak, W. F. W., & Alenezi, R. (2017). Production, characterization and performance of biodiesel as an alternative fuel in diesel engines–A review. *Renewable and Sustainable Energy Reviews*, *72*, 497–509.

McInerney, M. J., Bryant, M. P., & Pfennig, N. (1979). Anaerobic bacterium that degrades fatty acids in syntrophic association with methanogens. *Archives of Microbiology*, *122*(2), 129–135.

Melikoglu, M., Lin, C. S. K., & Webb, C. (2013a). Analyzing global food waste problem: Pinpointing the facts and estimating the energy content. *Central European Journal of Engineering*, *3*(2), 157–164.

Melikoglu, M., Lin, C. S. K., & Webb, C. (2013b). Stepwise optimization of enzyme production in solid state fermentation of waste bread pieces. *Food and Bioproducts Processing*, *91*(4), 638–646.

Mondal, S., & Palit, D. (2019). Effective role of microorganism in waste management and environmental sustainability. In *Sustainable Agriculture, Forest and Environmental Management* (pp. 485–515). Singapore: Springer.

Morita, M., & Sasaki, K. (2012). Factors influencing the degradation of garbage in methanogenic bioreactors and impacts on biogas formation. *Applied Microbiology and Biotechnology*, *94*(3), 575–582.

Mozejko-Ciesielska, J., & Kiewisz, R. (2016). Bacterial polyhydroxyalkanoates: Still fabulous? *Microbiological Research*, *192*, 271–282.

Muhammad Nasir, I., Mohd Ghazi, T. I., & Omar, R. (2012). Production of biogas from solid organic wastes through anaerobic digestion: A review. *Applied Microbiology and Biotechnology*, *95*(2), 321–329.

Muñoz, L. E. A., & Riley, M. R. (2008). Utilization of cellulosic waste from tequila bagasse and production of polyhydroxyalkanoate (PHA) bioplastics by *Saccharophagus degradans*. *Biotechnology and Bioengineering*, *100*, 882–888.

Nagao, N., et al. (2012). Maximum organic loading rate for the single-stage wet anaerobic digestion of food waste. *Bioresource Technology*, *118*, 210–218.

Nasir, I. M., Mohd Ghazi, T. I., & Omar, R. (2012). Anaerobic digestion technology in livestock manure treatment for biogas production: A review. *Engineering in Life Sciences*, *12*(3), 258–269.

Obruca, S., Benesova, P., Petrik, S., Oborna, J., Prikryl, R., & Marova, I. (2014a). Production of polyhydroxyalkanoates using hydrolysate of spent coffee grounds. *Process Biochemistry*, *49*, 1409–1414.

Ohkouchi, Y., & Inoue, Y. (2007). Impact of chemical components of organic wastes on L(+)-lactic acid production. *Bioresource Technology*, *98*(3), 546–553.

Okino-Delgado, C. H., Prado, D. Z. D., Facanali, R., Marques, M. M. O., Nascimento, A. S., Fernandes, C. J. D. C., & Fleuri, L. F. (2017). Bioremediation of cooking oil waste using lipases from wastes. *PLoS ONE*, *12*(10), e0186246.

Paritosh, K., Kushwaha, S. K., Yadav, M., Pareek, N., Chawade, A., & Vivekanand, V. (2017). Food waste to energy: An overview of sustainable approaches for food waste management and nutrient recycling. *BioMed Research International*.

Poomipuk, N., Reungsang, A., & Plangklang, P. (2014). Poly-β-hydroxyalkanoates production from cassava starch hydrolysate by *Cupriavidus* sp. KKU38. *International Journal of Biological Macromolecules*, *65*, 51–64.

Sedhai, P. (2022). Agricultural and food waste treatment using microbial enzyme extracted from thermophiles. *Acta Scientific Biotechnology*, *3*(2), 25–28.

Qi, J., Zhu, F., Wei, X., Zhao, L., Xiong, Y., Wu, X., & Yan, F. (2016). Comparison of biodiesel production from sewage sludge obtained from the A2/O and MBR processes by *in situ* transesterification. *Waste Management*, *49*, 212–220.

Qi, D. H., Chen, B., Zhang, D., & Lee, C. F. (2016). Optical study on the combustion characteristics and soot emissions of diesel–soybean biodiesel–butanol blends in a constant volume chamber. *Journal of the Energy Institute*, *89*(4), 807–820.

Qureshi, N., Saha, B. C., & Cotta, M. A. (2007). Butanol production from wheat straw hydrolysate using *Clostridium beijerinckii*. *Bioprocess and Biosystems Engineering*, *30*, 419–427.

Qureshi, N., Saha, B. C., Dien, B. S., Hector, R., & Cotta, M. A. (2010). Production of butanol from agricultural residues: Use of barley straw hydrolysate. *Biomass and Bioenergy*, *34*, 559–565.

Qureshi, N., Saha, B. C., Hector, R. E., Hughes, S., & Cotta, M. A. (2008). Butanol production from wheat straw by simultaneous saccharification and fermentation using *Clostridium beijerinckii* Part 1: Batch fermentation. *Biomass and Bioenergy*, *32*, 168–175.

Qureshi, N., Singh, V., Liu, S., Ezeji, T. C., Saha, B. C., & Cotta, M. A. (2014). Process integration for simultaneous saccharification, fermentation, and recovery (SSFR): Production of butanol from corn stover using *Clostridium beijerinckii* P260. *Bioresource Technology*, *154*, 222–228.

Rikame, S. S., Mungray, A. A., & Mungray, A. K. (2012). Electricity generation from acidogenic food waste leachate using dual chamber mediator-less microbial fuel cell. *International Biodeterioration & Biodegradation*, *75*, 131–137.

Seidavi, A. R., Zaker-Esteghamati, H., & Scanes, C. G. (2019). Present and potential impacts of waste from poultry production on the environment. *World's Poultry Science Journal*, *75*(1), 29–42.

Show, K. Y., Lee, D. J., & Tay, J. H. (2012). Aerobic granulation: Advances and challenges. *Applied Biochemistry and Biotechnology*, *167*(6), 1622–1640.

Siddiquee, M. N., & Rohani, S. (2011). Lipid extraction and biodiesel production from municipal sewage sludges: A review. *Renewable and Sustainable Energy Reviews*, *15*(2), 1067–1072.

Silva, L. F., Taciro, M. K., Raicher, G., Piccoli, R. A. M., Mendonça, T. T., Lopes, M. S. G., & Gomez, J. G. C. (2014). Perspectives on the production of polyhydroxyalkanoates in biorefineries associated with the production of sugar and ethanol. *International Journal of Biological Macromolecules*, *71*, 2–7.

Sindhu, R., Gnansounou, E., Binod, P., & Pandey, A. (2016). Bioconversion of sugarcane crop residue for value added products–An overview. *Renewable Energy*, *98*, 203–215.

Sindhu, R., Silviya, N., Binod, P., & Pandey, A. (2013). Pentose-rich hydrolysate from acid pretreated rice straw as a carbon source for the production of poly-3-hydroxybutyrate. *Biochemical Engineering Journal*, *78*, 67–72.

Singh, S. P., & Singh, D. (2010). Biodiesel production through the use of different sources and characterization of oils and their esters as the substitute of diesel: A review. *Renewable and Sustainable Energy Reviews*, *14*(1), 200–216.

Tao, L., He, X., Tan, E. C., Zhang, M., & Aden, A. (2014). Comparative techno-economic analysis and reviews of n-butanol production from corn grain and corn stover. *Biofuels, Bioproducts and Biorefining*, *8*, 342–361.

Tuck, C. O., Pérez, E., Horváth, I. T., Sheldon, R. A., & Poliakoff, M. (2012). Valorization of biomass: Deriving more value from waste. *Science*, *337*(6095), 695–699.

Ujor, V., Bharathidasan, A. K., Cornish, K., & Ezeji, T. C. (2014). Feasibility of producing butanol from industrial starchy food wastes. *Applied Energy*, *136*, 590–598.

Valta, K., Damala, P., Panaretou, V., Orli, E., Moustakas, K., & Loizidou, M. (2017). Review and assessment of waste and wastewater treatment from fruits and vegetables processing industries in Greece. *Waste and Biomass Valorization*, *5*, 1629–1648.

Van-Thuoc, D., Quillaguaman, J., Mamo, G., & Mattiasson, B. (2008). Utilization of agricultural residues for poly(3-hydroxybutyrate) production by *Halomonas boliviensis* LC1. *Journal of Applied Microbiology*, *104*, 420–428.

Vohra, M., Manwar, J., Manmode, R., Padgilwar, S., & Patil, S. (2014). Bioethanol production: Feedstock and current technologies. *Journal of Environmental Chemical Engineering*, *2*(1), 573–584.

Vyas, A. P., Verma, J. L., & Subrahmanyam, N. (2010). A review on FAME production processes. *Fuel*, *89*(1), 1–9.

Wang, L., & Chen, H. (2011). Increased fermentability of enzymatically hydrolyzed steam-exploded corn stover for butanol production by removal of fermentation inhibitors. *Process Biochemistry*, *46*, 604–607.

Wu, Q., Wang, Y., & Chen, G. Q. (2009). Medical application of microbial biopolyesters polyhydroxyalkanoates. *Artificial Cells, Blood Substitutes, and Immobilization Biotechnology*, *37*, 1–12.

Yu, J., & Stahl, H. (2008). Microbial utilization and biopolyester synthesis of bagasse hydrolysates. *Bioresource Technology*, *99*, 8042–8048.

Zhang, Y., & Ezeji, T. C. (2014). Elucidating and alleviating impacts of lignocellulose-derived microbial inhibitors on *Clostridium beijerinckii* during fermentation of *Miscanthus giganteus* to butanol. *Journal of Industrial Microbiology and Biotechnology*, *41*, 1505–1516.

Zhang, Y., Sun, W., Wang, H., & Geng, A. (2013). Polyhydroxybutyrate production from oil palm empty fruit bunch using *Bacillus megaterium* R11. *Bioresource Technology*, *147*, 307–314.

Zhu, Y., Romain, C., & Williams, C. K. (2016). Sustainable polymers from renewable resources. *Nature*, *540*(7633), 354–362.

Index

B

C

G

N

P

X

Y

Z

For Product Safety Concerns and Information please contact our EU representative GPSR@taylorandfrancis.com Taylor & Francis Verlag GmbH, Kaufingerstraße 24, 80331 München, Germany